Information Optics and Photonics

Thierry Fournel • Bahram Javidi
Editors

Information Optics and Photonics

Algorithms, Systems, and Applications

Springer

Editors

Prof. Thierry Fournel
Université Jean Monnet
UMR CNRS-UJM 5516
Laboratoire Hubert Curien
rue du Pr.Benoît Lauras 18
42000 St.-Etienne
France
fournel@univ-st-etienne.fr

Prof. Dr. Bahram Javidi
University of Connecticut
Electrical and Computer Engineering Department
371 Fairfield Rd. Unit 2157
Storrs, CT 06269-2157
USA
Bahram.Javidi@UConn.edu

ISBN 978-1-4419-7379-5 e-ISBN 978-1-4419-7380-1
DOI 10.1007/978-1-4419-7380-1
Springer New York Dordrecht Heidelberg London

Library of Congress Control Number: 2010937647

Printed on acid-free paper

Springer is part of Springer Science+Business Media (www.springer.com)

In memory of my mother Maryam

Contents

Part I Optical Devices and Systems

1 **General Solution of Two-Dimensional Beam-Shaping with Two Surfaces** 3
K.-H. Brenner

2 **Nanophotonics for Information Systems** 13
Y. Fainman, D.T.H. Tan, K. Ikeda, and M. Abashin

3 **Liquid Crystal Light-Valves for Slow-Light and Applications** 39
U. Bortolozzo, S. Residori, and J.-P. Huignard

4 **Diversity of Optical Signal Processing Led by Optical Signal Form Conversion** 55
T. Konishi

5 **Dynamic Wavefront Sensing and Correction with Low-Cost Twisted Nematic Spatial Light Modulators** 63
V. Durán, V. Climent, J. Lancis, E. Tajahuerce, S. Bará, J. Arines, J. Ares, P. Andrés, and Z. Jaroszewicz

6 **Nanoinjection Detectors and Imagers for Sensitive and Efficient Infrared Detection** 77
O.G. Memis and H. Mohseni

7 **Biological Applications of Stimulated Parametric Emission Microscopy and Stimulated Raman Scattering Microscopy** 89
S. Kajiyama, Y. Ozeki, K. Fukui, and K. Itoh

Part II 3D Passive/Active Imaging and Visualization

8 **Novel Approaches in 3D Sensing, Imaging, and Visualization** 101
R. Schulein, M. DaneshPanah, M. Cho, and B. Javidi

9 **Overview of Free-viewpoint TV (FTV)** 115
M. Tanimoto

10 **Presentation and Perception of Digital Hologram Reconstructions of Real-World Objects on Conventional Stereoscopic Displays** 129
R. Näsänen, T.M. Lehtimäki, T. Pitkäaho, and T.J. Naughton

11 **Parallel Phase-Shifting Digital Holography Based on the Fractional Talbot Effect** 143
L. Martínez-León, M. Araiza-E, B. Javidi, P. Andrés, V. Climent, J. Lancis, and E. Tajahuerce

12 **Improvement of Viewing-Zone Angle and Image Quality of Digital Holograms** 155
T. Nomura, Y. Teranishi, E. Nitanai, and T. Numata

13 **Shot Noise in Digital Holography** 163
F. Joud, F. Verpillat, M. Atlan, P.-A. Taillard, and M. Gross

14 **Deformation of Digital Holograms for Full Control of Focus and for Extending the Depth of Field** 177
M. Paturzo and P. Ferraro

15 **Zoom Algorithms for Digital Holography** 187
B.M. Hennelly, D.P. Kelly, D.S. Monaghan, and N. Pandey

Part III Polarimetric Imaging

16 **Partial Polarization of Optical Beams: Temporal and Spectral Descriptions** 207
T. Setälä, F. Nunziata, and A.T. Friberg

17 A Degree of Freedom and Metrics Approach for Nonsingular Mueller Matrices Characterizing Passive Systems 217
V. Devlaminck and P. Terrier

18 Resolution-Enhanced Imaging Based upon Spatial Depolarization of Light 225
D. Sylman, Z. Zalevsky, V. Micó, and J. García

Part IV Algorithms for Imaging and Analysis

19 Hybrid Imaging Systems for Depth of Focus Extension With or Without Postprocessing 235
F. Diaz, F. Goudail, B. Loiseaux, and J.-P. Huignard

20 Multispectral Image Pansharpening Based on the Contourlet Transform 247
I. Amro and J. Mateos

21 Faces Robust Image Hashing by ICA: An Attempt to Optimality 263
T. Fournel and D. Coltuc

22 Minkowski Metrics: Its Interaction and Complementarity with Euclidean Metrics 273
J.-M. Becker and M. Goeb

Contributors

M. Abashin Department of Electrical and Computer Engineering, University of California, San Diego, 9500 Gilman Drive, La Jolla, CA 92093-0407, USA

Israa Amro Universidad de Granada, Granada, Spain
and
Al-Quds Open University, Hepron, Palestine, iamro@correo.ugr.es; isamro@qou.edu

Pedro Andrés Departamento de Óptica, Universitat de València, E46100 Burjassot, Spain, pedro.andres@uv.es

María Araiza-E Laboratorio de Procesamiento Digital de Señales, Universidad Autónoma de Zacatecas, Zacatecas, Mexico

Jorge Ares Departamento de Física Aplicada, Universidad de Zaragoza, Zaragoza, Spain

Justo Arines Departamento de Física Aplicada, Universidad de Zaragoza, Zaragoza, Spain

Michael Atlan Fondation Pierre-Gilles de Gennes, Institut Langevin: UMR 7587 CNRS, U 979 INSERM, ES-PCI ParisTech, Université Paris 6, Université Paris 7, 10 rue Vauquelin, 75 231 Paris Cedex 05, France, atlan@optique.espci.fr

Salvador Bará Departamento de Física Aplicada, Universidade de Santiago de Compostela, 15782 Compostela, Galiza, Spain

Jean-Marie Becker CPE Lyon, 43 Bd du 11 Nov. 1918, 69616 Villeurbanne Cedex, France
and
Université de Lyon/CNRS, UMR 5516, Laboratoire Hubert Curien/Université de Saint-Etienne, Jean Monnet, 18 rue du Prof. B. Lauras, 42023 Saint-Etienne, France, becker@cpe.fr

Umberto Bortolozzo INLN, Université de Nice Sophia-Antipolis, CNRS, 1361 route des Lucioles 06560 Valbonne, France, Umberto.Bortolozzo@inln.cnrs.fr

K.-H. Brenner Chair of Optoelectronics, University of Heidelberg, B6, 23-29, D 68131 Mannheim, Germany, brenner@ziti.uni-heidelberg.de

Myungjin Cho Department of Electrical and Computer Engineering, University of Connecticut, U-2157, Storrs, CT 06269-2157, USA

Vicent Climent GROC·UJI, Departament de Física, Universitat Jaume I, 12071 Castelló, Spain
and
INIT (Institute of New Imaging Technologies), Universitat Jaume I, 12071 Castelló, Spain

Daniela Coltuc Faculty of Electronics, Telecommunications and Information Technology, Politehnica University of Bucharest, 1-3 Blv. Iuliu Maniu, Bucharest 6, Romania, daniela.coltuc@upb.ro

Mehdi DaneshPanah Department of Electrical and Computer Engineering, University of Connecticut, U-2157, Storrs, CT 06269-2157, USA

Vincent Devlaminck LAGIS, FRE CNRS 3303, Université Lille 1, Sciences et Technologies 59655 Villeneuve d'Ascq, France, vincent.devlaminck@univ-lille1.fr

Frédéric Diaz Thales Research and Technology, Campus Polytechnique, 1 Avenue Augustin Fresnel, 91767 Palaiseau Cedex, France
and
Laboratoire Charles Fabry de l'Institut d'Optique, CNRS, Université Paris-Sud, Campus Polytechnique, RD 128, 91127 Palaiseau Cedex, France, frederic.diaz@thalesgroup.com

Vicente Durán GROC.UJI, Physics Department-INIT, Universitat Jaume I, E12080 Castelló, Spain, vduran@sg.uji.es

Y. Fainman Department of Electrical and Computer Engineering, University of California, San Diego, 9500 Gilman Drive, La Jolla, CA 92093-0407, USA, fainman@ece.ucsd.edu

Pietro Ferraro CNR – Istituto Nazionale di Ottica, via Campi Flegrei 34, 80078-Pozzuoli (NA), Italy, pietro.ferraro@inoa.it

Thierry Fournel Université de Lyon/CNRS, UMR 5516, Laboratoire Hubert Curien/Université de Saint-Etienne, Jean Monnet, 18 rue du Prof. B. Lauras, 42023 Saint-Etienne, France, fournel@univ-st-etienne.fr

Ari T. Friberg Department of Applied Physics, Aalto University, P. O. Box 13500, FI-00076 Aalto, Finland, ari.friberg@tkk.fi
and
Department of Physics and Mathematics, University of Eastern Finland, P. O. Box 111, FI-80101 Joensuu, Finland, ari.friberg@joensuu.fi
and
Department of Microelectronics and Applied Physics, Royal Institute of Technology, Electrum 229, SE-164 40 Kista, Sweden, atf@kth.se

Kiichi Fukui Department of Biotechnology, Graduate School of Engineering, Osaka University, 2-1 Yamadaoka, Suita, Osaka 565-0871, Japan

Javier García Departamento de Óptica, Universitat de Valencia, c/Dr. Moliner, 50, 46100 Burjassot, Spain

Michel Goeb Université de Lyon/CNRS, UMR 5516, Laboratoire Hubert Curien/Université de Saint-Etienne, Jean Monnet, 18, rue du Prof. B. Lauras, 42023, Saint-Etienne, France

François Goudail Laboratoire Charles Fabry de l'Institut d'Optique, CNRS, Université Paris-Sud, Campus Polytechnique, RD 128, 91127 Palaiseau Cedex, France

Michel Gross Laboratoire Kastler Brossel École Normale Supérieure, UMR 8552, UPMC, CNRS 24 rue Lhomond, 75231 Paris Cedex 05, France, gross@lkb.ens.fr

Bryan M. Hennelly National University of Ireland Maynooth, Ireland, bryanh@cs.nuim.ie

Jean-Pierre Huignard Thales Research and Technology, Campus Polytechnique, 1 Avenue A. Fresnel, 91767 Palaiseau Cedex, France, jeanpierre.huignard@thalesgroup.com

K. Ikeda Department of Electrical and Computer Engineering, University of California, San Diego, 9500 Gilman Drive, La Jolla, CA 92093-0407, USA

Kazuyoshi Itoh Department of Material and Life Science, Graduate School of Engineering, Osaka University, 2-1 Yamadaoka, Suita, Osaka 565-0871, Japan, itoh@mls.eng.osaka-u.ac.jp

Zbigniew Jaroszewicz Institute of Applied Optics, Kamionkowska 18, 03-805 Warsaw, Poland
and
National Institute of Telecommunications, Szachowa 1, 04-894 Warsaw, Poland

Bahram Javidi University of Connecticut Electrical and Computer Engineering, Department 371 Fairfield Rd. Unit 2157 Storrs, CT 06269-2157 USA
Bahram.Javidi@UConn.edu

Fadwa Joud Laboratoire Kastler Brossel École Normale Supérieure, UMR 8552, UPMC, CNRS 24 rue Lhomond, 75231 Paris Cedex 05, France, joud@lkb.ens.fr

Shin'ichiro Kajiyama School of Biology-oriented Science and Technology, Kinki University, 930 Nishimitani, Kinokawa, Wakayama, 649-6493, Japan
and
Department of Material and Life Science, Graduate School of Engineering, Osaka University, 2-1 Yamadaoka, Suita, Osaka 565 0871, Japan

Damien P. Kelly National University of Ireland, Maynooth, Ireland, damienk@cs.nuim.ie

Tsuyoshi Konishi Graduate School of Engineering, Osaka University, Osaka, Japan, konishi@mls.eng.osaka-u.ac.jp

Jesus Lancis GROC·UJI, Departament de Física, Universitat Jaume I, 12071 Castelló, Spain
and
INIT (Institute of New Imaging Technologies), Universitat Jaume I, 12071 Castelló, Spain

Taina M. Lehtimäki RFMedia Laboratory, University of Oulu, Vierimaantie 5, FI-84100 Ylivieska, Finland

Brigitte Loiseaux Thales Research and Technology, Campus Polytechnique, 1 avenue Augustin Fresnel, 91767 Palaiseau Cedex, France

Lluís Martínez-León GROC·UJI, Departament de Física, Universitat Jaume I, 12071 Castelló, Spain
and
INIT (Institute of New Imaging Technologies), Universitat Jaume I, 12071 Castelló, Spain

Javier Mateos Universidad de Granada, Granada, Spain, jmd@decsai.ugr.es

Omer G. Memis Bio-Inspired Sensors and Optoelectronics Lab (BISOL), EECS Department, Northwestern University, Evanston, IL 60208, USA

Vicente Micó Departamento de Óptica, Universitat de Valencia, c/Dr. Moliner, 50, 46100 Burjassot, Spain

Hooman Mohseni Bio-Inspired Sensors and Optoelectronics Lab (BISOL), EECS Department, Northwestern University, Evanston, IL 60208, USA, hmohsen@northwestern.edu

David S. Monaghan National University of Ireland, Maynooth, Ireland, david.monaghan@dcu.com

Risto Näsänen RFMedia Laboratory, University of Oulu, Vierimaantie 5, FI-84100 Ylivieska, Finland

Thomas J. Naughton RFMedia Laboratory, Oulu Southern Institute, University of Oulu, Vierimaantie 5, FI-84100 Ylivieska, Finland
and
Department of Computer Science, National University of Ireland, Maynooth, Ireland, tomn@cs.nuim.ie

Eiji Nitanai Faculty of Systems Engineering, Wakayama University, 930 Sakaedani, Wakayama, Wakayama, 640-8510, Japan, nitanai@sys.wakayama-u.ac.jp

Takanori Nomura Faculty of Systems Engineering, Wakayama University, 930 Sakaedani, Wakayama, Wakayama, 640-8510, Japan, nom@sys.wakayama-u.ac.jp

Takuhisa Numata Faculty of Systems Engineering, Wakayama University, 930 Sakaedani, Wakayama, Wakayama, 640-8510, Japan, numata@sys.wakayama-u.ac.jp

Ferdinando Nunziata Dipartimento per le Tecnologie, Centro Direzionale, Universita di Napoli Parthenope, isola C4, I-80143 Napoli, Italy, ferdinando.nunziata@uniparthenope.it

Yasuyuki Ozeki Department of Material and Life Science, Graduate School of Engineering, Osaka University, 2-1 Yamadaoka, Suita, Osaka 565-0871, Japan

Nitesh Pandey National University of Ireland, Maynooth, Ireland, npandey@cs.nuim.ie

Melania Paturzo CNR – Istituto Nazionale di Ottica, via Campi Flegrei 34, 80078-Pozzuoli (NA), Italy

Tomi Pitkäaho RFMedia Laboratory, University of Oulu, Vierimaantie 5, FI-84100 Ylivieska, Finland

Stefania Residori INLN, Université de Nice Sophia-Antipolis, CNRS, 1361 route des Lucioles 06560 Valbonne, France, Stefania.Residori@inln.cnrs.fr

Robert Schulein Department of Electrical and Computer Engineering, University of Connecticut, 371 Fairfield Way, Unit 2157, Storrs CT 06269-2157, USA

Tero Setälä Department of Applied Physics, Aalto University, P. O. Box 13500, FI-00076 Aalto, Finland, tero.setala@tkk.fi

David Sylman School of Engineering, Bar-Ilan University, Ramat-Gan 52900, Israel

Pierre-André Taillard Conservatoire de musique neuchâtelois, Avenue Léopold-Robert 34, 2300 La Chaux de-Fonds, Switzerland, taillard@hispeed.ch

Enrique Tajahuerce GROC·UJI, Departament de Física, Universitat Jaume I, 12071 Castelló, Spain
and
INIT (Institute of New Imaging Technologies), Universitat Jaume I, 12071 Castelló, Spain, enrique.tajahuerce@uji.es

D.T.H. Tan Department of Electrical and Computer Engineering, University of California, San Diego, 9500 Gilman Drive, La Jolla, CA 92093-0407, USA

Masayuki Tanimoto Graduate School of Engineering, Nagoya University, Chikusa-ku, Furo-cho, Nagoya 464-8603, Japan, tanimoto@nuee.nagoya-u.ac.jp

Yusuke Teranishi Faculty of Systems Engineering, Wakayama University, 930 Sakaedani, Wakayama, Wakayama, 640-8510, Japan, s082039@sys.wakayama-u.ac.jp

Patrick Terrier LAGIS, FRE CNRS 3303, Université Lille 1, Sciences et Technologies 59655 Villeneuve d'Ascq, France

Frédéric Verpillat Laboratoire Kastler Brossel École Normale Supérieure, UMR 8552, UPMC, CNRS 24 rue Lhomond, 75231 Paris Cedex 05, France, verpillat@lkb.ens.fr

Zeev Zalevsky School of Engineering, Bar-Ilan University, Ramat-Gan 52900, Israel, zalevsz@macs.biu.ac.il

Part I
Optical Devices and Systems

Chapter 1
General Solution of Two-Dimensional Beam-Shaping with Two Surfaces

K.-H. Brenner

Abstract Beam shaping is a technique, by which a known input irradiance is transformed into a desired output irradiance by changing the local propagation vector of the wave front. Unlike one-dimensional beam-shaping, which leads to a simple differential equation which can be integrated in a straight forward manner, the two-dimensional beam shaping problem leads to a Monge–Ampere type equation, which is difficult to solve. In this paper, we generalize the problem to refractive and reflective systems and use the shifted-base-function approach to obtain a general solution.

1.1 Introduction

Beam shaping is used to transform a given intensity distribution into a different, desired intensity distribution. Applications of one-dimensional (rotationally symmetric) beam shaping [1–6] are typically in the area of high power laser operations, where the Gaussian laser beam shape leads to an inefficient usage of the available laser power due to loss at the focusing lens. Unlike one-dimensional beam-shaping, which leads to a simple differential equation which can be integrated in a straight-forward manner, the two-dimensional beam shaping problem leads to a nonlinear Monge–Ampere type equation [7–9], for which numerical solutions [10–15] are difficult to obtain. We will first generalize the equations for optical beam shaping with two surfaces, derive the nonlinear Monge–Ampere type differential equation, and present the shifted-base-function (SBF) approach [16] to model the surfaces and to solve this equation.

K.-H. Brenner (✉)
Chair of Optoelectronics, University of Heidelberg, B6, 23-29, D 68131 Mannheim, Germany
e-mail: brenner@ziti.uni-heidelberg.de

T. Fournel and B. Javidi (eds.), *Information Optics and Photonics: Algorithms, Systems, and Applications*, DOI 10.1007/978-1-4419-7380-1_1,

1.2 Optical Beam Shaping with Two Surfaces

Beam shaping can be realized by diffractive [17], reflective [18–20], or refractive optics. We consider refractive optics in a telescopic geometry, as illustrated in Fig. 1.1. The input medium and the output medium are assumed to have the same refractive index, here denoted by n_1. The intermediate medium, which has a thickness of D on the optical axis has a refractive index of n_2. In a notation similar to Shealy [6], the input coordinates are denoted by small letters (x, y) and the output coordinates are denoted by captial letters (X, Y). Any beam shaping problem can be decomposed into a sequence of two tasks: The mapping of the coordinates (x, y) to the coordinates (X, Y) and the determination of the optical surface, required to achieve this task. For the moment, we consider the mapping as given by the two functions $X(x, y)$ and $Y(x, y)$, which describe the exit position of a ray as function of the entrance coordinates (x, y).

The mapping determines the intensity $I_1(X, Y)$ in the exit plane as function of the brightness in the entrance plane $I_0(x, y)$ according to the local energy conservation law:

$$I_0\, \mathbf{s}_0 \cdot \mathrm{d}\mathbf{A}_0 = I_1\, \mathbf{s}_1 \cdot \mathrm{d}\mathbf{A}_1. \tag{1.1}$$

The vectors $\mathbf{s}_0$ and $\mathbf{s}_1$ represent the unit direction vectors of the source in the entrance and the exit plane and $\mathrm{d}\mathbf{A}_0$ and $\mathrm{d}\mathbf{A}_1$ the corresponding vectorial area elements. In the telescopic geometry, the direction vectors are equal and parallel to the z-axis.

Since the area elements in the entrance and the exit plane are also parallel to the z-axis, the equation can also be treated in a scalar form. The brightness in the entrance plane $I_0(x, y)$ is typically defined by the light source. For the brightness in the exit plane, a frequent choice is the flat-top profile or a smoothened version of it, which enables an analytic solution [21] of the one-dimensional beam shaping problem. For the second task, the determination of the optical surfaces we consider Fig. 1.2. The heights of the front surface are denoted by z and the heights of the back surface by Z. The origin is at $z = Z = 0$. Both surfaces are defined so that the heights are $h = 0$ and $H = D$ on axis. Thus, $z(0, 0) = 0$ and $Z(0, 0) = D$. The ray directions,

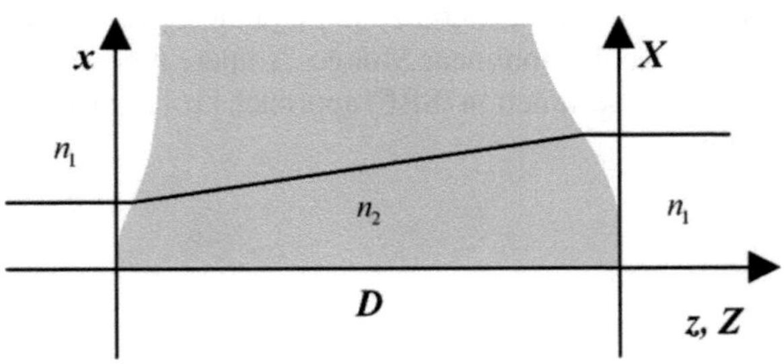

Fig. 1.1 Geometry for general two-dimensional beam shaping in the refractive telescopic arrangement

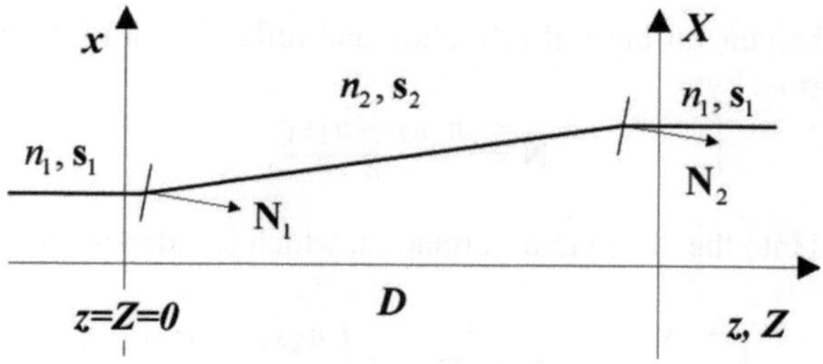

Fig. 1.2 Geometry and direction vectors for general two-dimensional beam shaping

expressed by the corresponding unit vectors $\mathbf{s}_{1,2}$ at the front surface are related to the surface normal by

$$n_2\mathbf{s}_2 = n_1\mathbf{s}_1 + \beta\mathbf{N}. \tag{1.2}$$

For a refractive system, the parameter β is determined by Snell's law of refraction:

$$\beta = \sqrt{n_2^2 - n_1^2 + (n_1\mathbf{s}_1\mathbf{N})^2} - n_1\mathbf{s}_1\mathbf{N}. \tag{1.3}$$

Likewise, for a reflective system, $n_1 = n_2$ and the parameter β is determined by the vectorial law of refraction:

$$\beta = -2(\mathbf{s}_1\mathbf{N}). \tag{1.4}$$

The relations for the ray directions at the back surface are identical to those in (1.2)–(1.4) with the indices 1 and 2 interchanged. In the following, we will observe, that the parameter β is completely irrelevant for the solution of the problem. If we assume that the index n_1 and the direction vector $\mathbf{s}_1$ are equal in the entrance and the exit surface, it is obvious from (1.2) to (1.4), that the surface normals are also equal on the entrance position and the mapped exit position:

$$\mathbf{N}_1(x, y) = \mathbf{N}_2[X(x), Y(y)]. \tag{1.5}$$

First, we will establish a general relation between the surface normal and the surface gradient. We start with the well known relation:

$$\mathbf{N} = -N_z \begin{pmatrix} z_x \\ z_y \\ -1 \end{pmatrix} \text{ with } N_z = \left(\sqrt{z_x^2 + z_y^2 + 1}\right)^{-1} \tag{1.6}$$

with the abbreviation $z_u = \partial z/\partial u$. Thus, we can write the two-dimensional surface gradient by:

$$\begin{pmatrix} z_x \\ z_y \end{pmatrix} = -\frac{1}{N_z}\begin{pmatrix} N_x \\ N_y \end{pmatrix}. \tag{1.7}$$

Referring to the general laws of refraction and reflection in (1.2), we can express the surface normal by:

$$\mathbf{N} = \frac{n_2\mathbf{s}_2 - n_1\mathbf{s}_1}{\beta}, \tag{1.8}$$

which inserted into the (1.7) yields a relation, which is independent of β:

$$\begin{pmatrix} z_x \\ z_y \end{pmatrix} = -\frac{1}{n_2 s_{2z} - n_1 s_{1z}} \begin{pmatrix} n_2 s_{2x} - n_1 s_{1x} \\ n_2 s_{2y} - n_1 s_{1y} \end{pmatrix}. \tag{1.9}$$

For the telescopic arrangement, the direction vectors are given by:

$$\mathbf{s}_1 = \begin{pmatrix} 0 \\ 0 \\ 1 \end{pmatrix}, \quad \mathbf{s}_2 = \frac{1}{W} \begin{pmatrix} \Delta x \\ \Delta y \\ \Delta z \end{pmatrix}, \tag{1.10}$$

with $\Delta x = X - x$, $\Delta y = Y - y$, $\Delta z = Z - z$ and $W = \sqrt{\Delta x^2 + \Delta y^2 + \Delta z^2}$. With this and the abbreviation $\nu = n_1/n_2$, we have:

$$\begin{pmatrix} z_x \\ z_y \end{pmatrix} = \frac{1}{\nu W - \Delta z} \begin{pmatrix} \Delta x \\ \Delta y \end{pmatrix}. \tag{1.11}$$

In this equation, Δx and Δy are given by the mapping, but Δz is still undetermined. We can solve for Δz by using the constant optical path length (OPL) condition, which in this geometry reads:

$$n_1 z + n_2 W + n_1(D - Z) = n_2 D. \tag{1.12}$$

Using (1.11) and (1.12) we can solve for Δz and obtain the solution:

$$\Delta z = \frac{\nu D + \sqrt{D^2 + [(\nu + 1)/(\nu - 1)]\Delta^2}}{\nu + 1}. \tag{1.13}$$

If we insert these two solutions into (1.11), we obtain: the general equation for 2D-beam shaping in the telescopic arrangement:

$$\begin{pmatrix} z_x \\ z_y \end{pmatrix} = \frac{1}{\sqrt{(\nu - 1)^2 D^2 + (\nu^2 - 1)\,\Delta^2}} \begin{pmatrix} \Delta x(x, y) \\ \Delta y(x, y) \end{pmatrix} \tag{1.14}$$

with $\Delta^2 = \Delta x^2 + \Delta y^2$. Thus, for a given mapping $\Delta x(x, y)$ and $\Delta y(x, y)$, the front surface can be obtained by integration of the gradient. Considering $z(0, 0) = 0$, the front surface is determined uniquely. For the determination of the back-surface, we can use (1.13) or alternatively (1.5).

1.3 SBF-Approximation of Functions

In the following, we will model the position displacement $\Delta x, \Delta y$ by shifted base functions (SBF). The "Shifted base function" approach (SBF), presented in [16], was inspired by radial basis function-interpolation, with two main differences: first: the approach is not an interpolation scheme but instead a least square approximation. Consequently, any measurement errors are reduced in the model. Second, this approach does not require the basis function to be rotationally symmetric. Thus, the extent of this function can be adapted to the degree of smoothness of the function to be modeled. We model the scalar function S by:

$$S(x, y) = \sum_{k=0}^{B-1} c_k \, \psi\left(\frac{x - x_k}{\sigma_x}, \frac{y - y_k}{\sigma_y}\right). \tag{1.15}$$

In this model, S is a superposition of a single base-function ψ, which is anisotropically scaled and shifted to the base points $\mathbf{r}_k$. For the base function ψ, we often choose the Lorenzian amplitude:

$$\psi(x, y) = \frac{1}{\sqrt{1 + x^2 + y^2}}, \tag{1.16}$$

because all its derivatives can be calculated easily. The corresponding first order partial derivatives are then given by $S_u = \sum_{k=0}^{B-1} c_k \, \psi_u[(x - x_k)/\sigma_x, (y - y_k)/\sigma_y]$ $(u \in \{x, y\})$ and the second order partial derivatives by $S_{u,v} = \sum_{k=0}^{B-1} c_k \, \psi_{u,v}$ $[(x - x_k)/\sigma_x, (y - y_k)/\sigma_y](u, v \in \{x, y\})$. In order to find the coefficient vector $\mathbf{c}$, which approximates a given function F best, one can determine the minimum of the cost function:

$$\Phi(\mathbf{c}) = \sum_{k=0}^{B-1} [F(x_k, y_k) - S(x_k, y_k)]^2, \tag{1.17}$$

which leads to a linear set of equations that can be solved by standard matrix methods. When the optimal coefficients c_k are determined, the scalar function S can be determined immediately from (1.15).

1.4 Transport Equation for Mapping Intensity

The mapping determines the brightness $I_1(X, Y)$ in the exit plane as function of the brightness in the entrance plane $I_0(x, y)$ according to the local energy conservation law $I_0 \, dA_0 = I_1 \, dA_1$ in the scalar form. Consequently, the intensity transformation (or beam shaping) is controlled by the Jacobian of the mapping:

$$I_0(x, y) = I_1(X, Y) \left| \begin{pmatrix} \frac{\partial X}{\partial x} & \frac{\partial X}{\partial y} \\ \frac{\partial Y}{\partial x} & \frac{\partial Y}{\partial y} \end{pmatrix} \right|. \tag{1.18}$$

In order to solve (1.18), we introduce a scalar function $S(x, y)$, which relates the entrance and exit coordinates by:

$$X = x + D\frac{\partial S}{\partial x}, \; Y = y + D\frac{\partial S}{\partial y}. \tag{1.19}$$

With this approach, we can guarantee that the integral $\int_0^{\mathbf{r}} \nabla S \, d\mathbf{r}'$ is path independent and the mapping, thus curl free [22]. With this assumption, (1.18) can be rewritten as:

$$I_1(x + S_x, y + S_y)\,[(1 + S_{xx})(1 + S_{yy}) - S_{xy}^2] - I_0(x, y) = 0, \tag{1.20}$$

where we have abbreviated the first order partial derivatives by $S_x = \partial S/\partial x$, $S_y = \partial S/\partial y$ and the second order partial derivatives by $S_{xx} = \partial^2 S/\partial x^2$, $S_{xy} = \partial^2 S/\partial x \partial y$, $S_{yy} = \partial^2 S/\partial y^2$. Equation (1.20) is a nonlinear Monge–Ampere type differential equation, which has been studied extensively in the mathematical literature in combination with the solution of optimal mass transport. We can establish a very simple relation to the surface gradient of (1.14), if we rewrite this equation in the form:

$$\begin{pmatrix} z_x \\ z_y \end{pmatrix} = \frac{1}{\sqrt{(\nu - 1)^2 + (\nu^2 - 1)(\Delta^2/D^2)}} \begin{pmatrix} \frac{x(x,y)}{D} \\ \frac{y(x,y)}{D} \end{pmatrix}. \tag{1.21}$$

By assuming that the lateral shift Δ is small compared to the separation D and using (1.19), we can write (1.21) in the form:

$$\nabla_\perp z \cong \frac{1}{|\nu - 1|} \nabla S, \tag{1.22}$$

which indicates, that the surface is a scaled version of the scalar function in this approximation. We want to point out, however, that this approximation only provides a visual interpretation but is not necessary to solve for the surface. Solving the beam shaping problem means finding a function S, which satisfies (1.20). For the general case, nonlinear optimization techniques are not suitable for finding a solution, since this function has many local minima and convergence is thus a main problem. Alternatively, one could use simulated annealing techniques, which are more likely to converge, but the speed of convergence is extremely slow. Therefore, we have applied a local optimization technique, which provided approximate solutions in typically less than eight iterations.

1.5 Examples

In the following two examples, we first consider a binary pattern and then a gray valued pattern as the design intensity. For the binary example, we chose a Gaussian intensity distribution in the entrance plane with $\sigma = 0.4$ as shown in Fig. 1.3.

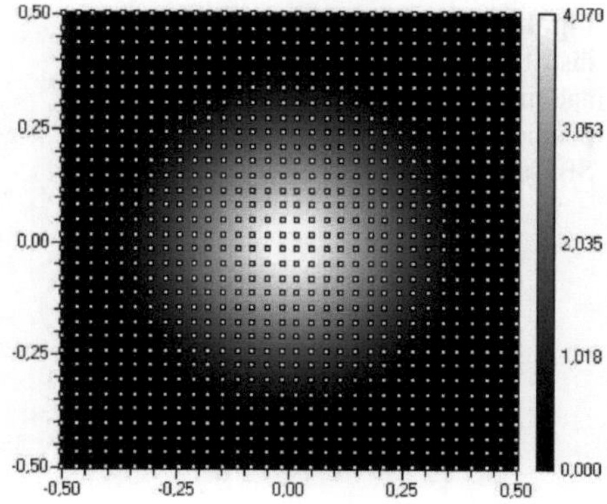

Fig. 1.3 Gaussian input distribution and marked reference positions of the SBF-approximation

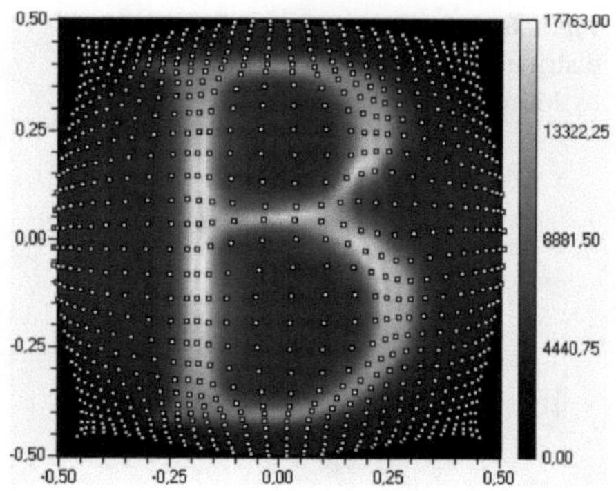

Fig. 1.4 Intensity obtained by Monte Carlo ray trace and mapped positions of the SBF-approximation

The figure also indicates the locations of the 32×32 base points chosen for the SBF-approximation. In Fig. 1.4, we show the intensity in the plane $z = D$, obtained by first calculating the optical surface and then applying a Monte Carlo ray trace using rays, which are distributed in the x–y plane according to the Gaussian distribution assumed in Fig. 1.3. The ray direction for all rays was assumed parallel to the optical axis. Then the law of refraction at the surface was applied to the rays and they were propagated to $z = D$. By accumulating approximately five million rays, the intensity image in Fig. 1.4 was obtained. We also show in Fig. 1.4, how the original base points are mapped to the output plane.

Because the Gaussian is very concentrated, the algorithm has to spread out the light in the center, concentrating some rays to those locations, where the character B has to be formed. In Fig. 1.5, we chose a wider Gaussian intensity distribution in the entrance plane with $\sigma = 0.8$ together with the locations of the 32×32 base points chosen for the SBF-approximation. The gray level picture in Fig. 1.6 again shows the Monte Carlo ray trace intensity obtained from this surface.

The picture in Fig. 1.6 shows a photograph of Adolf W. Lohmann, taken at the occasion of receiving the Emmett N. Leith medal in October 2008. The example clearly demonstrates that with this design technique, arbitrary intensity distributions

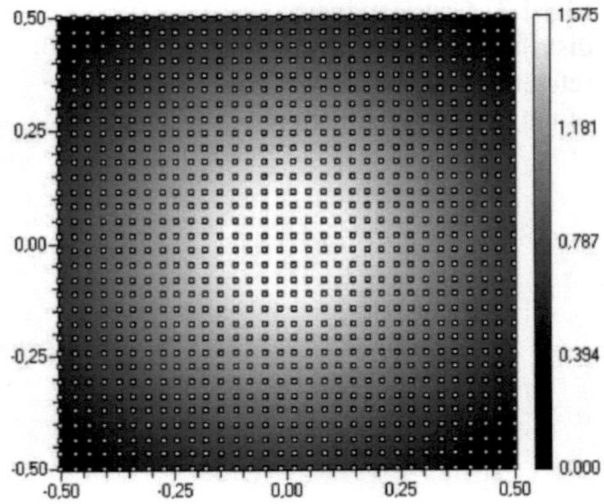

Fig. 1.5 Gaussian input distribution with sigma 0.8 and marked reference positions for the SBF-approximation

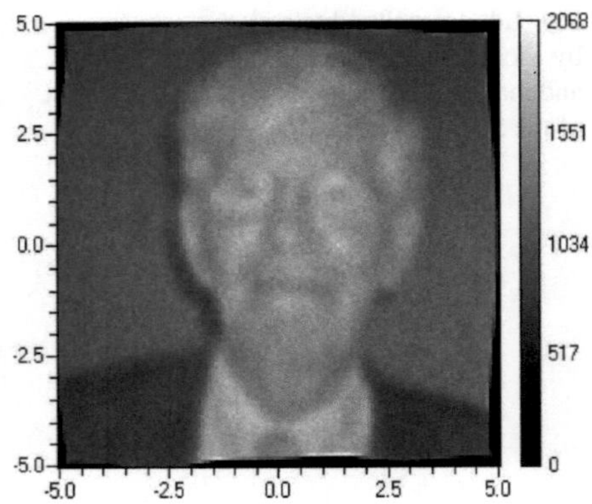

Fig. 1.6 Intensity distribution obtained by Monte Carlo ray trace

can be formed from any directed illumination. The quality of the images can be improved by increasing the number of base points. With 32×32 base points, the calculation time for both examples was in the order of 5 min.

1.6 Conclusion

We have presented an exact relationship between the coordinate mapping and the gradient of the surfaces. This relationship also has applications in generating a given coordinate mapping with optical components. We have shown that the solution can be applied to both, reflective and refractive beam shaping with two surfaces. As model function for the surfaces, we used with the shifted-base-function (SBF) approach. This approach provides a new and efficient method for solving the general two-dimensional beam shaping problem. The law of power conservation leads to a nonlinear Monge–Ampere type equation, which was solved in good approximation numerically.

References

1. L. A. Romero and F. Dickey, "Lossless laser beam shaping," J. Opt. Soc. Am. A 13, 751–760 (1996).
2. J. Hoffnagle and C. Jefferson, "Design and performance of a refractive optical system that converts a Gaussian to a flattop beam," Appl. Opt. 39, 5488–5499 (2000).
3. F. Dickey and S. Holswade, Laser Beam Shaping: Theory and Techniques (Marcel Dekker, New York, 2000).
4. J. Hoffnagle and C. Jefferson, "Beam shaping with a plano-aspheric lens pair," Opt. Eng. 42, 3090–3099 (2003).
5. D. Shealy and S. Chao, "Design and analysis of an elliptical Gaussian laser beam shaping system," Proc. SPIE 4443, 24–35 (2001).
6. D. Shealy and S. Chao, "Geometric optics-based design of laser beam shapers," Opt. Eng. 42, 1–16 (2003).
7. D. Gilbarg and N. Trudinger, Elliptic Partial Differential Equations of Second Order, second ed., Grundlehren Math.Wiss. [Fund. Princ. Math. Sci.], Vol. **224** (Springer-Verlag, Berlin, 1983).
8. J. Benamou and Y. Brenier, "A computational fluid mechanics solution to the Monge–Kantorovich mass transfer problem," Numer. Math. 84, 375–393 (2000).
9. E. J. Dean and R. Glowinski, Numerical analysis numerical solution of the two-dimensional elliptic Monge–Ampère equation with Dirichlet boundary conditions: an augmented Lagrangian approach, C. R. Acad. Sci. Paris, Ser. I 336, 779–784 (2003).
10. H. Ries and J. Muschaweck, "Tailored freeform optical surfaces," J. Opt. Soc. Am. A 19, 590–595 (2002).
11. V. I. Oliker and L. D. Prussner, "On the numerical solution of the equation $z_{xx}\, z_{yy} - z_{xy}^2 = f$ and its discretization. I," Numer. Math. 54, 271–293 (1988).
12. V. I. Oliker and L. Prussner, "A new technique for synthesis of offset dual reflector systems," In: 1994 Conference Proceeding, 10th Annual Review of Progress in Applied Computational Electromagnetics (Applied Computational Electromagnetics Society), Vol. I, pp. 45–52 (1994).
13. V. Oliker, "Mathematical aspects of design of beam shaping surfaces in geometrical optics," In: Trends in Nonlinear Analysis, M. Kirkilionis, S. Krömker, R. Rannacher, and F. Tomi, eds., pp. 191–222 (Springer-Verlag, Berlin, 2002).
14. T. Glimm and V. Oliker, "Optical design of two-reflector systems, the Monge–Kantorovich mass transfer problem and Fermat's principle," Indiana Univ. Math. J. 53, 1255–1278 (2004).
15. V. Oliker, "Optical design of two-mirror beam-shaping systems. Convex and non-convex solutions for symmetric and non-symmetric data," Proc. SPIE 5876, 203–214 (2005).
16. K.-H. Brenner "Shifted base functions: an efficient and versatile new tool in optics," J. Phys. Conf. Ser. 139, 012002, 11 pp, Workshop on Information Optics (WIO'08), 1–5August 2008 in Annecy (2008).
17. O. Bryngdahl, "Geometrical transformations in optics," J. Opt. Soc. Am. 64, 1092–1099 (1974).
18. P. Scott and W. Southwell, "Reflective optics for irradiance redistribution of laser beams: design," Appl. Opt. 20, 1606–1610 (1981).
19. P. H. Malyak, "Two-mirror unobscured optical system for reshaping irradiance distribution of a laser beam," Appl. Opt. 31 4377–4383, (1992).
20. K. Nemoto, T. Fujii, N. Goto, H. Takino, T. Kobayashi, N. Shibata, K. Yamamura, and Y. Mori, "Laser beam intensity profile transformation with a fabricated mirror," Appl. Opt. 36, 551–557 (1997).
21. D. L. Shealy, J. A. Hoffnagle, and K.-H. Brenner "Analytic beam shaping for flattened output irradiance profile," SPIE Proceedings, Vol. **6290**, No. 01, Annual Meeting of SPIE, Laser Beam Shaping Conference VII, San Diego (13–14 August 2006).
22. R. T. Frankot and R. Chellappa, "A method for enforcing integrability in shape from shading algorithms," IEEE Trans. Pattern. Anal. Mach. Intell. 10(4), 439–451 (1988).

Chapter 2
Nanophotonics for Information Systems

Y. Fainman, D.T.H. Tan, K. Ikeda, and M. Abashin

Abstract The field of photonics finds applications in information technology, health care, lighting, and sensing. This chapter explores the role of nanotechnology with focus on nanophotonics in dielectric and inhomogeneous metamaterials for optical communications, computing, and information and signal processing.

2.1 Introduction

Optics has the potential to solve some of the most exciting problems in communication and computing hardware. It promises crosstalk-free interconnects with essentially unlimited bandwidth, long-distance data transmission without skew and without power- and time-consuming regeneration, miniaturization, parallelism, and efficient implementation of important algorithms such as Fourier transforms. In the past, when the speed of digital computers was able to support only relatively small information processing throughput, the optical information processing techniques were developed and used to construct processors and systems in support of numerous applications that required high throughput for real time operation. These methods exploited the parallelism of optics supported by the richness of the modal continuum of free space and a variety of optoelectronic devices that were developed in support of these applications and systems. The constructed information processing systems and concepts were used for image processing [3–6], pattern recognition [7, 8], neural networks [9], and linear algebra calculus [10] to name a few. However with rapid advancements of the speed and, therefore, the information processing throughput of digital computers, optical signal processing systems were not able to support these applications in a broad sense due to high cost, lesser accuracy, and lack of user-friendly interfaces. Later, the optical information processing

Y. Fainman (✉)
Department of Electrical and Computer Engineering, University of California, San Diego, 9500 Gilman Drive, La Jolla, CA 92093-0407, USA
e-mail: fainman@ece.ucsd.edu

T. Fournel and B. Javidi (eds.), *Information Optics and Photonics: Algorithms, Systems, and Applications*, DOI 10.1007/978-1-4419-7380-1_2,

transformed from parallelism in space domain to parallelism in optical frequency domain in support of processing information carried by ultrashort optical pulses in the femtosecond range. Such waveforms vary too rapidly for even the fastest photodetectors to resolve, leading to the need to develop optical information processing methods. Time-domain and spectral-domain processors utilized linear [11] and nonlinear [12] processes and found useful applications for ultrafast waveform synthesis, detection and processing [12–14].

It is evident that optical processing in space and time has so far failed to move out of the lab. The free-space and guided-wave devices are costly, bulky, and fragile in their alignment. They are also difficult to integrate with electronic systems, both in terms of the fabrication process and in terms of delivery and retrieval of the massive volumes of data the optical elements can process. However, with the most recent emphases on construction of *chip-scale integration* using advanced lithographic tools employed in surrounding electronics, things may be changing. Experts predict lithographic resolution as fine as 16 nm by 2013 [65] which is about 100 times smaller than the telecommunication wavelength of 1,550 nm. These techniques can be used to create deeply subwavelength features that act as a *metamaterial* with optical properties controlled by the *density* and *geometry* of the pattern and its *constituent materials.* In this chapter, we focus on metamaterials composed primarily of dielectric materials that are engineered on the nanometer scale so as to have emergent optical properties not otherwise present. The increased localization of the optical field as a result of these engineered materials brings about phenomena such as form-birefringence, structural dispersion, and enhanced optical nonlinear interactions. Equivalently, characteristics such as the local polarizability of the metamaterial and the dispersion may be controlled by the geometry, properties of constituent materials, and their composition. The introduction of periodicity in these engineered materials result in a modification in the dispersion relation that can be used to create an artificial bandgap [15–17]. The manipulation and modification of this periodicity allows the bandgap to shift and parts of the bandgap to be accessed by propagating modes. Photonic crystal (PhC) waveguides rely on this exact concept – a line of defects is introduced into the otherwise periodic structure so as to guide light [18–20]. The confinement of light within the PhC lattice is also used to realize devices such as super collimators, super prisms, super lenses, omnidirectional filters, modulators, and lasers through proper design and optimization of Bloch modes [21–30].

Similar to the PhC in its periodic nature is a class of metamaterials which exploits the advantages of both continuous free space and discrete guided wave modes. The simplest example involves the propagation of light in a waveguiding slab, where confinement occurs only in the vertical direction; the free space propagation occurs in the plane of the slab. This configuration, aptly termed "free space optics on a chip" (FSOC) enables interaction with discrete optical components which are located along the propagation direction. Functionalizing devices for such integrated systems would require free space implementations of focusing, beam steering, and wavelength selectivity [31]. Realization of these functionalities can exploit periodic, quasi-periodic, or even random nanostructured composites. By altering at nanometer scales, the surface morphology of a dielectric using nanolithography and advanced etching techniques, we can realize these complex structures and control the material's local polarizability. As we shall see in this chapter, these structures fall into the deeply sub-wavelength regime with spatial features $\ll \lambda/2n$ and require metamaterials engineering with very high spatial resolution.

The engineering of composite dielectrics can continue to larger scales creating metamaterials that involve feature sizes on the sub-wavelength scale, e.g., just $<\lambda/2n$. The common themes of periodicity/quasi-periodicity and enhanced effects due to light confinement of guided modes will still remain. Continuing the simplification of PhC with periodicity in two dimensions that has been used for planar confinement, an alternative system whereby light confinement in 2D is achieved by total internal reflection and periodicity in 1D is introduced to create a bandgap in the third dimension. One method of implementation involves using a channel waveguide instead of a slab waveguide in the FSOC case for guiding light, and periodically modulating the effective index of the channel waveguide along the propagation direction. This results in a periodic photonic nano-wire akin to a 1D PhC. Thereafter, interesting properties may be engineered by once again, introducing defects in the periodic structure to access forbidden modes or slightly changing the periodicity to alter the effective bandgap and dispersive properties. The strong confinement of fields in these engineered nano-wire structures will also enable exploring interaction with other overlapping fields or discrete components despite the highly guided nature of these modes.

Silicon on insulator (SOI) materials will be used for most of the discussion in this chapter because SOI is compatible with the well established CMOS fabrication process. In addition, a large index difference between silicon and its oxide exists, which leads to highly confined modes and enables the miniaturization of on-chip silicon based photonic circuits. Furthermore, silicon is optically transparent and has a very low material absorption coefficient at the wavelengths region around 1.55 μm which is also used for telecommunications. Waveguiding loss in SOI platforms has a state of the art value of less than 1 dB/cm. In terms of the impact for future systems applications, it is evident that next generation computing would benefit greatly from all-optical data transfer and processing on a chip. Electrical interconnections inherent in today's computing cannot measure up in terms of both speed and bandwidth. Researchers in the field are aware of the need to bypass any sort of electro-optic process in order to take computing speeds to the next level. Much work is being done in creating both passive and active devices in SOI. Discrete device components such as filters, modulators, and resonators have been demonstrated. Active devices utilizing Raman gain and hybrid silicon lasers which achieve gain from a bonded III–V material have been demonstrated. The momentum of research in this area is the best evidence that silicon photonics is set to revolutionize the field of computing and communications.

In this chapter, we will divide the analysis of the dielectric metamaterials into two categories, namely those in the deeply sub-wavelength scale and the sub-wavelength scale. As we shall see, interesting material emergent properties arise when the materials are engineered to sizes smaller than or comparable to the wavelength of light in the said medium. Optical field concentrators, compact sources, polarizers, chromatic dispersers, diffractive structures, and other optical processing devices can now be implemented on-chip using metamaterials wherever natural materials with similar properties either do not exist, or (more frequently) would not be compatible with lithographic fabrication processing. Moreover, there exists an opportunity to develop a new family of optical devices exploiting near field interactions in a much broader sense than that which has been done to date. In summary, there exists an opportunity in using advanced *lithography* and materials composition geometry to "lithographically write and assemble" optical materials and devices with *novel optical functionalities* into circuits and subsystems on a chip.

2.2 Nanophotonics Process

To advance this technology, investigations of nanostructures and their interaction with electromagnetic field are critical. Engineers also need appropriate modeling and design tools, new fabrication recipes, and test instruments capable of characterizing on-chip components (see Fig. 2.1). The nanophotonics process will help to establish near field optical systems science and underlying technologies to advance future integrated information systems.

The design of integrated photonic systems is a challenging task as it not only involves the accurate solution of electromagnetic equations, but also the need to incorporate the material and quantum physics equations to enable the investigation and analysis of near field interactions. These studies need to be integrated with device fabrication and characterization to verify device concepts and optimize device designs. In this chapter we discuss examples of the SOI metamaterials and devices that can be realized using CMOS-compatible fabrication process which we demonstrated recently in our lab. These examples include birefringent elements that utilize a combination of geometry and material properties to separate light into orthogonal polarizations, graded-index lenses, frequency-selective resonators and Bragg gratings, and metal-dielectric nanostructures that can achieve extremely tight field confinement. Some of these example devices are tested using our near field characterization tool, the heterodyne near-field scanning optical microscope (H-NSOM).

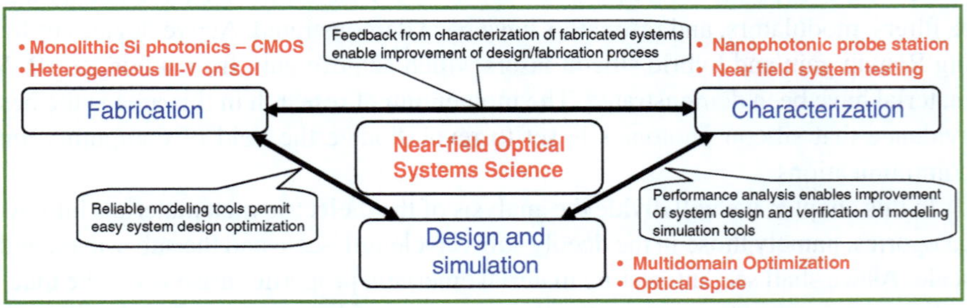

Fig. 2.1 Nanophotonics process

This microscope uses a fiber probe tapered to about 200–500 nm diameter with an aperture of about 50–200 nm which is brought close enough to the nanostructure under test to pick up its evanescent electromagnetic fields. The idea to improve resolution of optical measurements by bringing a subwavelength aperture close to the object of interest was first introduced by Synge in 1928 and experimentally realized only in 1983 by two independently working research groups: Dieter W. Pohl and his colleagues working in IBM laboratories [32] and Aaron Lewis with colleagues in Cornell University [33]. The efficiency of light transmission through a small aperture ($d \ll \lambda$) rapidly drops as the aperture size decreases: $T \sim (d/\lambda)^4$ [33], thus for realistic aperture of 100 nm and visible wavelengths the transmission efficiency only reach 10^{-6} to 10^{-8} [34]. Such small transmission coefficients demands using powerful optical sources (often lasers), efficient detection schemes and detectors. Heterodyne detection, as one of the examples, an interferometric technique which not only improves the detection efficiency but also allows measurement of optical phase.

In addition, heterodyne setup is ideal for studying pulses propagation in the nanophotonic circuits [1], which is of particular importance for information systems. It was also shown that dynamic response (e.g., pulse propagation velocity) can be measured as well by using continuous but spectrally broad optical sources [2].

The concept of heterodyne detection is to mix the signal of interest with a coherent reference beam possessing slightly shifted optical frequency. This can be done by implementing Mach–Zehnder interferometer scheme with one arm (signal arm) including NSOM and the other arm (reference arm) providing a frequency-shifted reference. The two fields are added coherently, yielding the desired interference signal of interest oscillating at the heterodyne frequency. The coherent gain of the heterodyne detection significantly improves the sensitivity of the instrument. An example of the H-NSOM [35] arrangement is shown on the Fig. 2.2. The system is composed of the readily available telecom components.

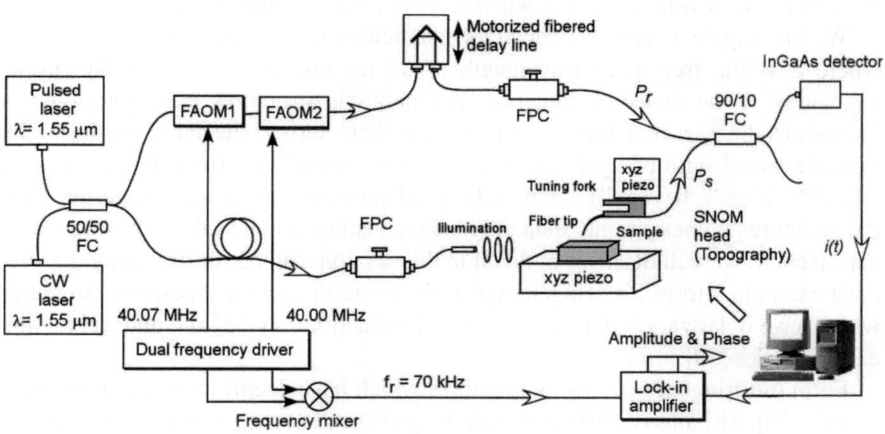

Fig. 2.2 Schematic diagram of the H-NSOM setup with the collection system for effect of the tip analysis [35]

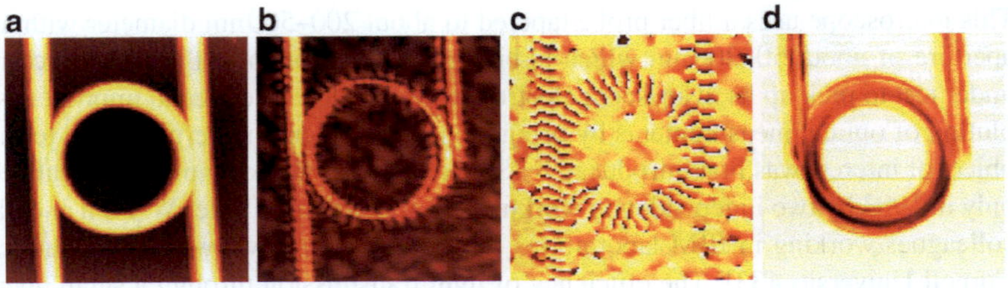

Fig. 2.3 Four images obtained simultaneously: (**a**) topography detected with the AFM, (**b**) near field amplitude, (**c**) near field optical phase, and (**d**) measurement of the effect of the tip at the output [66]

This in-fiber realization provides better interferometric stability; polarization maintaining fibers can also be used for maximizing interference term. The scanning process provides simultaneously three images: sample topography, deducted from the AFM feedback system which keeps the probe at the constant height above the sample surface; amplitude and phase distributions of the evanescent optical fields. Example of H-NSOM characterization of microring resonator with 10 μm diameter is given in Fig. 2.3. The mapping of evanescent fields has proven to be a powerful aid in understanding the performance of nano-scale optical materials, devices and circuits.

2.3 Dielectric Metamaterials

In this section, we will describe the analysis of the deeply sub-wavelength scale dielectric metamaterials and also their interesting emergent behavior in the sub-wavelength scale limit as well as when they are perturbed into aperiodic composites.

We investigate a class of dielectrics, characterized by feature sizes $\delta \ll \lambda/2n$, where λ is the free-space wavelength of device operation and n is the refractive index of the dielectric material. The photonic structures having periodic or quasi-periodic refractive index variation with characteristic distances much smaller than the wavelength of light can be called "metamaterials" (from the Greek word "μετά"="after," "beyond") – materials that determine their optical properties from the structure geometry rather than only from constituent material composition. This approach, as we will discover in detail in this section, can be illustrated by the simplest example – form-birefringent materials, a one-dimensional periodic structures which have polarization-dependent index of refraction [36–38] and unusual nonlinear properties [39].

Form birefringence occurs in structures which have deeply subwavelength periodicity [20]. The altered surface morphology of the dielectric used to construct such structures results in a large difference between the effective indices of the TE and TM polarized optical fields, since they need to satisfy different boundary conditions.

Form-birefringent nanostructures (FBN) are advantageous to naturally birefringent materials in that (1) the strength of birefringence, $\Delta n/n$ (where Δn and n are the difference and average refractive indices for the two orthogonal polarizations) is larger in the former; (2) the extent of form birefringence, Δn may be adjusted by varying the duty ratio as well as the shape of the microstructures; and (3) FBNs may be used to modify the reflection properties of the dielectric boundaries [40]. These features are useful for constructing polarization-selective beam splitters and general-purpose polarization-selective diffractive optical elements such as birefringent computer-generated holograms (BCGHs) [38]. Extending this concept to the 2D geometry or implementing aperiodicity enables other useful functionalities such as converting a linear polarization state to radial or azimuthal polarization [40, 41] and creating graded-index medium [41, 42]. It was also shown that the metamaterial approach can help to overcome fabrication difficulties and create Fresnel lens analogue using binary lithographic fabrication with deeply sub-wavelength feature size of less than 60 nm [31].

2.3.1 Inhomogeneous Dielectric Metamaterials with Space-Variant Polarizability

It is also possible to mold the light flow in the planar configuration using the metamaterial approach. Bringing functionality of the table-top optical information processing components to a chip will create compact devices, which can benefit from fast data transfer, small form-factor, parallel processing, and low power consumption. Implementing free-space-like propagation for planar optics means that while the light is confined by index difference in the vertical direction (chip plane), the horizontal beam size is regulated by the phenomena similar to that in 3D free-space optics such as diffraction, reflection and refraction. This can be seen as a direct and more natural transition of the conventional free space bulk optical components and devices to chip-scale photonic integrated circuits.

To create a dielectric metamaterial we use a subwavelength structure that can be fabricated in the high refractive index slab (see Fig. 2.4a). The slab has an index of refraction, n_1 and the gaps in the etched subwavelength structure can be filled with a material possessing a low index of refraction n_2 such as for example air with $n_2 = 1$. This slab structure is constructed on the cladding with an overall lower index of refraction $n_c < n_1$, to ensure confinement in the vertical direction. For some material systems such as SOI, the cladding with the guiding slab is located on the top of a thick substrate with the refractive index n_s. Consider a grating with a period, Λ with $F\Lambda$ being a fraction of the unit cell filled with high-index material. It can be shown that the second-order effective medium theory approximation [43, 44] is accurate for small grating periods $\Lambda < \lambda/n$ [20] and for grating thickness larger than $\lambda/3$ [45]. Other approaches in design and analysis of these subwavelength grating metamaterial structures include numerical methods such as RCWA, finite element method (FEM), and the finite-difference time-domain (FDTD) approach.

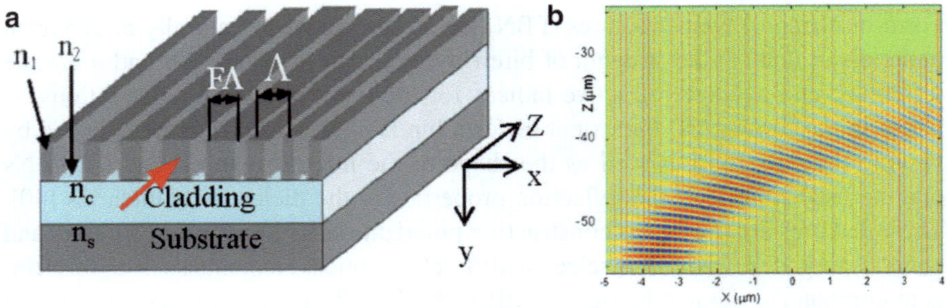

Fig. 2.4 Schematic diagram of a subwavelength planar metamaterial: (**a**) description of a deeply subwavelength approach to dielectric metamaterial structure; (**b**) Numeric simulation results showing light propagation in a subwavelength SOI gratings with spatial chirp introduced by linearly varying the filling factor (increasing from left to right) and initial periods of $\Lambda = 150\,\text{nm}$

This concept can be used for example in creating new materials with refractive indices different from those of the constituent materials. For example, the SOI material system includes silicon with index of refraction of $n_{\text{Si}} = 3.48$ and silicon dioxide with $n_{\text{SiO}_2} = 1.46$ as the only materials available for structure design. In the table-top free-space optics we, on the other hand, have a variety of materials such as different glasses, crystals, polymers, etc. This fact makes it difficult to directly transfer table-top optical setups for on-chip implementations. Metamaterials can provide a solution to overcome this difficulty. For example, by implementing the scheme of subwavelength gratings, the index achievable for SOI material system varies from 1.5 to 3.4, thus covering almost fully the range between high-index silicon and low-index oxide. This range was estimated for TE-polarized fields in structures with a period of $\Lambda = 400\,\text{nm}$ satisfying $\Lambda < \lambda/n = 1{,}500\,\text{nm}/3.5 \approx 400\,\text{nm}$ condition, with filling factors varying from 0.1 to 0.9 to comply with the state of the art nanofabrication capabilities (e.g., feature size ~40 nm). We examine subwavelength structures with a linearly varying filling factor, corresponding to introducing a linear spatial chirp. Such a slab will equivalently act as a graded index metamaterial media, where the effective index of refraction in the transverse direction decreases or increases linearly. It is well known that in such a "graded" index material, the incident beam of light will bend towards a higher index of refraction. We performed numerical simulations of light propagation in such spatially chirped subwavelength grating structures with initial periods of $\Lambda = 150\,\text{nm}$. The numeric simulation result for SOI material platform is shown in Fig. 2.4b.

2.3.2 Graded Index Structures

The simplest example of non-resonant metamaterials is a subwavelength periodic nanostructure that can dramatically affect the polarization of propagating light due to anisotropic polarizability of the medium, an effect known as form-birefringence

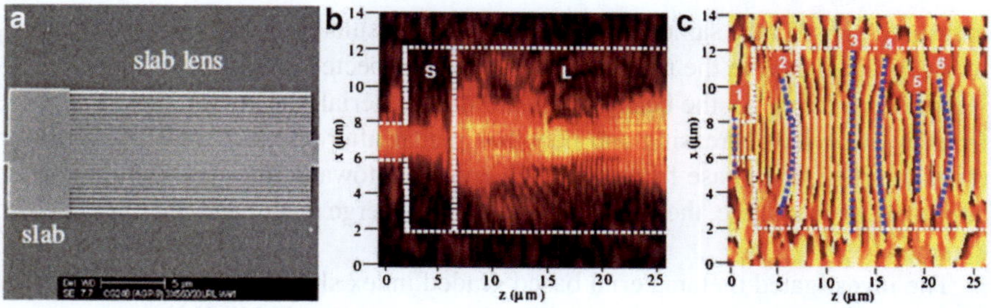

Fig. 2.5 Artificial dielectric slab lens. **(a)** Scanning electron micrograph. **(b)** Amplitude distribution. In-plane focusing effect visible inside lens. **(c)** Phase distribution with quadratic phase fronts

[20, 36, 38, 46], where the effective properties can be engineered through the choice of constituent materials and their geometry. For a specific eigen polarization, a mode matching device can be realized using the nonresonant metamaterial with space-variant polarizability to guide and manipulate the modes of light propagating on a chip. The device is realized by lithographically defining and etching subwavelength features into a high refractive index slab waveguide [31], thus modifying its local effective index of refraction. We use a sub-wavelength periodic structure and locally modulate its duty cycle in the transverse direction x, to achieve modulation of the index of refraction, i.e., $n(x) = n_0\left[1 - (1/2)\alpha x^2\right]$, where n_0 and α are constants representing the maximal effective index and the gradient strength, respectively. To validate our approach, we designed and fabricated a graded index slab element that focuses light into a 2-μm wide Si ridge waveguide. We used a SOI geometry with a Si slab thickness of 250 nm and an oxide thickness of 3 μm. The fabricated element in Fig. 2.5a shows the layout of the device. A grating period of 400 nm is chosen to assure the validity of the effective medium approximation on one hand, and to avoid the need for fabricating ultra small features. For compatibility with CMOS fabrication, the minimal air gap is chosen to be 100 nm, imposing a maximal duty cycle of 75%.

Typically, characterization of nanophotonic devices is performed by analyzing the light intensity measured at the output of the device. Unfortunately, this approach lacks the ability to probe the amplitude and, even more importantly, the phase profile of the optical beam as it propagates within the structure. To overcome this deficiency, the fabricated samples are characterized using our H-NSOM [35], capable of measuring both amplitude and phase of the propagating optical field with a resolution of about 100 nm. Figure 2.5b, c shows the measured amplitude and phase of the optical field propagating through the device at a wavelength of $\lambda = 1{,}550$ nm. Figure 2.5b shows the amplitude of the optical field in the region that includes the input waveguide, the non-patterned slab ("S") and large portion of the slab lens section ("L"). The dashed vertical white lines mark the boundaries between the various sections of the device. Light propagates from left to right. Figure 2.5c shows the measured phase in the same region. The obtained results clearly show the expanding of the optical beam in the region of the slab. As expected, the phase front is diverging in this section. As the beam enters the metamaterial, the curvature of the phase front is gradually decreasing and becomes planar after about 5 μm propagation in the slab. Then, the phase front begins to converge towards the focus. As the beam continues to propagate, the phase front starts to diverge again, and the optical beam expands.

The investigated metamaterial based graded index slab "lens" device is the first step towards the realization of the FSOC concept. Our H-NSOM measurements clearly demonstrate the focusing effect. This experimental demonstration opens new possibilities in the field of on chip integrated photonic devices, as the demonstrated component can be integrated with other building blocks (some of these yet to be developed) to create future devices and systems based on the concept of FSOC. We believe that this new concept may become essential for applications such as optical interconnections, information processing, spectroscopy and sensing on a chip.

2.4 Photonic Nano-wires: Sub-Wavelength Inhomogeneous Dielectrics

We refer to sub-wavelength structures as structures with features that are smaller but comparable to the wavelength of light. An example of such periodic structures is a general family of resonant structured materials such as Photonic Crystal (PhC) lattices and the whole family of devices that can be implemented in the PhC lattice. In this section we explore a novel practical approach that we call photonic nano-wire metamaterials which can be used to implement both sub-wavelength as well as deeply sub-wavelength nanostructures. It should be noted that sub-wavelength scale devices are characterized by feature sizes $\Lambda < \lambda/2n$, where λ is the vacuum wavelength of operation of the device, and n is the effective index of the specific mode in the device. The effective index of the mode in the material, $n_{\text{eff}} = \beta/k$, where β is the propagation constant of the waveguide mode and k is the wave number in vacuum. For example, for SOI materials system with silicon refractive index, $n_{\text{Si}} = 3.48$ and silicon oxide refractive index, $n_{\text{SiO}_2} = 1.46$, we can construct a typical single-mode silicon waveguide with n_{eff} of around 2.5 for operation at the telecommunications wavelength of 1.55 μm. In the following we will mainly focus on the sub-wavelength regime to demonstrate the unique capabilities of this approach and demonstrate experimentally example devices. We first investigate a periodic 1D PhC nano-wire, and extend the investigation to a quasi-periodic nano-wire. Finally, we study the characteristics of a filter created by coupling two such photonic nano-wires together.

2.4.1 Photonic Crystal-Based Resonant Device

Photonic crystal (PhC) lattice [15–17] is a well known resonant inhomogeneous material, which found a number of realizations during the past decade [21, 22, 37, 47]. Photonic crystal lattice may play a unique role as an integration platform of nanophotonic devices. Numerous devices with various functionalities have been utilized in a PhC lattice including modulators, detectors, filters, lasers, superprisms, and elements with negative refraction. However, fabrication of 3D resonant inhomogeneous materials and devices remains challenging, and frequently, 1D and 2D topologies are used due to ease in their design, fabrication, and integration.

A simple example of a 1D resonant structure is a distributed Bragg reflector (DBR) used in planar waveguide technology to perform and enhance various functionalities of optical elements such as a single-mode selector in semiconductor lasers, optical filters, switches, modulators, couplers, detectors, and sensors. The DBRs conventionally fabricated on the surface of the waveguide involves an additional fabrication procedure separate from the waveguide. Recently, in contrast to traditional approaches, we exploit single step fabrication method to define the DBRs and other nanostructured resonant devices using corrugation of waveguide sidewalls. In this approach, both the period and the modulation strength of the DBRs are lithographically assigned on the waveguide sidewalls [48, 49]. We term this class of devices as "vertical gratings" to distinguish them from conventional surface DBRs.

The designed devices are Fabry–Perot (FP) type filters made of a pair of identical Bragg reflectors each having reflection (r) and transmission (t) amplitude coefficients, and separated by a spacer of length ($d = \Lambda/2$) causing a phase shift ($\phi = \pi/2$), as shown in Fig. 2.6a. The transmission amplitude of the resonant filter (t_{RF}) as a function of λ is given by $t_{\mathrm{RF}} = [t^2 \exp(i\phi)]/1 - r^2 \exp(2i\phi)$, where t is transmittance of the Bragg structure. These filters are fabricated on a piece ($\sim 1 \times 0.5\,\mathrm{cm}^2$) of 6-inch SOI wafer consisting of a silicon top layer with a mean thickness of 252.2 nm with distribution of $6\sigma \sim 18.3$ nm on a ~3-μm-thick silicon dioxide layer. Figure 2.6b, c shows SEM micrographs of the fabricated device. For the testing process of the device, a linearly polarized tunable laser source is coupled into a polarization maintaining (PM) tapered fiber with an output mode field diameter of ~2.5 μm, producing ~20 dB polarization extinction. Another PM tapered fiber is used to collect light from the fabricated devices and its relative power transmission over the wavelength is analyzed. All measurements are performed using TE polarization. Figure 2.7 shows the measured transmission spectrum. The measured transmission spectrum of the fabricated device shows a wide stopband of ~19 nm bandwidth with a narrow transmission band with linewidth ($\Delta\lambda$) of ~0.5 nm in the center of the stop band. The developed fabrication procedures show very good surface quality which is also indicated by the measured cavity $Q \sim \lambda_{\mathrm{B}}/\Delta\lambda$ of about 3,000.

The described resonant photonic crystal based devices and components implemented using photonic nano-wires are essential for future integration of various information systems on a chip. A resonant nanophotonic device utilizing vertical gratings has been realized on SOI wafer, demonstrating fabrication procedures that

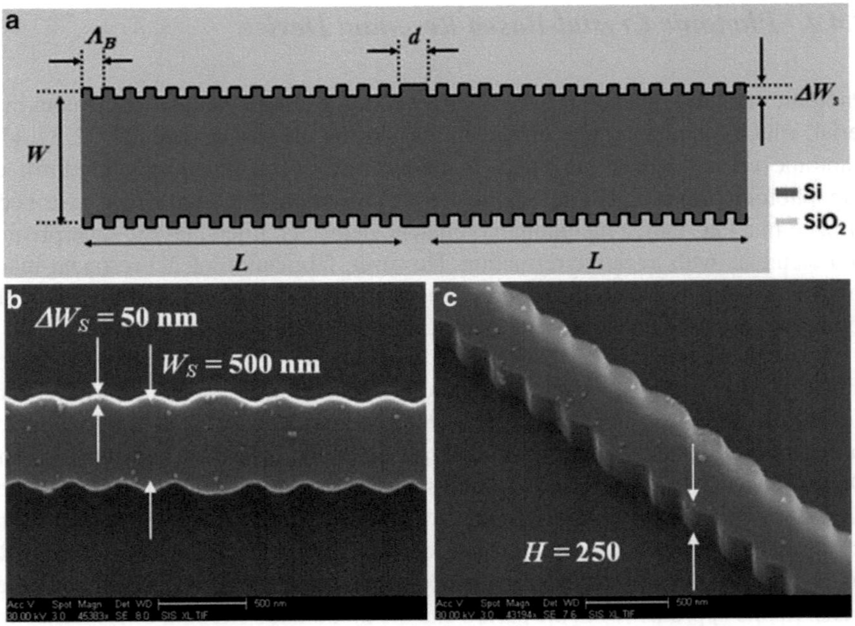

Fig. 2.6 (a) Top view of resonant transmission filter in sidewall modulated nano-wire, where W_S is the average width of the waveguide, and ΔW_S is the full depth of a single sidewall of the waveguide, L and Λ_B are the total length and the period of a single Bragg reflector, respectively. For SOI realization we use $d = \Lambda_B/2$, $\Lambda_B = 305$ nm, $W = 500$ nm and $\Delta W_S = 50$ nm; scanning electron micrographs of a dry etched silicon nano-wire filter device: (b) *top view*; (c) *tilted view*

Fig. 2.7 Measured spectra from the fabricated devices with $W_S = 500$ nm and $\Lambda = 305$ nm for $\Delta W_S = 50$ nm, $2 \times L = 70\,\mu$m [48,50]

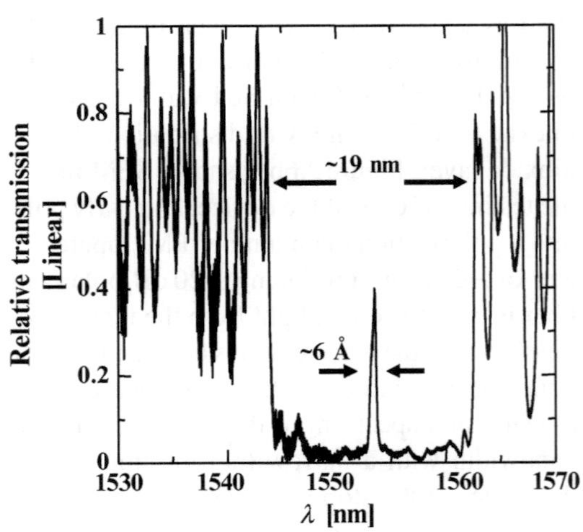

show very good surface quality. These experimental demonstrations open new possibilities in the field of on chip integrated photonic devices, as the demonstrated component can be integrated with other building blocks to create future devices and systems based on the concept of FSOC. Finally, the introduced approach on using lithography for side wall modulation opens up numerous opportunities in constructing various resonant devices for detection, modulation, generation, and manipulation of light on a chip.

2.4.2 Aperiodic Chirped Photonic Nano-Wires

The sidewall modulated photonic nano-wire gratings can be extended to engineer dispersion for chip-scale integration by introducing a chirp into its period. The device is similar to chirped fiber Bragg gratings (CFBG), used to compensate for group velocity dispersion (GVD) in fiber-optic applications. In this device, the sidewall modulating gratings are quasi-periodic, in contrast to the previous section where they had a fixed period [49]. Our GVD device uses the same SOI technology that was demonstrated in the previous section. A schematic diagram of the device is shown in Fig. 2.8. The refractive indices of Si and SiO_2 are assumed to be 3.48 and 1.46 respectively for the purposes of this study. The mean width, $W = 500$ nm and height, $H = 250$ nm of the waveguide result in an effective mode index, $n_{\text{eff}} = 2.63$ and the chosen Bragg wavelength, $\lambda_o = 1.55\,\mu$m. The corresponding mean Bragg grating period, $\Lambda_o = \lambda_o/(2n_{\text{eff}}) = 295$ nm. Next, a linear chirp is applied to the Bragg grating to introduce a quadratic phase across the optical spectrum centered at the resonant Bragg frequency. Consequently, a linear group delay will appear across the reflection band – this leads to the term, chirped Bragg grating (CBG). The CBG period at any point, z is given by $\Lambda(z) = \Lambda_o + (F/2\pi)\,(\Lambda_o^2/L)\,(2z/L)$, where F is the chirp parameter and L is the device length [51]. The total chirp in period for a fixed F is $\Delta\Lambda = (F/\pi)\,(\Lambda_o^2/L)$.

The large sidewall modulation of 50 nm on our CBG devices implies that in contrast to weakly coupled CFBGs in silica, they are operating in a strong coupling regime. Moreover, the relatively short length of strongly chirped device may not meet the slowly varying envelope approximation inherent in coupled mode theory (CMT). Therefore, we perform FDTD simulations to study their characteristics. The effective index of a slab of height, H infinite in the y-direction was used in place of the material refractive index of silicon to reduce the problem to two dimensions (2D). Although the reduction of the 3D problem to a 2D one is approximate, this

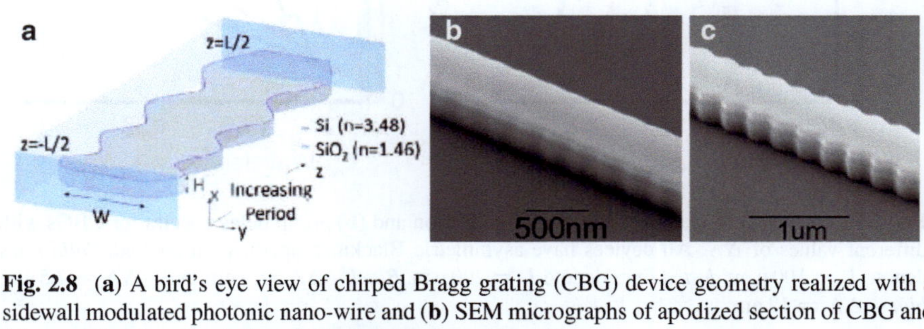

Fig. 2.8 (**a**) A bird's eye view of chirped Bragg grating (CBG) device geometry realized with a sidewall modulated photonic nano-wire and (**b**) SEM micrographs of apodized section of CBG and (**c**) unapodized section before SiO_2 cladding deposition [49]

method retains the salient spectral characteristics [52] and has shown good experimental agreement for similar structures [53].

A well known phenomenon which occurs in Bragg filters is the spectral ripple which arises from the abrupt start and end of the periodic structures in the filter. The windowing effect which leads to the degradation in the filter response may be alleviated by gradually increasing the amplitude of index modulation, and equivalently κ, from 0 at the edge to its maximum value at the center of the grating. Therefore, the effectiveness of different apodization filters [54, 55] in achieving a flat-top response and in-band ripple suppression is first investigated. While our study shows that various apodization filters are effective in suppressing group delay ripple, asymmetric Blackman apodization [54] is the most effective in maximizing the CBG bandwidth while maintaining a flat response. The apodization applied is asymmetric in that the coupling strength of the input half of the CBG increases from 0 at the CBG edge to its maximum value at the center of the CBG. The rear half of the CBG has a constant κ. We adopt asymmetric Blackman apodization for our CBGs for further studies of the effect of $\Delta\Lambda$ on the CBG devices. We expect that the group delay scales with the device length, and the group velocity dispersion (GVD), which is the derivative of the group delay with respect to wavelength relates inversely to $\Delta\Lambda$. Increasing $\Delta\Lambda$ leads to a larger range of periods and hence, larger bandwidth over which a fixed group delay is distributed. Therefore, the GVD decreases for increasing $\Delta\Lambda$. Referring to Fig. 2.9a, we observe an increase in bandwidth from 38 to 88 nm for $L = 100\,\mu\text{m}$ as $\Delta\Lambda$ increases from 4 to 12 nm, and a corresponding drop in the GVD (Fig. 2.9b) from 7.0×10^5 to 2.1×10^5ps/nm/km. In addition, longer device lengths are desirable to achieve reflectivities close to 100% even for

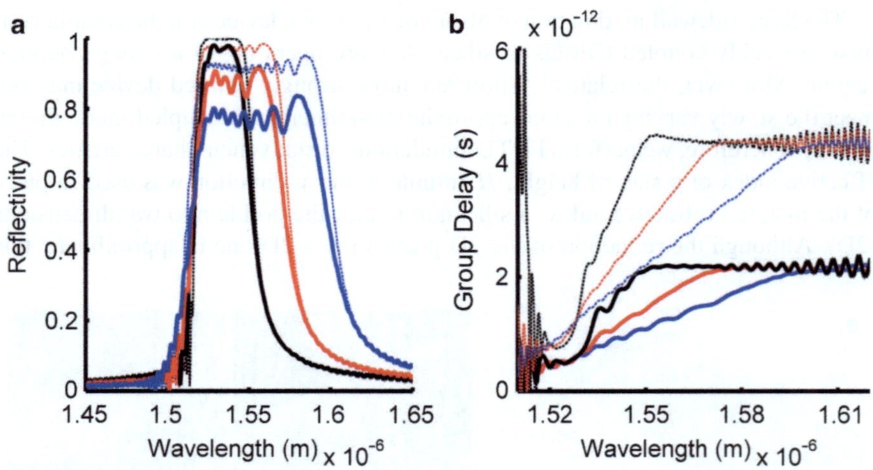

Fig. 2.9 2D FDTD modeling results for (**a**) reflection and (**b**) group delay spectra for CBGs with different values of $\Delta\Lambda$. All devices have asymmetric Blackman apodization applied. *Solid lines* denote $L = 100\,\mu\text{m}$; *dotted lines* denote $L = 200\,\mu\text{m}$. *Black* – $\Delta\Lambda = 4$ nm, *red* – $\Delta\Lambda = 7.5$ nm, *blue* – $\Delta\Lambda = 12$ nm

large values $\Delta\Lambda$. Referring to Fig. 2.9, doubling the CBG length to 200 μm results in higher reflectivity and significant ripple reduction in both the reflection and group delay spectra.

The designed CBG device was fabricated on a SOI wafer with a 250 nm layer of silicon on top of a 3 μm buried oxide layer on a silicon wafer. We use e-beam lithography followed by reactive ion etching. SiO_2 cladding was deposited over the fabricated Si structure using plasma-enhanced chemical vapor deposition (PECVD) SEM micrographs of fabricated structures before PECVD deposition of SiO_2 are shown in Fig. 2.8b, c.

We fabricated test samples with an access waveguide around 200 μm in length in front of the CBG structure. For the dispersion characterization, we use Fabry–Perot (FP) resonance oscillations in the measured reflection spectrum of the device. The resonator is defined by two reflectors, the cleaved input facet of the Si waveguide and the start of the CBG. We use a method similar to that reported in [56]. The length traversed by each wavelength component from the cleaved input facet to the point of reflection in the CBG may be experimentally determined using the relation, $l_c = \lambda^2 / [2n_g(\Delta\lambda_{FP})]$, where $\Delta\lambda_{FP}$ is the free spectral range of the FP resonator and the value of n_g is fixed at 4.2. The total group delay is subsequently found using $(2l_c \times n_g)/c = \lambda^2/[c(\Delta\lambda)]$, which is independent of the value of n_g. Finally, the GVD is obtained using the derivative of the group delay with respect to wavelength. Since the GVD of the access waveguide is two orders of magnitude lower than that of our device [57], the measured dispersion properties are dominated by these of our CBG devices. Typical experimental data of the reflection spectrum is compared with the numeric modeling result obtained for the same CBG geometry. The modeled result shows good agreement with the envelope and bandwidth of the measured spectrum.

Three fabricated CBG devices with $L = 100$ μm and $\Delta\Lambda = -4$, -7.5 and 12 nm were characterized in terms of their group delay (Fig. 2.10). The GVD is extracted from the slope of a linear fit applied to the experimental data. The expected

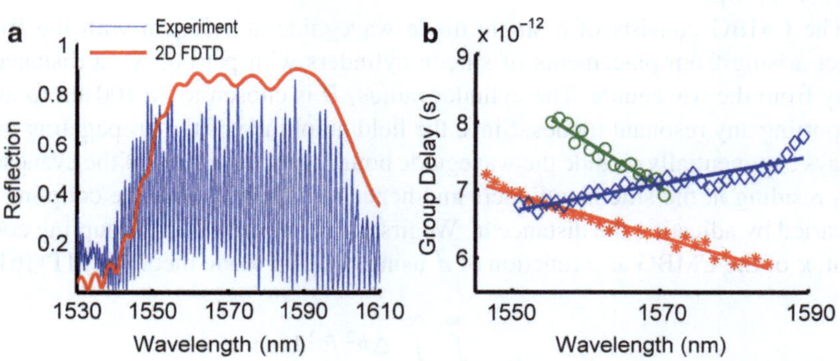

Fig. 2.10 (**a**) Measured reflection spectrum showing FP oscillations (*blue*) and 2D FDTD modeling result for $L = 100$ μm and $\Delta\Lambda = 7.5$ nm. (**b**) Measured group delay for CBGs with $L = 100$ μm for different values of $\Delta\Lambda$. *Solid lines* denote linear fit to measured data [49]

GVD values for $L = 100\,\mu\text{m}$, $\Delta\Lambda = -4$, -7.5 and 12 nm from the simulations are −0.070, −0.033 and 0.021 ps/nm, respectively; the measured GVD values are −0.067, −0.032 and 0.020 ps/nm, respectively, showing good agreement between the numeric and experimental results. Note that as expected, changing the sign of the chirp changes the sign of the dispersion (Fig. 2.10).

The CBG device provides a platform to engineer normal or anomalous dispersion for photonic systems applications including GVD compensation in silicon photonic structures on a chip. Asymmetric Blackman apodization was found to be most effective in suppressing group delay ripple while maximizing the useable bandwidth. The experiments demonstrated a dispersion value of $7.0 \times 10^5 \text{ps/nm/km}$ in a wide spectral range of about 40 nm in the near infrared spectral range of 1.55 μm. Both normal and anomalous dispersion has been demonstrated. The operating bandwidth may be adjusted and arbitrary amounts of dispersion achieved by adjusting the sign and magnitude of the chirp. The tunability of the CBG bandwidth makes it highly suitable for accommodating ultrashort pulses with high spectral content.

2.4.3 Cladding-Modulated Photonic Nano-wires

Periodic and aperiodic structures are fundamental in implementing various optical devices for switching, wavelength division multiplexing (WDM) and sensing applications [54, 58]. We have previously demonstrated on-chip sidewall modulated Bragg gratings (SMBG) realized by single-step lithography [49,59], in which the strength of the coupling coefficient is determined by the modulation amplitude which in turn is limited by the resolution of electron beams used to pattern these structures. Narrow bandwidth WDM components require Bragg gratings which have weak coefficients and which can be easily controlled. To this end, we introduce a cladding-modulated Bragg grating (CMBG) on SOI which overcomes the limitations inherent in SMBGs in fabricating devices with weak coupling (see Fig. 2.11) [60].

The CMBG consists of a single mode waveguide at 1.55 μm with the Bragg effect arising from placements of silicon cylinders with period, Λ_B a distance, d away from the waveguide. The cylinder radius, R is chosen to be 100 nm to avoid supporting any resonant modes. Since the field amplitude of the propagating mode decays exponentially outside the waveguide boundaries, the extent of the evanescent tails residing in the silicon cylinders and hence the strength of mode coupling can be varied by adjusting the distance d. We first calculate the CMBG coupling coefficient, κ of the CMBG as a function of d using coupled mode theory (CMT) [61]:

$$\kappa = \frac{k_o}{4n_\text{eff}} \frac{\int\limits_{-\infty}^{\infty}\int\limits_{-\infty}^{\infty} \Delta n^2 E^2 \text{d}x\text{d}y}{\int\limits_{-\infty}^{\infty}\int\limits_{-\infty}^{\infty} E^2 \text{d}x\text{d}y} \tag{2.1}$$

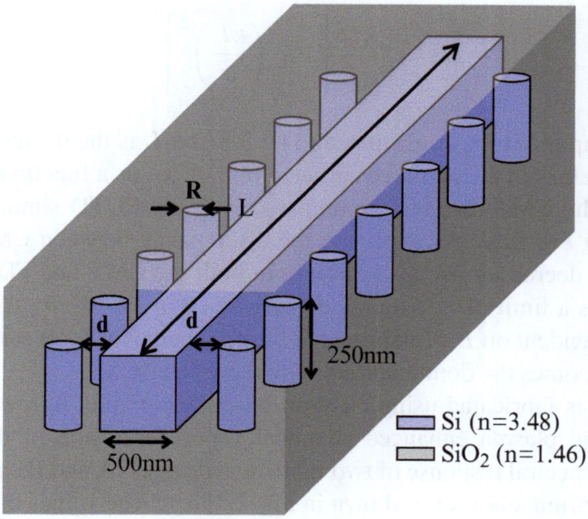

Fig. 2.11 Schematic diagram of the CMBG with R = 100 nm [60]

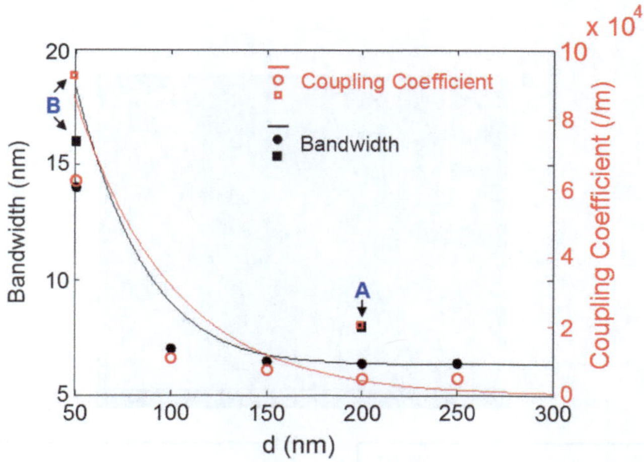

Fig. 2.12 κ and $\Delta\lambda$ calculated using CMT (*solid lines*) and 2D FDTD (*circles*). Parameters used: $L = 100\,\mu\text{m}$, $n_{\text{Si}} = 3.48$ and $n_{\text{SiO}_2} = 1.46$. Experimentally measured values are denoted with *squares*

The spatial distribution of the E-field and effective index is found using a fully vectorial beam propagation method. For small values of d, the coupling is large and CMT is no longer valid. The inverse logarithmic relationship between κ and d is apparent in Fig. 2.12 and is a direct result of the exponential nature of the decay in modal amplitude away from the core-cladding interface. From CMT, we expect that the bandwidth (defined here as the width between zeros of the central lobe) of the device increases as the coupling strength increases [62, 63]:

$$\Delta\lambda = \frac{\lambda^2}{n_g L}\left[1 + \left(\frac{\kappa L}{\pi}\right)^2\right]^{1/2} \tag{2.2}$$

where the group index, n_g is estimated to be 3.8 and L is the device length. Using (2.2) and the values of κ calculated earlier, a plot of $\Delta\lambda$ as a function of d is shown in Fig. 2.12. The CMT results are compared with 2D FDTD simulations marked with circles on Fig. 2.12. As expected, the discrepancy between CMT and FDTD increases as d decreases. We also observe in both the CMT and FDTD plots, that $\Delta\lambda$ approaches a limit of $\sim$ 5 nm as d increases. For $\kappa L \ll \pi$, $\Delta\lambda$ becomes increasingly dependent on L explaining the limiting behavior for weak coupling. For $\kappa L \gg \pi$, κ becomes the dominant factor on the value of $\Delta\lambda$.

The device is fabricated using electron-beam lithography followed by reactive ion etching and plasma enhanced chemical vapor deposition of the SiO_2 overcladding. The spectral response of two fabricated devices (A and B) measured using an optical spectrum analyzer is shown in Fig. 2.13b, c. The values of $\Delta\lambda$ and κ obtained experimentally are marked on Fig. 2.12. Even though $L = 70\,\mu$m for device B, $\kappa L \gg \pi$ and therefore κ is the dominant factor on $\Delta\lambda$. κ calculated from (2.2) should therefore be similar to that for $L = 100\,\mu$m. $\Delta\lambda$ measured for devices A

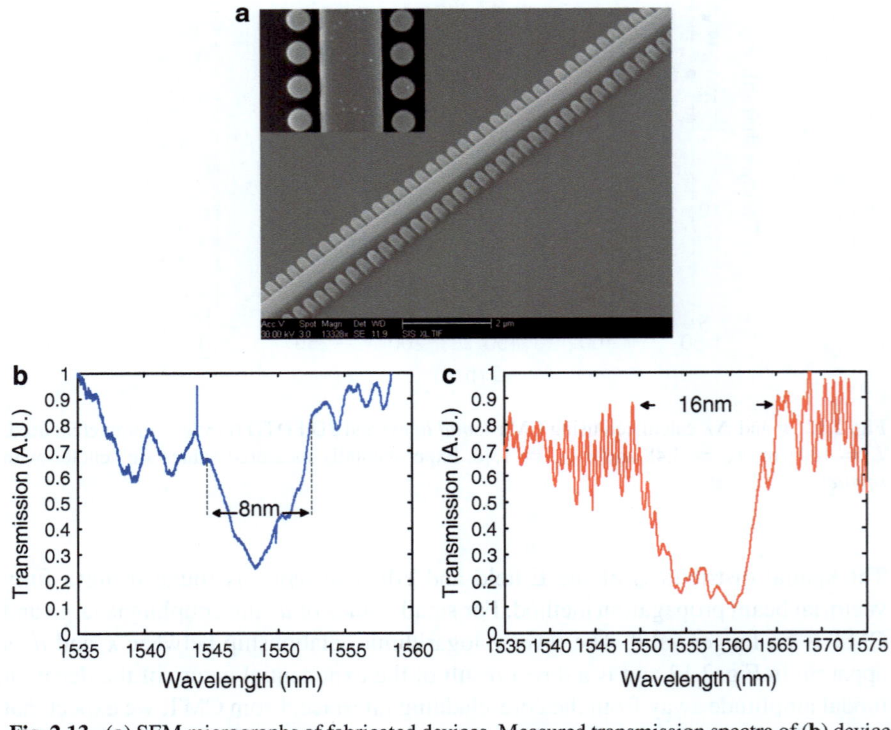

Fig. 2.13 (**a**) SEM micrographs of fabricated devices. Measured transmission spectra of (**b**) device A ($L = 100\,\mu$m, $d = 200$ nm) and (**c**) device B ($L = 70\,\mu$m, $d = 50$ nm)

and B are 8 and 16 nm respectively, close to the expected values of 7 and 13 nm respectively from 2D FDTD. The slightly higher value of $\Delta\lambda$ measured for device B may be attributed to the proximity effect during e-beam writing resulting in a value of d smaller than the target. For small values of d, small changes in d result in large changes in κ owing to the inverse exponential relation between κ and d observed in Fig. 2.12.

The demonstrated CMBG device may be useful for on-chip switching and add-drop filtering applications. By integrating several CMBGs with slightly different values of Λ_B, the device could be used to route data at different carrier frequencies (i.e., wavelengths). For example, smaller values of κ with sufficiently long L to achieve ~100% efficiency/extinction will allow more narrow band data channels to be accommodated for in dense WDM applications, whereas a large value of κ accommodates fewer channels but with much higher bandwidth. To circumvent the limiting behavior of the bandwidth, devices with small values of κ can be made longer such that $\kappa L \gg \pi$. For example, for $d = 200$ nm and $L = 2.3$ mm, $\kappa L = 10$ and $\Delta\lambda = 0.9$ nm from (2.2).

The designed, fabricated and experimentally validated novel CMBG using SOI material platform allows large dynamic range, resolution and precise control of the coupling strengths of the device. Devices with weak and strong coupling strengths differing by an order of magnitude were demonstrated experimentally confirming the numeric predictions. This method of creating Bragg gratings enables weak coupling coefficients to be realized to meet narrow bandwidth requirements in switching and add-drop filtering applications.

2.5 Nanophotonic Devices and Circuits: Wavelength Selective Add Drop Filter with Vertical Gratings on a Silicon Chip

As a further application of the sidewall modulated photonic nano-wire demonstrated in previous sections, we developed a wavelength selective add/drop filter. The device provides a viable platform for realizing next generation integrated WDM functionalities in order to bypass the aforementioned limitations in electro-optic conversions. The wavelength selective add/drop filter is ideally suited to optical interconnects, and will be implemented using two coupled vertical gratings on the same SOI platform used to study the other vertical grating devices.

Figure 2.14a shows a schematic diagram of the add/drop filter. Two vertical gratings with different widths, W_1 and W_2 are placed in parallel such that three Bragg conditions arise [64]: $2\beta_1 = 2\pi/\Lambda$ (backward coupling in waveguide 1 (WG1)), $2\beta_2 = 2\pi/\Lambda$ (backward coupling in WG2) and $\beta_1 + \beta_2 = 2\pi/\Lambda$ (cross coupling), where β_1 and β_2 are the propagation constants of WG1 and WG2 respectively, L is the device length and Λ is the grating period. A raised cosine apodization filter is applied to both gratings to minimize out of band ripple. Here, we allocate the cross coupling wavelength to $\lambda_B = 1.55\,\mu$m by setting the period as $\Lambda_B = 2\pi/[\beta_1(\lambda_B) + \beta_2(\lambda_B)]$, and backward coupling wavelengths in WG1 and

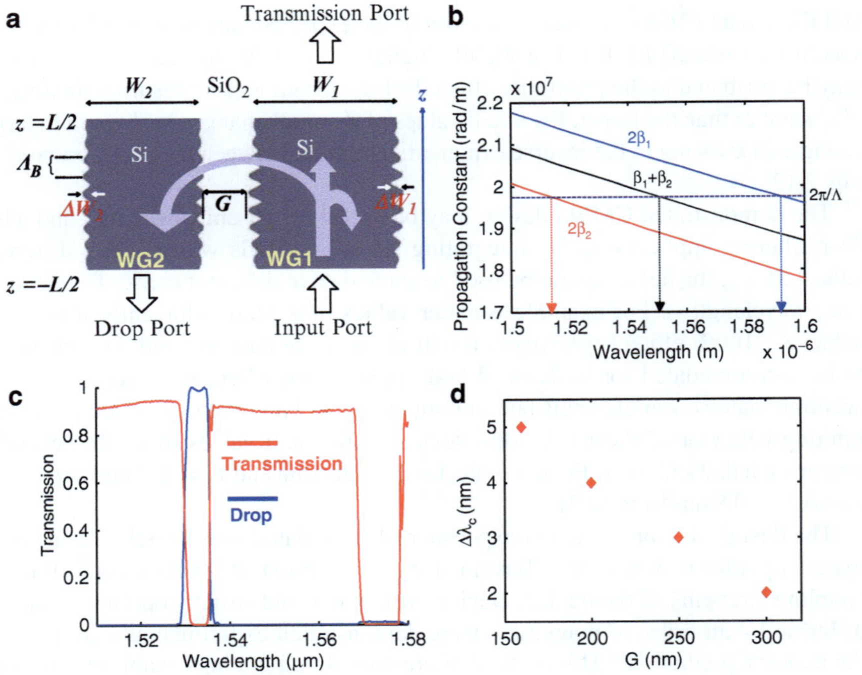

Fig. 2.14 (**a**) Schematic diagram of wavelength selective add-drop filter (*top view and cross-section*). (**b**) Propagation constants, $\beta_{1,2}$ calculated for TE polarized optical fields using a fully vectorial beam propagation method in 2D, assuming a silicon slab thickness of 250 nm and both sides of the silicon slab surrounded by silica cladding; the intersection between the solid lines and the dotted vertical lines correspond to the three Bragg conditions; (**c**) transmission and drop port spectral response for light incident at the input port for $L = 400\,\mu\text{m}$ and $G = 160\,\text{nm}$ and (**d**) plot showing inverse trend between $\Delta\lambda_c$ and G obtained using 2D FDTD

WG2 to be outside the C-band by setting $2\beta_1(\lambda > 1.57\,\mu\text{m}) = 2\pi/\Lambda_B$ and $2\beta_2(\lambda < 1.53\,\mu\text{m}) = 2\pi/\Lambda_B$. Figure 2.14b shows the dispersion plots of $2\beta_1$, $2\beta_2$, and $\beta_1 + \beta_2$, calculated using a fully vectorial beam propagation method for the quasi-TE mode. Based on the calculated propagation constants, we obtain $\Lambda_B = 316\,\text{nm}$ for $\lambda_B = 1.55\,\mu\text{m}$. An approximation to the full 3D problem is obtained by performing 2D FDTD simulations using the effective index method for $L = 400\,\mu\text{m}$ and $G = 160\,\text{nm}$, as shown in Fig. 2.14a. The obtained transmission and drop port spectra is summarized in Fig. 2.14c. The stop band centered close to $1.57\,\mu\text{m}$ in the transmission plot arises from the self coupling in WG1 whereas the stop band centered at $1.535\,\mu\text{m}$ is the cross coupling band which brings about the device's add/drop functionality. Another important feature of the device is the tunability of the cross coupling bandwidth, $\Delta\lambda_c$ using G. A larger G implies a smaller overlap integral between the forward mode in WG1 and the backward mode in WG2, and therefore a lower cross coupling coefficient, κ_c. According to coupled mode theory [59, 64] and (2.2), the coupling coefficient is inversely related to the stop-band of the resonant device. $\Delta\lambda_c$ obtained using 2D FDTD for several values of G is plotted in Fig. 2.14d, and confirms the inverse relation between $\Delta\lambda_c$ and G.

For device characterization, we fabricated several add/drop filters using an SOI wafer with 250 nm of Si on a 3 μm buried oxide layer on a Si substrate. E-beam lithography was performed followed by reactive ion etching and plasma-enhanced chemical vapor deposition of the top layer of SiO_2 cladding. Four add/drop filters with varying G from 60 to 200 nm were fabricated. The length of the fabricated add/drop filters was fixed at 400 μm for the three smallest values of G. All access waveguides are terminated in inverse tapers to improve coupling and to minimize Fabry–Perot oscillations. For the filter with the weakest coupling ($G = 200$ nm), the length had to be increased to 1 mm in order to obtain a sufficiently large extinction at the drop port. Figure 2.15a shows SEM micrographs at the center of the add/drop filter, where the sidewall modulation amplitude is at its maximum. The measured transmission spectra for the transmission and drop ports are shown in Fig. 2.15b. The drop port's bandwidth is also observed to decrease as G is increased from 60 to 200 nm. The inverse relation is shown in Fig. 2.15c, and shows that $\Delta\lambda_c$ decreases

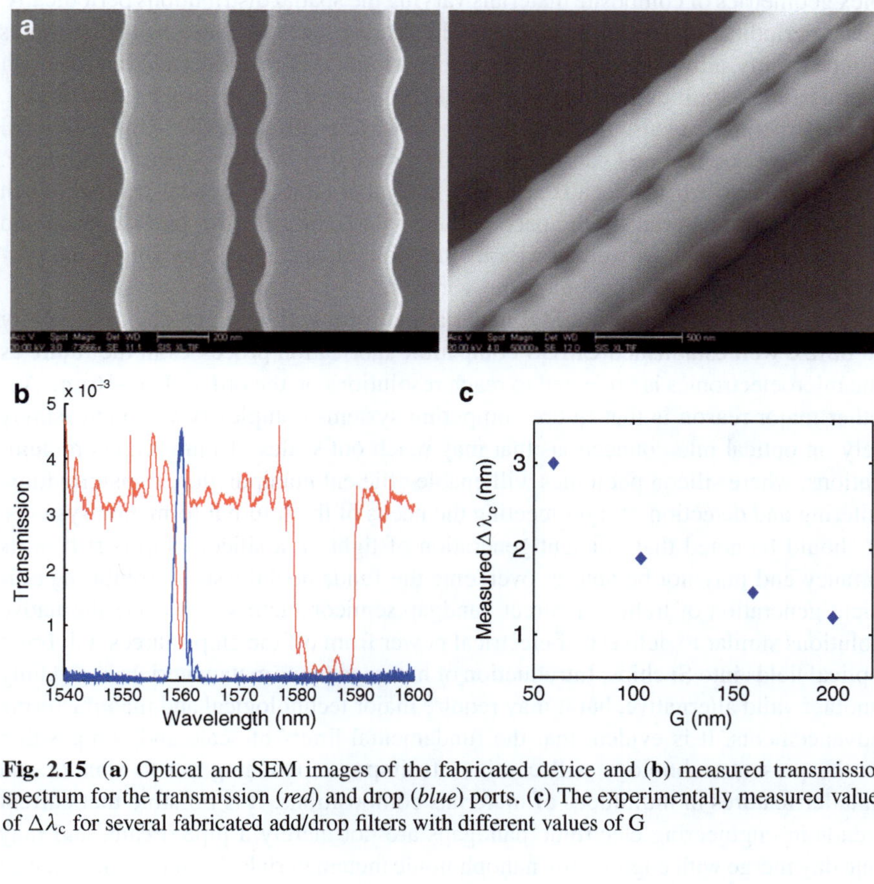

Fig. 2.15 (**a**) Optical and SEM images of the fabricated device and (**b**) measured transmission spectrum for the transmission (*red*) and drop (*blue*) ports. (**c**) The experimentally measured values of $\Delta\lambda_c$ for several fabricated add/drop filters with different values of G

from 3 nm in the most strongly coupled case of $G = 60$ and 1.2 nm in the device with weakest coupling ($G = 200$ nm).

To incorporate the add/drop filter as a building block of a WDM system, several such coupled vertical grating add/drop filters may be combined to multiplex different wavelengths on a chip. Several additional vertical gratings, each with a slightly different width and hence, different propagation constant and dispersion relation, may be coupled with WG1 in order to mutiplex/drop a different cross coupling wavelength. Consequently, the device will be able to serve as a WDM system to be integrated with next generation on-chip computing and communication networks.

2.6 Discussions and Future Perspectives

This chapter has given an introductory view on the emerging field of nanophotonics and its potential for chip-scale integration of information systems. The main points include our ability to use standard and novel nanofabrication tools to create complex geometries of composite materials varying the spatial distributions periodically, quasi-periodically or even at random. The feature sizes of these nanocomposites can be deeply subwavelength, which enables creation of materials and devices with unique functionalities that is not possible to achieve with existing technologies. These structures can also be created on a sub-wavelength scale where the individual features are coupled and can be operated in spatial as well as spectral resonance. We have described the localized effects and alterations in optical behavior such dielectric metamaterials with space variant polarizability (form birefringence and metamaterials lens), as well as various functionalities achieved by single and spatially coupled resonant photonic nano-wires.

The current trend of using SOI material platform will continue due to its ability to utilize well established CMOS compatible fabrication processes in the future as the microelectronics is projected to reach resolutions on the order of 10–16 nm. Another major reason is that future computing systems complexity will increasingly rely on optical interconnections that may reach out scales of intra-chip communications, where silicon photonics will enable efficient manipulation, transformation, filtering and detection of light meeting the needs of the future information systems. It should be noted that efficient generation of light on a silicon chip is still in its infancy and may not be able to overcome the fundamental issues prohibiting efficient generation of light in indirect bandgap semiconductors; however, alternative solutions similar to delivery of electrical power from off the chip sources will bring optical fields into Si chips. Introduction of heterogeneous metamaterials is certainly another valid alternative, but it may require major technological and manufacturing advancements. It is evident that the fundamental limits of scale and composition in the present technology will continue to improve, reaching out to smaller and smaller features in multitudes of materials compositions. It is possible that current trends in engineering electronic bandgaps are not merely a pipe dream, and may one day merge with engineering nanophotonic metamaterials, leading to new device concepts that will serve as the backbone of future information systems technologies. Engineering the polarizability of dielectric nanostructures is still in its nascent stages and numerous novel functionalities are still to come.

References

1. M. L. M. Balistreri, H. Gersen, J. P. Korterik, L. Kuipers and N. F. van Hulstdagger, "Tracking femtosecond laser pulses in space and time," *Science* 294, 1080–1082, 2001
2. M. Abashin, K. Ikeda, R. Saperstein and Y. Fainman, "Heterodyne near-field scanning optical microscopy with spectrally broad sources," *Opt. Lett.* 34, 1327–1329, 2009
3. Y. Fainman, E. Klancnik and S. H. Lee, "Optimal coherent image amplification by two beam coupling in photorefractive $BaTiO_3$," *Opt. Eng.* 25, 228–234, 1986
4. Y. Fainman, C. C. Guest and S. H. Lee, "Optical digital logic operations by two beam coupling in photorefractive material," *Appl. Opt.* 25, 1598–1603, 1986
5. P. Ambs, S. H. Lee, Q. Tian and Y. Fainman, "Optical implementation of the Hough transform by a matrix of holograms," *Appl. Opt.* 25, 4039–4045, 1986
6. P. Ambs, Y. Fainman, S. H. Lee and J. Gresser, "Computerized design and generation of space-variant holographic filters. Part I: System design considerations and applications of space-variant filters to image processing," *Appl. Opt.* 27, 4753–4760, 1988
7. Q. Tian, Y. Fainman, Z. H. Gu and S. H. Lee, "Comparison of statistical pattern recognition algorithms for hybrid processing. Part I: Linear mapping algorithms," *J. Opt. Soc. Am. A* 5, 1655–1669, 1988
8. Q. Tian, Y. Fainman and S. H. Lee, "Comparison of statistical pattern recognition algorithms for hybrid processing. Part II: Eigenvector-based algorithms," *J. Opt. Soc. Am. A* 5, 1670–1682, 1988
9. J. Y. Jau, Y. Fainman and S. H. Lee, "Comparison of artificial neural network with pattern recognition and signal processing," *Appl. Opt.* 28, 302–305, 1989
10. H. Rajbenbach, Y. Fainman and S. H. Lee, "Optical implementation of an iterative algorithm for matrix inversion," *Appl. Opt.* 26, 1024–1031, 1987
11. P. C. Sun, Y. Mazurenko, W. Chang, P. Yu and Y. Fainman, "All-optical parallel-to-serial conversion by holographic spatial-to-temporal frequency encoding," *Opt. Lett.* 20, 1728–1730, 1995
12. P. C. Sun, Y. Mazurenko and Y. Fainman, "Femtosecond pulse imaging: ultrafast optical oscilloscope," *J. Opt. Soc. Am. A* 14, 1159–1169, 1997
13. D. M. Marom, D. Panasenko, P. C. Sun and Y. Fainman, "Spatial-temporal wave mixing for space-time conversion," *Opt. Lett.* 24, 563–565, 1999
14. R. E. Saperstein, D. Panasenko and Y. Fainman, "Demonstration of a microwave spectrum analyzer based on time-domain optical processing in fiber," *Opt. Lett.* 29, 501–503, 2004
15. J. D. Joannopoulos, S. G. Johnson, J. N. Winn and R. D. Meade, "Photonic Crystals: Molding the Flow of Light," 2nd ed., Princeton University Press, Princeton, 2008
16. A E. Yablonovitch, "Inhibited spontaneous emission in solid-state physics and electronics," *Phys. Rev. Lett.* 58, 2059–2062, 1987
17. S. John, "Strong localization of photons in certain disordered dielectric superlattices," *Phys. Rev. Lett.* 58, 2486–2489, 1987
18. M. Abashin, P. Tortora, I. Marki, U. Levy, W. Nakagawa, L. Vaccaro, H. Herzig and Y. Fainman, "Near-field characterization of propagating optical modes in photonic crystal waveguides," *Opt. Express* 14, 1643–1657, 2006
19. R. D. Meade, A. Devenyi, J. D. Joannopoulos, O. L. Alerhand, D. A. Smith and K. Kash, "Novel applications of photonic band gap materials: Low-loss bends and high Q cavities," *J. Appl. Phys.* 75, 4753, 1994

20. I. Richter, P. C. Sun, F. Xu and Y. Fainman, "Design considerations of form birefringent microstructures," *Appl. Opt.* 34, 2421–2429, 1995
21. R. Tyan, P. C. Sun, A. Scherer and Y. Fainman, "Polarizing beam splitter based on the anosotropic spectral reflectivity characteristic of form-birefringent multilayer grating," *Opt. Lett.* 21, 761–763, 1996
22. W. Nakagawa, P. C. Sun, C. H. Chen and Y. Fainman, "Wide field of view narrow band spectral filter base on photonic crystal nanocavities," *Opt. Lett.* 27, 191–193, 2002
23. S. Lin, E. Chow, V. Hietala, P. R. Villeneuve and J. D. Joannopoulos, "Experimental demonstration of guiding and bending of electromagnetic waves in a photonic crystal," *Science* 282, 274–276, 1998
24. H. Kosaka, T. Kawashima, A. Tomita, M. Notomi, T. Tamamura, T. Sato and S. Kawakami, "Self-collimating phenomena in photonic crystals," *Appl. Phys. Lett.* 74, 1212, 1999
25. E. Schonbrun, Q. Wu, W. Park, T. Yamashita, C. J. Summers, M. Abashin and Y. Fainman, "Wave front evolution of negatively refracted waves in a photonic crystal," *Appl. Phys. Lett.* 90, 041113, 2007
26. H. Kosaka, T. Kawashima, A. Tomita, M. Notomi, T. Tamamura, T. Sato and S. Kawakami, "Superprism phenomena in photonic crystals," *Phys. Rev. B* 58, R10096, 1998
27. E. Schonbrun, M. Abashin, J. Blair, Q. Wu, W. Park, Y. Fainman and C. J. Summers, "Total internal reflection photonic crystal prism," *Opt. Express* 15, 8065–8075, 2007
28. C. Luo, S. G. Johnson and J. D. Joannapoulos, "Subwavelength imaging in photonic crystals," *Phys. Rev. B* 68, 045115, 2003
29. Y. Fink, J. N. Winn, S. Fan, C. Chen, J. Michel, J. D. Joannopoulos and E. L. Thomas, "A dielectric omnidirectional reflector," *Science* 282, 1679–1682, 1998
30. O. Painter, R. K. Lee, A. Scherer, A. Yariv, J. D. O'Brien, P. D. Dapkus and I. Kim, "Two-dimensional photonic band-gap defect mode laser," *Science* 284, 1819–1821, 1999
31. U. Levy, M. Abashin, K. Ikeda, A. Krishnamoorthy, J. Cunningham and Y. Fainman, "Inhomogenous dielectric metamaterials with space-variant polarizability," *Phys. Rev. Lett.* 98, 243901, 2007
32. D. W. Pohl, W. Denk and M. Lanz, "Optical stethoscopy: Image recording with resolution $\lambda/20$," *Appl. Phys. Lett.* 44, 651, 1984
33. A. Lewis, M. Isaacson, A. Harootunian, et al., "Development of a 500Å spatial resolution light microscope I. light is efficiently transmitted through $\lambda/16$ diameter apertures," *Ultramicroscopy* 13, 227, 1984
34. G. A. Valaskovic, M. Holton and G. H. Morrison, "Parameter control, characterization, and optimization in the fabrication of optical fiber near-field probes," *Appl. Opt.* 34, 1215–1228, 1995
35. A. Nesci and Y. Fainman, "Complex amplitude of an ultrashort pulse with femtosecond resolution in a waveguide using a coherent NSOM at 1550 nm," In Wave Optics and Photonic Devices for Optical Information Processing II, P. Ambs and F. R. Beyette, Jr., eds., Proc. SPIE 5181, 62–69, 2003
36. F. Xu, R. Tyan, P. C. Sun, C. Cheng, A. Scherer and Y. Fainman, "Fabrication, modeling, and characterization of form-birefringent nanostructures," *Opt. Lett.* 20, 2457–2459, 1995
37. R.-C. Tyan, A. Salvekar, C.-C. Cheng, A. Scherer, F. Xu, P. C. Sun and Y. Fainman, "Design, fabrication, and characterization of form-birefringent multilayer polarizing beam splitter," *J. Opt. Soc. Am. A* 14, 1627–1636, 1997
38. F. Xu, R.-C. Tyan, P.-C. Sun, Y. Fainman, C.-C. Cheng and A. Scherer, "Form-birefringent computer-generated holograms," *Opt. Lett.* 21(18), 1513, 1996
39. W. Nakagawa, R. Tyan and Y. Fainman, "Analysis of enhanced second-harmonic generation in periodic nanostructures using modified rigorous coupled-wave analysis in the undepleted-pump approximation," *J. Opt. Soc. Am. A* 19, 1919–1928, 2002
40. U. Levy, C. H. Tsai, L. Pang and Y. Fainman, "Engineering space-variant inhomogeneous media for polarization control," *Opt. Lett.* 29, 1718–1720, 2004
41. U. Levy, M. Nezhad, H.-C. Kim, C.-H. Tsai, L. Pang and Y. Fainman, "Implementation of a graded-index medium by use of subwavelength structures with graded fill factor," *J. Opt. Soc. Am. A* 22, 724–733, 2005

42. J. N. Mait, A. Scherer, O. Dial, D. W. Prather and X. Gao, "Diffractive lens fabricated with binary features less than 60 nm," *Opt. Lett.* 25, 381–383, 2000
43. S. M. Rytov, "Electromagnetic properties of a finely stratified medium," *Sov. Phys. JETP* 2, 466–475, 1956
44. E. N. Glytsis and T. K. Gaylord, "High-spatial-frequency binary and multilevel stairstep gratings: polarization-selective mirrors and broadband antireflection surfaces," *Appl. Opt.* 31, 4459–4470, 1992
45. P. Lalanne and D. L. Lalanne, "Depth dependence of the effective properties of subwavelength gratings," *J. Opt. Soc. Am. A* 14, 450–458, 1997
46. U. Levy and Y. Fainman, "Dispersion properties of inhomogeneous nanostructures," *J. Opt. Soc. Am. A.* 21, 881–889, 2004
47. C. C. Cheng, A. Scherer, R. C. Tyan, Y. Fainman, C. Witzgall and E. Yablonovitch, "New fabrication techniques for high quality photonic crystals," *J. Vac. Sci. Technol. B* 15, 2764–2767, 1997
48. H. C. Kim, K. Ikeda and Y. Fainman, "Resonant waveguide device with vertical gratings," *Opt. Lett.* 32, 539–541, 2007
49. D. T. H. Tan, K. Ikeda, R. E. Saperstein, B. Slutsky and Y. Fainman, "Chip-scale dispersion engineering using chirped vertical gratings," *Opt. Lett.* 33, 3013–3015, 2008
50. H. C. Kim, K. Ikeda and Y. Fainman, "Tunable transmission resonant filter and modulator with vertical gratings," *J. Lightwave Technol.* 25, 1147–1151, 2007
51. H. Kogelnik, "Filter response of nonuniform almost-periodic structures," *Bell Sys. Tech. J.* 55, 109–126, 1975
52. M. Gnan, G. Bellanca, H. M. H. Chong, P. Bassi and R. M. De la Rue, "Modeling of photonic wire Bragg gratings," *Opt. Quantum Electron.* 38, 133, 2006
53. V. R. Almeida, R. R. Panepucci and M. Lipson, "Nanotaper for compact mode conversion," *Opt. Lett.* 28, 1302–1304, 2003
54. R. Kashyap, Fiber Bragg Gratings. Academic, New York, 1999
55. K. O. Hill, F. Bilodeau, B. Malo, T. Kitagawa, S. Thériault, D. C. Johnson, J. Albert and K. Takiguchi, "Chirped in-fiber Bragg gratings for compensation of optical-fiber dispersion," *Opt. Lett.* 19, 1314–1316, 1994
56. M. Notomi, K. Yamada, A. Shinya, J. Takahashi, C. Takahashi and I. Yokohama, "Extremely large group velocity dispersion of line-defect waveguides in photonic crystal slabs," *Phys. Rev. Lett.* 87, 253902, 2001
57. A. C. Turner, C. Manolatou, B. S. Schmidt and M. Lipson, "Tailored anomalous group-velocity dispersion in silicon channel waveguides," *Opt. Express* 14, 4357–4362, 2006
58. A. D. Kersey, M. A. Davis, H. J. Patrick, M. LeBlanc, L. P. Koo, C. G. Askins, M. A. Putnam and E. J. Friebele, "Fiber grating sensors," *J. Lightwave Technol.* 15, 1442–1463, 1997
59. K. Ikeda, M. Nezhad and Y. Fainman, "Wavelength selective coupler with vertical gratings on silicon chip," *Appl. Phys. Lett.* 92, 201111, 2008
60. D. T. H. Tan, K. Ikeda and Y. Fainman, "Cladding-modulated Bragg gratings in silicon wave-guides," *Opt. Lett.* 34, 1357–1359, 2009
61. A. Yariv, "Coupled-mode theory for guided-wave optics," *IEEE J. Quantum Electron.* 9, 919–933, 1973
62. D. C. Flanders, H. Kogelnik, R. V. Schmidt and C. V. Shank, "Grating filters for thin-film optical-waveguides," *Appl. Phys. Lett.* 24, 194–196, 1974
63. V. R. Almeida, C. A. Barrios, R. R. Panepucci and M. Lipson, "All-optical control of light on a silicon chip," *Nature* 431, 1081–1084, 2004
64. P. Yeh and H. F. Taylor, "Contradirectional frequency-selective couplers for guided-wave optics," *Appl. Opt.* 19, 2848–2855, 1980
65. C. Wagner and N. Harned, "EUV lithography: Lithography gets extreme," *Nature Photonics* 4, 24–26, 2010
66. M. Abashin, U. Levy, K. Ikeda and Y. Fainman, "Effects produced by metal-coated near-field probes on the performance of silicon waveguides and resonators," *Opt. Lett.* 32, 2602–2604, 2007

42. J. N. Mait, A. Scherer, O. Dial, D. W. Prather and X. Gao, "Diffractive lens fabricated with binary features less than 60 nm," Opt. Lett. 25, 381–383, 2000.
43. S. M. Rytov, "Electromagnetic properties of a finely stratified medium," Sov. Phys. JETP 2, 466–475, 1956.
44. E. N. Glytsis and T. K. Gaylord, "High-spatial-frequency binary and multilevel stairstep gratings: polarization-selective mirrors and broadband antireflection surfaces," Appl. Opt. 31, 4459–4470, 1992.
45. P. Lalanne and D. Lemercier-Lalanne, "Depth dependence of the effective properties of subwavelength gratings," J. Opt. Soc. Am. A 14, 450–458, 1997.
46. U. Levy and Y. Fainman, "Dispersion properties of inhomogeneous nanostructures," J. Opt. Soc. Am. A 21, 881–889, 2004.
47. C. C. Cheng, A. Scherer, R. C. Tyan, Y. Fainman, G. Witzgall and E. Yablonovitch, "New fabrication techniques for high quality photonic crystals," J. Vac. Sci. Technol. B 15, 2764–2767, 1997.
48. H. C. Kim, K. Ikeda and Y. Fainman, "Resonant waveguide device with vertical gratings," Opt. Lett. 32, 539–541, 2007.
49. D. T. H. Tan, K. Ikeda, R. E. Saperstein, B. Slutsky and Y. Fainman, "Chip-scale dispersion engineering using chirped vertical gratings," Opt. Lett. 33, 3013–3015, 2008.
50. H. C. Kim, K. Ikeda and Y. Fainman, "Tunable transmission resonant filter and modulator with vertical gratings," J. Lightwave Technol. 25, 1147–1151, 2007.
51. H. Kogelnik, "Filter response of nonuniform almost-periodic structures," Bell Sys. Tech. J. 55, 109–126, 1976.
52. M. Chen, G. Bellanca, H. M. H. Chong, P. Bassi and R. M. De la Rue, "Modelling of photonic wire Bragg gratings," Opt. Quantum Electron. 38, 133, 2006.
53. V. R. Almeida, R. R. Panepucci and M. Lipson, "Nanotaper for compact mode conversion," Opt. Lett. 28, 1302–1304, 2003.
54. R. Kashyap, Fiber Bragg Gratings. Academic, New York, 1999.
55. K. O. Hill, F. Bilodeau, B. Malo, T. Kitagawa, S. Theriault, D. C. Johnson, J. Albert and K. Takiguchi, "Chirped in-fiber Bragg gratings for compensation of optical-fiber dispersion," Opt. Lett. 19, 1314–1316, 1994.
56. M. Notomi, K. Yamada, A. Shinya, J. Takahashi, C. Takahashi and I. Yokohama, "Extremely large group-velocity dispersion of line-defect waveguides in photonic crystal slabs," Phys. Rev. Lett. 87, 253902, 2001.
57. A. C. Turner, C. Manolatou, B. S. Schmidt and M. Lipson, "Tailored anomalous group-velocity dispersion in silicon channel waveguides," Opt. Express 14, 4357–4362, 2006.
58. A. D. Kersey, M. A. Davis, H. J. Patrick, M. LeBlanc, K. P. Koo, C. G. Askins, M. A. Putnam and E. J. Friebele, "Fiber grating sensors," J. Lightwave Technol. 15, 1442–1463, 1997.
59. K. Ikeda, M. Nezhad and Y. Fainman, "Wavelength selective coupler with vertical gratings on silicon chip," Appl. Phys. Lett. 92, 201111, 2008.
60. D. T. H. Tan, K. Ikeda and Y. Fainman, "Cladding-modulated Bragg gratings in silicon waveguides," Opt. Lett. 34, 1357–1359, 2009.
61. A. Yariv, "Coupled-mode theory for guided-wave optics," IEEE J. Quantum Electron. 9, 919–933, 1973.
62. D. C. Flanders, H. Kogelnik, R. V. Schmidt and C. V. Shank, "Grating filters for thin-film optical waveguides," Appl. Phys. Lett. 24, 194–196, 1974.
63. V. R. Almeida, C. A. Barrios, R. R. Panepucci and M. Lipson, "All-optical control of light on a silicon chip," Nature 431, 1081–1084, 2004.
64. P. Yeh and H. F. Taylor, "Contradirectional frequency-selective couplers for guided-wave optics," Appl. Opt. 19, 2848–2855, 1980.
65. C. Wagner and N. Harned, "EUV lithography: Lithography gets extreme," Nature Photonics 4, 24–26, 2010.
66. M. Abashin, U. Levy, K. Ikeda and Y. Fainman, "Artifacts produced by metal coated near-field probes on the performance of silicon waveguides and resonators," Opt. Lett. 32, 2602–2604, 2007.

Chapter 3
Liquid Crystal Light-Valves for Slow-Light and Applications

U. Bortolozzo, S. Residori, and J.-P. Huignard

Abstract By performing two-wave mixing experiments in a liquid crystal light valve we obtain slow and fast-light effects with a deceleration of light pulses down to group velocities as small as a few tenths of mm/s. The wave-mixing process is characterized by the presence of multiple-order output beams, each experiencing a different group delay. On each output order, the group delay can be controlled by changing the frequency detuning between the pump and probe as well as by varying the external voltage applied to the light-valve. The large group delay provided by the beam coupling in the light-valve corresponds to a large group index, which, on the other hand, is associated to a narrow frequency bandwidth of the two-wave-mixing gain. These properties can be used to realize applications in high precision interferometry and adaptive holography, with systems that allows the detection of subpicometer displacements.

3.1 Introduction

Slow and fast-light have recently received a large interest, both from the fundamental point of view and for potential applications, in pulse buffering/multiplexing, high sensitivity interferometers, precision metrology and optical sensing [1,2]. After the early achievements obtained by exploiting electromagnetically induced transparency in ultracold atoms [3–5], slow-light has been realized at room temperature by using a mechanism called coherent population oscillation in solid crystals [6] and, nowadays, different methods to realize slow and fast-light are currently employed. Among them, the processes of nonlinear wave mixing offer the advantage of large and tunable dispersion properties, which can be used to obtain controllable group velocities in small-sized experiments. These properties have been exploited in photorefractive crystals [7–10], in optical fibers through stimulated Brillouin

J.-P. Huignard (✉)
Thales Research and Technology, 1 Avenue A. Fresnel, 91767 Palaiseau, France
e-mail: jean-pierre.huignard@thalesgroup.com

T. Fournel and B. Javidi (eds.), *Information Optics and Photonics: Algorithms, Systems, and Applications*, DOI 10.1007/978-1-4419-7380-1_3,

scattering [11,12] and in semiconductor structures [13,14]. Recently, we have shown that by performing two-wave mixing experiments in a liquid crystal light-valve (LCLV), we are able to obtain fast and slow-light down to group velocities as slow as a few tenths of mm/s [15]. In the LCLV, the wave-mixing process is characterized by the presence of multiple-order output beams, each experiencing a different group delay. On each output order, the group delay can be controlled by changing the frequency detuning between the pump and probe as well as by varying the external voltage applied to the light-valve. The large transverse size of the LCLV allows also to delay whole images, a property that can be exploited for optical storage and wavefront sensing.

The LCLV and its nonlinear optical properties are presented in Sect. 3.2. Section 3.3 provides a detailed description of the TWM process occurring in LCLV. Section 3.4 shows the slow-light behavior of the LCLV, with the possibility of obtaining large group delays. Given the large group delay and the slow response time provided by the liquid crystal media, the natural field of applications of slow-light in LCLV is that of precision interferometry and adaptive holography. Two examples showing these two types of potential applications are reported in Sect. 3.5.

3.2 Liquid Crystal Light-Valves as Nonlinear Optical Media

A liquid crystal light-valve is made by the association of a liquid crystal layer with a photoconductive material. A schematic representation of the LCLV is depicted in Fig. 3.1a. In our case, the photoconductive material is a photorefractive $Bi_{12}SiO_{20}$ (BSO) crystal, cut in the form of a thin plate (1 mm thickness, $20 \times 30\,mm^2$ lateral size), that acts as one of the cell wall [17, 18]. The other wall is a glass window.

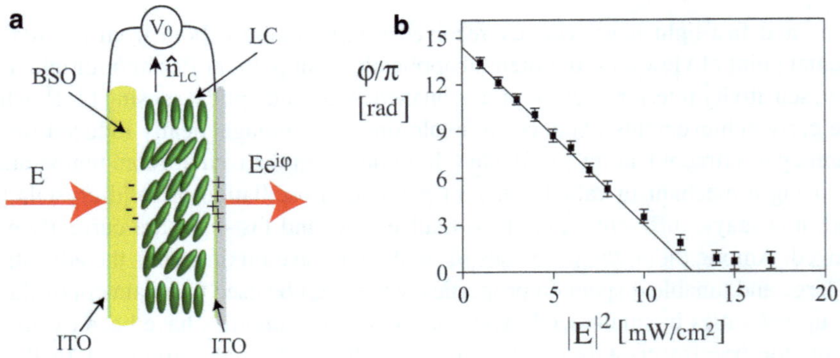

Fig. 3.1 (color online). (**a**) Schematic representation of the liquid crystal light-valve (LCLV): the liquid crystal (LC) layer is inserted in between a photoconductor (BSO) and a glass wall; transparent ITO electrodes allow the application of an external voltage V_0. (**b**) Typical response of the LCLV: the phase shift φ of the output beam is plotted as a function of the input intensity $|E|^2$ (from [16])

In between the two walls there is a gap of $d = 14\,\mu\text{m}$, which is filled with the nematic liquid crystal E48. Transparent electrodes (Indium Tin Oxide, ITO, layers) are deposited over the BSO and the glass wall and allow the application of an external voltage V_0 across the LC layer. The voltage applied is a.c., with a rms value from 2 to 20 V and a frequency from 50 Hz to 20 kHz. While liquid crystals are here used for their large birefringence, the BSO is used for its large photoconductivity and transparency in the visible range [19].

The nematic phase is characterized by a long range orientational order for which all the molecules are aligned, in average, along a preferential direction, so called the nematic director n_{LC} [20]. The molecules have a different polarizability along their long and short axis, hence, when an electric field, or a voltage V_0, is applied across the nematic layer, an induced dipole moment arises and all the molecules reorient towards the direction of the applied field. Because of the LC birefringence, the nematic layer as a whole behaves like a strongly birefringent material, characterized by a different refractive index for a beam polarized along the long or short molecular axis, so called, respectively, the extraordinary n_e and the ordinary n_o index. Typical values for nematics are $n_e = 1.7$ and $n_o = 1.5$, which gives a large birefringence $\Delta n = n_e - n_o = 0.2$.

When the LC molecules reorient under the action of an applied field, their collective motion implies a change of the principal axis of the nematic layer and, hence, an incoming light field experiences a large refractive index change. When a light beams impinges onto the LCLV, photo-generation of charges occurs at the BSO surface because of its photoconductive properties. Hence, the local voltage across the LC layer increases inducing a further molecular reorientation and, thus, an additional refractive index change. As a result, at the exit of the LCLV the light beam acquires a phase shift φ that is a function of the applied voltage and of the beam intensity.

A typical response of the LCLV is shown in Fig. 3.1b, where the phase shift φ acquired by the light beam when passing through the valve is plotted against the input beam intensity $I = |E|^2$. The response is linear up to intensities of the order of $10\,\text{m W/cm}^2$. Then, it saturates to a value that corresponds to the maximum LC birefringence, $\Delta n = 0.2$, for which all the LC molecules are aligned along the direction of the applied field. In between, that is, from the initial planar alignment and the final orthogonal alignment, the reorientation angle of the LC molecules varies from 0 to $\pi/2$ producing a phase shift φ of several π. In the linear region of its response, the LCLV behaves as a Kerr-like nonlinear medium, providing a refractive index change proportional to the input intensity $n = n_0 + n_2 I$, where n_0 is the value fixed by the applied voltage and n_2 is the nonlinear coefficient $\partial n/\partial I$. Thanks to the large LC birefringence, the nonlinear coefficient, which is the slope of the linear part of the response curve, is as large as $n_2 = -6\,\text{cm}^2/\text{W}$, the minus sign accounting for the defocusing character of the nonlinearity (the refractive index changes from n_e to n_o, with $n_e > n_o$).

The response time of the LCLV is dictated by the time τ_{LC} required by the collective motion of the LC molecules to establish over the whole thickness d of the nematic layer. This is given by, $\tau_{LC} = (\gamma/K)d^2$, where γ is the LC rotational viscosity and K the splay elastic constant [20]. For $d = 14\,\mu\text{m}$ and typical values of the LC constants, τ_{LC} is of the order of 100 ms. The spatial resolution, which is the minimal size of an independently addressed area, is given by the electric coherence length of the LC, $l_{LC} = (\Delta\varepsilon/K)^{1/2}\, d/V_0$, where $\Delta\varepsilon$ is the dielectric anisotropy of the LC, and, for the usual values of V_0, it is typically of the order of $10\,\mu\text{m}$.

3.3 Two-Wave Mixing and Optical Amplification in LCLV

Two-wave mixing in the LCLV takes place inside the thin layer of the nematic liquid crystals. Therefore, the beam coupling occurs in the Raman-Nath regime of optical diffraction [21], at difference with the Bragg regime that governs all beam coupling phenomena in bulk crystals [7–10]. The experimental setup for performing two-wave mixing experiments in the LCLV is sketched in Fig. 3.2. A pump beam E_p is sent onto the LCLV together with a weaker signal beam E_s. The total electric field at the input of the LCLV can be written as $E_{in}(\mathbf{r},t) = E_s e^{i[\mathbf{k}_s\mathbf{r}-\omega_s t]} + E_p e^{i[\mathbf{k}_p\mathbf{r}-\omega_p t]} +$ c.c., where E_p and E_s are the amplitudes of the pump, respectively, the signal wave, $\mathbf{k}_p$ and $\mathbf{k}_s$ their respective propagation vectors and ω_p, ω_s their frequencies. The two beams produce an intensity fringe pattern

$$|E_{in}(\mathbf{r},t)|^2 = I_T\left[1 + 2\frac{E_p E_s}{I_T}\cos\left(\mathbf{K}_g\mathbf{r} - \Delta\omega t\right)\right], \tag{3.1}$$

where $I_T \equiv |E_s|^2 + |E_p|^2 = I_s + I_p$ is the total input intensity, $\mathbf{K}_g = \mathbf{k}_p - \mathbf{k}_s$ is the grating wave vector and $\Delta\omega = \omega_p - \omega_s$ the frequency detuning between the pump and signal. The ratio between the pump and signal intensity $\beta \equiv I_p/I_s$, is usually much larger than one.

The fringe pattern induces, on its turn, a photo-induced space charge distribution, hence a molecular reorientation pattern in the LC layer, which creates a refractive index grating with the same wave vector $\mathbf{K}_g$. The spatial period of the grating

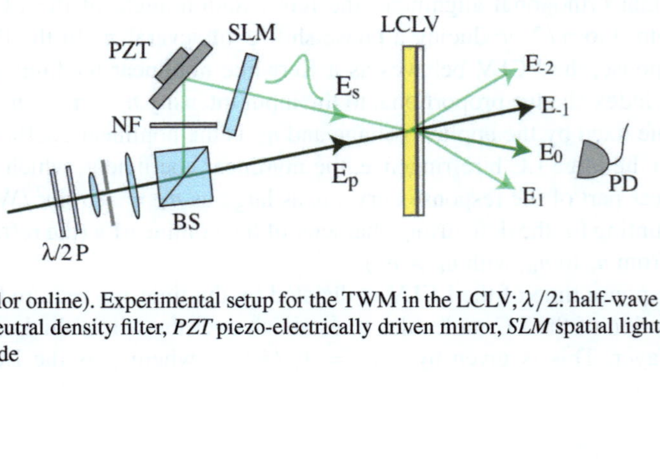

Fig. 3.2 (color online). Experimental setup for the TWM in the LCLV; $\lambda/2$: half-wave plate. P polarizer, *NF* neutral density filter, *PZT* piezo-electrically driven mirror, *SLM* spatial light-modulator, *PD* photodiode

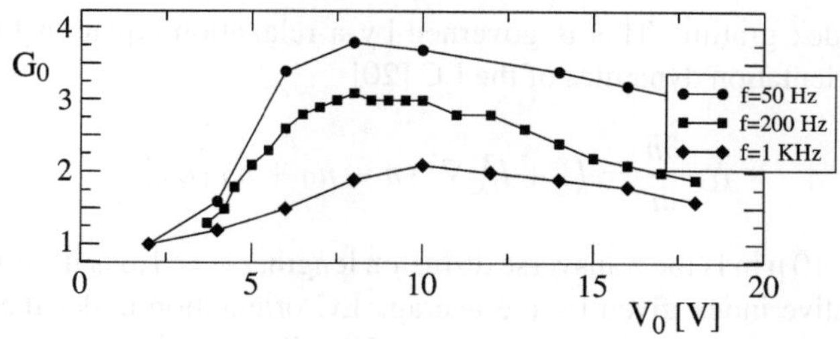

Fig. 3.3 Gain G_0 measured on the output order $m = 0$ as a function of the voltage V_0 applied to the LCLV; $I_p = 2.0\,\mathrm{mW/cm^2}$ and $\beta = 80$

$\Lambda \equiv 2\pi/\mathbf{K}_g$ is usually in between 50 and 300 μm. Because the thickness of the LC layer is much less than Λ, the LC grating acts as a thin hologram, thus, beam diffraction occurs in the Raman–Nath regime. Several diffracted beams, distinguished by the numbers $0, \pm 1, \pm 2, \ldots, \pm m$, are observed at the output of the LCLV. In Fig. 3.2 the output orders are identified as E_m. Due to self-diffraction, photons from the pump are transferred into the different output orders. The $m = 0, +1, \pm 2, \ldots$ orders are amplified, that is, they receive from the pump more photons than they are losing due to the scattering on the other orders. The $m = -1$ order is the pump beam that, even though depleted, remains of much higher intensity than the other beams [22].

In the experiment, both the pump and signal beams originate from a cw solid state laser, $\lambda = 532$ nm. They are enlarged and collimated, the beam diameter on the LCLV is 18 mm. The light polarization is linear and parallel to the LC nematic director n_{LC}. When the pump and signal are both continuous beams, we characterize the TWM as an optical amplification process of the input signal. By placing a photodiode on the zero order output, we measure the signal gain as $G_0 = I_0/I_s$, where I_s is the input signal intensity and $I_0 = |E_0|^2$. In Fig. 3.3 the measured gain G_0 is plotted as a function of the rms amplitude V_0 of the voltage applied to the LCLV and for different values of the frequency. An optical amplification up to a factor 4 is obtained.

By considering that the beam coupling occur over the thin LC layer, this corresponds to a huge exponential gain Γ, that is, by writing $I_0 = I_s e^{\Gamma d}$, for $G_0 = 4$ we have $\Gamma = 924\,\mathrm{cm}^{-1}$. Talbot effect has been used to further enhance the 2WM gain in a cascade of two or more light-valves in sequence [22]. Moreover, the TWM gain in the LCLV has been recently exploited to realize an optical cavity where the large number of competing modes leads to a complex dynamics with the formation of spatio-temporal pulses of light [23].

3.3.1 General Theoretical Description of the TWM in LCLV

To derive the full expression for the output field, that is, to account for all the scattered orders, we have to consider the evolution of the amplitude $n(\mathbf{r}, t)$ of the refractive index grating. This is governed by a relaxation equation following the molecular orientation dynamics of the LC [20]

$$\tau_{\mathrm{LC}}\frac{\partial n}{\partial t} = \left(1 + l_{\mathrm{LC}}^2\nabla^2\right) n + n_0 + n_2 \left|E_{\mathrm{in}}\right|^2 , \tag{3.2}$$

where $l_{\mathrm{LC}} = 10\,\mu\mathrm{m}$ is the transverse diffusion length, $n_0 = 1.6$ is the constant value of the refractive index given by the average LC orientation under the application of the voltage V_0, and n_2 is the equivalent Kerr-like coefficient of the LCLV. By coupling the above (2) with (1), it can been shown that the m output order field can be written as [24]

$$\widehat{E}_m = E_m \mathrm{e}^{i(\mathbf{k}_m\mathbf{r}-\omega_m t)} + \mathrm{c.c.}, \tag{3.3}$$

where $\omega_m = \omega_{\mathrm{s}} - m\Delta\omega$ is the frequency, $\mathbf{k}_m = \mathbf{k}_{\mathrm{s}} - m\mathbf{K}_{\mathrm{g}}$ the wave vector and the amplitude is given by

$$E_m = \left[E_{\mathrm{s}}J_m(\rho) + iE_{\mathrm{p}}J_{m+1}(\rho)\mathrm{e}^{-i\Psi}\right]\mathrm{e}^{i[k(n_0+n_2I_{\mathrm{T}})z+m((\pi/2)-\Psi)]}, \tag{3.4}$$

where J_m is the Bessel function of the first kind and of order m,

$$\rho = \frac{2kn_2E_{\mathrm{p}}E_{\mathrm{s}}}{\sqrt{\left(1+l_{\mathrm{LC}}^2K_{\mathrm{g}}^2\right)^2 + (\Delta\omega\tau_{\mathrm{LC}})^2}}d, \tag{3.5}$$

and

$$\tan\Psi = \frac{\Delta\omega\tau_{\mathrm{LC}}}{1+l_{\mathrm{LC}}^2K_{\mathrm{g}}^2} \tag{3.6}$$

From the above expression (4), we see that each order m receive two contributions, one is the scattering of the signal and the other is the scattering of the pump onto the refractive index grating.

It is now useful to write each output order field in the form

$$\widehat{E}_m = \sqrt{G_m}E_{\mathrm{s}}\mathrm{e}^{i\Phi_m}\mathrm{e}^{i(\mathbf{k}_m\mathbf{r}-\omega_m t)} + \mathrm{c.c.}, \tag{3.7}$$

where we define $G_m = |E_m|^2/|E_{\mathrm{s}}|^2$ as the gain amplification factor for each output order m and Φ_m is the associated nonlinear phase shift. Both G_m and Φ_m can be calculated from (4). In the $m = 0$ case, we have $E_0 = \sqrt{G_0}E_{\mathrm{s}}\mathrm{e}^{i\Phi_0}\mathrm{e}^{i(\mathbf{k}_{\mathrm{s}}\mathbf{r}-\omega_{\mathrm{s}}t)}$. When $\rho \ll 1$ the 2WM gain $G_0 = |E_0|^2/|E_{\mathrm{s}}|^2$ is given by

$$G_0 = 1 + 2g\sin(\Psi)\,z + g^2z^2, \tag{3.8}$$

where

$$g = \frac{kn_2\left|E_{\mathrm{p}}\right|^2}{\sqrt{\left(1+l_{\mathrm{LC}}^2K_{\mathrm{g}}^2\right)^2 + (\Delta\omega\tau_{\mathrm{LC}})^2}}, \tag{3.9}$$

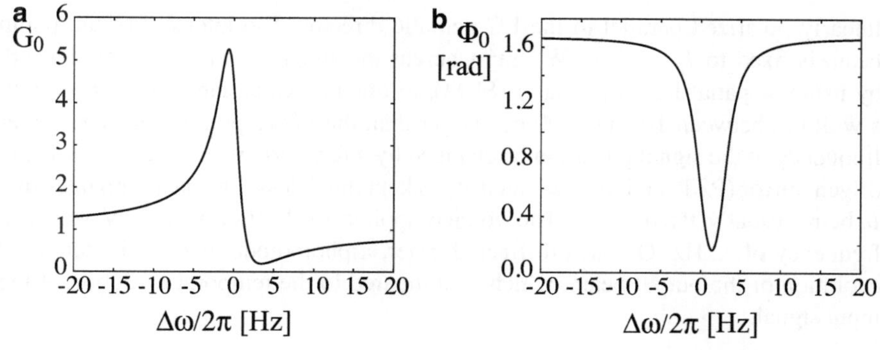

Fig. 3.4 (**a**) Gain G_0 and (**b**) phase shift Φ_0 for the $m = 0$ output order beam as a function of the frequency detuning $\Delta\omega$ between the pump and signal, $\beta = 30$ (from [16])

and the phase shift due to the nonlinear coupling is

$$\Phi_0 = \tan^{-1}\left(\frac{kn_2\left|E_{\mathrm{p}}\right|^2 E_{\mathrm{s}}\cos(\Psi)z}{E_{\mathrm{s}}\sqrt{\left(1+l_{\mathrm{LC}}^2K_{\mathrm{g}}^2\right)^2+(\Delta\omega\tau_{\mathrm{LC}})^2}\,kn_2\left|E_{\mathrm{p}}\right|^2+kn_2\left|E_{\mathrm{p}}\right|^2 E_{\mathrm{s}}\sin(\Psi)z}\right) + k\left[n_0+n_2I_{\mathrm{T}}\right]z. \quad (3.10)$$

In the limit $\delta = 0$, and for $l^2{}_{\mathrm{LC}}|K_{\mathrm{g}}|^2 \ll 1$, the above expression for the gain reduces to the already known formula [18]

$$G_0 = 1 + \left(\frac{2\pi n_2 I_{\mathrm{p}}}{\lambda}\right)^2 d^2, \quad (3.11)$$

which accounts for optical amplification in the LCLV. Note that the condition $l^2{}_{\mathrm{LC}}|K_{\mathrm{g}}|^2 \ll 1$ is usually satisfied, l_{LC} being of the order of ten microns while the spatial period of the grating is of the order of hundred microns.

The theoretical gain and associated phase retardation for the zero order of diffraction are numerically calculated and plotted in Fig. 3.4a, b, respectively, as a function of the frequency detuning δ between the pump and signal and by taking $\tau_{\mathrm{LC}} = 100$ ms, $l_{\mathrm{LC}} = 15$ μm, $\Lambda = 150$ μm and $n_2 = -6\,\mathrm{cm}^2/\mathrm{W}$. The maximum gain is obtained for $\delta \sim 0$, which agrees with the experimental observations. We can also note that the gain is highly selective in frequency.

3.4 Slow and Fast Light in LCLV: Tuning the Group Velocity of Light Pulses

Slow and fast-light in the LCLV are obtained by performing TWM experiments with a continuous pump and a time modulated signal, as represented in Fig. 3.2. As described above, the two beams, $\lambda = 532$ nm, are enlarged and collimated and linearly polarized parallel to the LC nematic director. The intensity of the pump beam is fixed to $I_p = 1.8\,\mathrm{mW/cm^2}$ whereas the signal beam is time-modulated, by using a spatial-light modulator (SLM), to obtain a Gaussian wave packet with a width in between 100 and 200 ms, larger than the LC response time. The center frequency of the signal pulse can be changed by a few hertz with a piezoelectrically driven mirror (PZT) and its peak intensity is kept much less than the pump intensity, β being usually fixed to 30. The voltage applied to the LCLV is 20 V rms at a frequency of 1 kHz. On each diffracted order, a photodiode records the temporal evolution of the output beam, which is compared to the temporal evolution of the input signal.

3.4.1 Theoretical Background

A light pulse, that is, a wave packet, is composed by many optical frequencies, each corresponding to a distinct Fourier component propagating with its own phase velocity. For the pulse to propagate without distortion, all the frequency components have to keep in phase, which, mathematically, can be expressed by requiring that the wave packet as a whole propagate with the group velocity [1]

$$v_g = \frac{\mathrm{d}\omega}{\mathrm{d}k} = \frac{c}{n + \omega \mathrm{d}n/\mathrm{d}\omega}, \tag{3.12}$$

where $\mathrm{d}n/\mathrm{d}\omega$ is the dispersion of the traversed medium. Formally, we can also express (13) as $v_g = c/n_g$, where n_g is the group index $n_g \equiv n + \omega\, \mathrm{d}n/\mathrm{d}\omega$. Controlling the group velocity of light pulses can thus be realized by controlling the dispersion properties of the medium. For $\mathrm{d}n/\mathrm{d}\omega$ large and positive, the group index becomes very large and, correspondingly, the group velocity very small, $0 < v_g \ll c$, so that slow-light is achieved. On the other side, when $\mathrm{d}n/\mathrm{d}\omega$ is negative and large, the group velocity becomes negative, $v_g < 0$ and fast-light is achieved. Correspondingly, the group delay $\tau_g = \mathrm{d}/v_g$, where d is thickness of the traversed medium, is large and becomes positive or, respectively, negative.

As we have seen above, TWM in the LCLV produces multiple output order beams. Associated to each order there is a gain and a dispersion curve that can be calculated from the model, (4). Then, the group delay Δt_m can be calculated for each output order m, and it is simply given by the slope of the dispersion curves

$$\Delta t_m = \frac{\partial \Phi_m}{\partial \omega_m}. \tag{3.13}$$

In Fig. 3.5a, the dispersion curves Φ_m, calculated for the same parameter values given above, are plotted for different orders m against $\omega_m - \omega_p$. Note that $\omega_m = \omega_s - \Delta\omega$, thus the plot of Φ_m against $\Delta\omega$ would change sign for odd m. We have chosen to display Φ_m as a function of $\omega_m - \omega_p$, so that the slopes of the dispersion

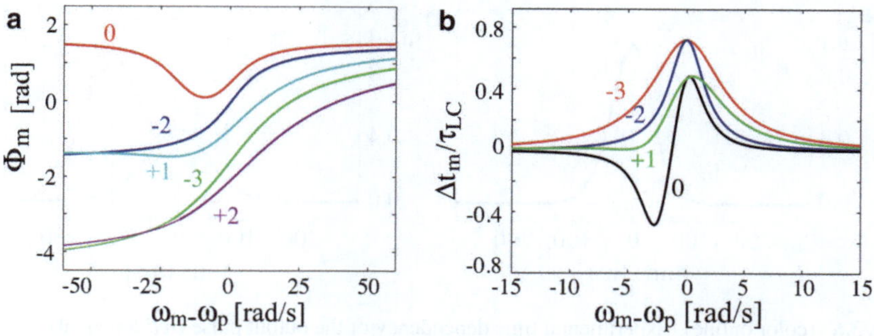

Fig. 3.5 (color online). (**a**) Dispersion curves Φ_m and (**b**) corresponding group delay Δt_m as a function of the frequency $(\omega_m - \omega_p)$ of the m output order; $\beta = 30$ (from [16])

curves in the graph give directly the correct sign for the group delay. In Fig. 3.5b, the group delay Δt_m is plotted against $\omega_m - \omega_p$. From these curves, the group velocity of each output pulse can be calculated as $v_m \equiv d/\Delta t_m$ with d the thickness of the LC layer. We can note that fast-light is obtained on the $m = 0$ order. Moreover, for every m the slope of the Φ_m curves is maximum for $\omega_m = \omega_p$, hence the maximum group delay is obtained when the pump and signal have the same frequency.

The theoretical limit for the maximum group delay can be calculated analytically and it is given

$$\Delta t_m^{(\max)} = \frac{m}{m+1}\tau_{LC} + \frac{1}{m+1}\left[\frac{\beta\left([J_{m+1}(\rho)]/J_m(\rho)\right)^2}{1+\beta\left([J_{m+1}(\rho)]/J_m(\rho)\right)^2}\right]\tau_{LC}, \tag{3.14}$$

for each order m. Given the structure of the Bessel functions, it is straightforward to verify that the theoretical absolute maximum delay is obtained for the $m = -2$ order and it is equal to $2\tau_{LC}$ [15]. Correspondingly, the theoretical limit for the maximum fractional delay, which is the pulse delay compared to the pulse width, is one. Indeed, in order to avoid distortions, the width of the input pulse has to be within the frequency bandwidth of the two-wave-mixing, which requires a minimum temporal width of the order of $2\tau_{LC}$.

3.4.2 Experimental Results

Different group delays can be obtained depending on the output order considered and on the frequency detuning $\Delta\omega$ between the pump and signal. Two representative data taken, respectively, on the $m = -2$ and $m = 0$ order are shown in Fig. 3.6, showing an anticipated, fast-light, and a delayed, slow-light, pulse. Fast-light occurs for $\Delta\omega/2\pi = 3$ Hz, which is close to the minimum of the gain, therefore the pulse is attenuated, whereas slow-light is obtained for $\Delta\omega = 0$, which is close to the

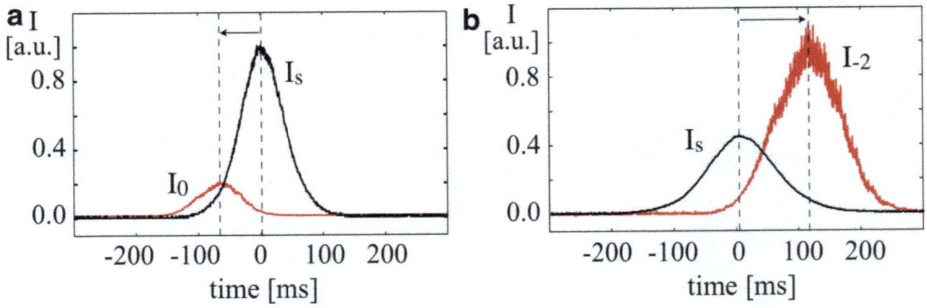

Fig. 3.6 (color online). Experimental time dependency of the output pulse (*red lines*) taken on the (**a**) $m = 0$ and (**b**) $m = -2$ diffraction order of the input pulse (*black line*) showing, respectively, an anticipated (fast-light) and a delayed (slow-light) pulse (from [16])

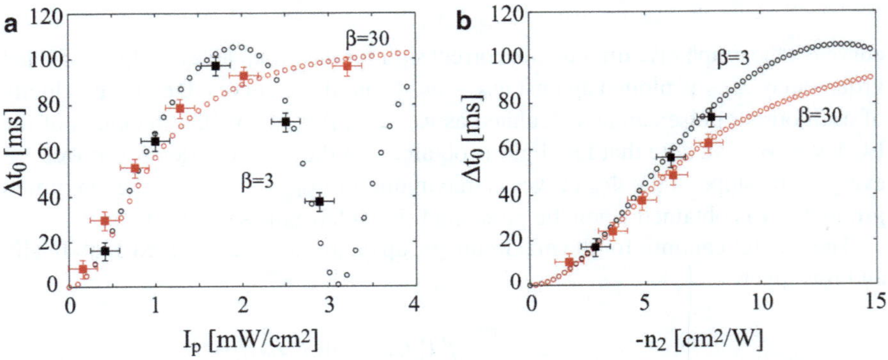

Fig. 3.7 (color online). Group delay Δt_0 as a function of (**a**) the pump intensity I_p and (**b**) the nonlinear coefficient n_2; *black dots* $\beta = 3$, *red dots* $\beta = 30$. *Black* and *red squares* are experimental points measured for $\beta = 3$ and $\beta = 30$, respectively (from [16])

maximum of the gain. Therefore, the slow-light pulse is not only delayed but also amplified. By fitting each pulse with a Gaussian, we have evaluated the time anticipation as $\Delta t_0 = -65$ ms for the fast pulse and the time retardation as $\Delta t_{-2} = 110$ ms for the slow pulse. The effective group velocity of each pulse can be determined as $v_m = d/\Delta t_m$, where d is the thickness of the traversed LC layer and Δt_m the group delay of the order m. We obtain $v_0 = -0.21$ mm/s for the fast pulse and $v_{-2} = 0.13$ mm/s for the slow pulse.

The group delay and, therefore, the group velocity can be tuned by changing the experimental parameters, such as the ratio of the pump and signal peak intensity β, the spacing of the refractive index grating Λ, and the nonlinear coefficient n_2. This last one can be changed by varying either the voltage V_0 applied to the LCLV or the pump intensity I_p. As an example, we show in Fig. 3.7 the behavior of the $m = 0$ order as a function of the pump intensity I_p and of the nonlinear coefficient n_2. The theoretical group delay Δt_0 is plotted together with the experimental points, which are measured for $\Delta\omega = 0, \beta = 3$ and $\beta = 30$, a grating spacing $\Lambda = 250\,\mu$m and a pump intensity $I_p = 1.5\,\text{mW/cm}^2$.

In Fig. 3.7a, we observe that for $\beta = 3$ there is a decrease of Δt_0 for $I_p > 2\,\mathrm{mW/cm^2}$. This is due to the dependency of the maximum group delay from the Bessel functions, which, depending on the respective value of I_p and β, can provide a decreasing contribution, as can be seen from (14). Note that also the argument of the Bessel functions, ρ, depends on β since it is proportional to $I_p I_s$. On the other side, we see in Fig. 3.7b that Δt_0 increases monotonically with n_2 for both values of β. Correspondingly, the group velocity can be finely tuned by changing the voltage V_0 applied to the LCLV.

3.5 Interferometry with the LCLV as a Slow-Light Medium

The slow-light phenomenon is associated to a large group index n_g, which allows to increase the spectral sensitivity, or resolution, of interferometric systems up to large factors, depending on the maximum group index achievable in the slow-light medium [2]. In the LCLV, the large group delay provided by the strong dispersion properties of the liquid crystals allows obtaining a group index of several orders of magnitude greater than in other media. At the same time, with the LCLV the optical amplification prevents losses and provides a gain that is tunable with the applied voltage, so that the slow-light interferometric systems can be adjusted to work in optimal conditions. As example of LCLV applications in the field of precision interferometry, we have realized a Mach–Zehnder interferometer, MZI, where a LCLV in inserted in one of the arms. In this system, we use the $m = 0$ order, so that the signal is present even when the LCLV is not operating and we can make a direct comparison between the sensitivity with and without the slow-light effect. Then, as a further example, we report an adaptive holographic system, where the gain of the LCLV provides a highly selective frequency filter, thus allowing reaching the detection of sub-picometer displacements.

3.5.1 *High Sensitivity LCLV-Based Interferometer*

The experimental setup for the LCLV-based MZI is shown in Fig. 3.8a. A pump beam is incident on the LCLV and the interferometer is aligned along the direction of the $m = 0$ output order [25]. If we introduce a small frequency perturbation Ω on the input signal, the transmission of the interferometer is given by $T = 1/2\,[1 + \cos\Delta\varphi]$ where the total phase difference $\Delta\varphi$ is given by the optical path difference ΔL between the two arms of the interferometer and by the slow-light nonlinear phase retardation Φ_0, which, as we have seen before, is a function of the frequency detuning Ω. If the frequency perturbation is small with respect to the bandwidth of the two-wave-mixing, $\Omega < 1/(2\tau_{LC})$, we can write

$$\Phi_0(\Omega) \cong \Phi_0\,(0) + \left[\frac{\partial\Phi_0}{\partial\Omega}\right]\Omega = \Phi_0(0) + \Delta t_0\Omega\,. \tag{3.15}$$

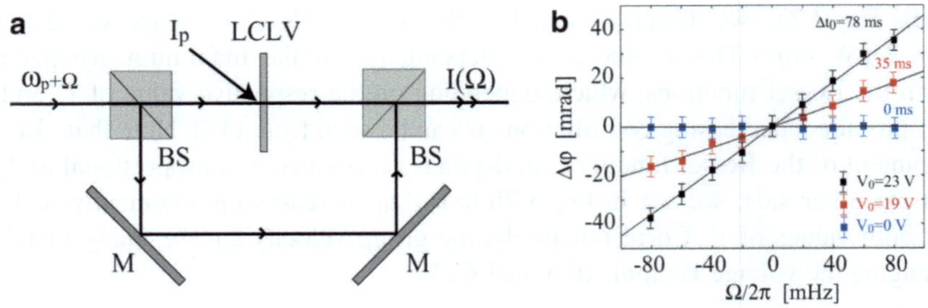

Fig. 3.8 (color online). (**a**) Schematic diagram of a Mach–Zehnder interferometer with the LCLV operating as a slow-light medium; BS beam-splitters, *M* mirrors. (**b**) Phase shift detected by the interferometer as a function of the perturbation frequency Ω for different voltages V_0 applied to the LCLV and correspondingly increasing group delays (from [16])

Since the group delay is much larger than $\Delta L/c$ (for example, for $\Delta L = 1$ m we have $\Delta L/c \sim 10^{-7}$s), we can neglect the linear contribution and the total phase shift can be written as

$$\Delta\varphi \cong \Phi_0(0) + \Delta t_0 \Omega. \tag{3.16}$$

Therefore, the spectral sensitivity of the interferometer is directly related to the group delay

$$T = \frac{1}{2}\left[1 + \cos\left(\Delta t_0 \Omega + \Phi_0(0)\right)\right], \tag{3.17}$$

and the group delay being very large, it is hugely enhanced. Thanks to the tunability of $\Phi_0(0)$ through V_0, in the experiment we have fixed $\Phi_0(0) = \pi/2$, so that $T \propto \sin \Delta t 0 \Omega$ and for small frequency perturbations the detection becomes linear, $T \approx \Delta t_0 \Omega$, with Δt_0 the large group delay provided by the slow-light process.

In Fig. 3.8b, we report the phase change detected at the output of the interferometer when a small frequency perturbation Ω, created by a piezoelectrically driven mirror, is added on the signal beam. While the interferometer shows no sensitivity when the LCLV is switched off, $V_0 = 0$, we see that the sensitivity largely increases when increasing V_0. The slopes of the detected phase curves are the corresponding group delays provided by the slow-light process in the LCLV.

3.5.2 Picometer Detection by Adaptive Holography in the LCLV

As we have seen in Sect. 3.4, the slow-light process in the LCLV is induced by a rapid change of the refractive index, which occurs due to the narrow frequency bandwidth of the gain provided by the two-beam coupling. While slow and fast-light exploit the dispersion properties, the gain counterpart acts as an optical filter and can be used to realize an adaptive holographic interferometer (AHI). The main objective of the adaptive interferometry is to detect frequency phase modulations of optical beams with complex wavefronts, such as speckle beams, characterized by

Fig. 3.9 Schematic diagram of the adaptive holographic system using the LCLV as a narrow frequency bandwidth medium; *NF* neutral density filters, *PZT* piezo-electrically driven mirror (from [16])

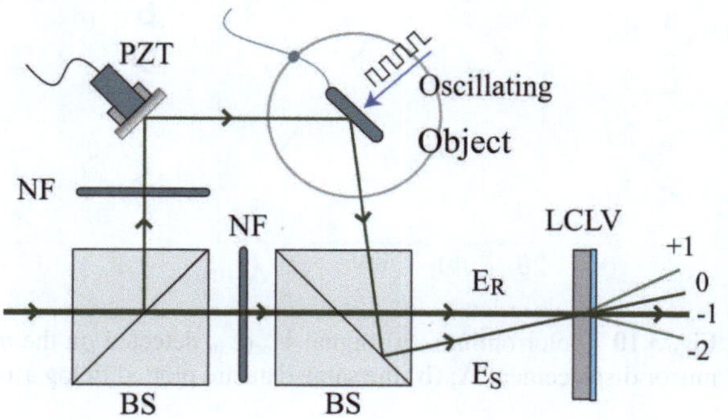

low frequency noise fluctuations, such as given by environmental disturbances when the beam passes, for instance, through the atmosphere, or turbid media or biological tissues. Adaptive holography addresses the problems related to these types of measurements by using an interferometer scheme where the signal beam, carrying the phase modulations, is made to interfere with a reference beam, thus producing an adaptive hologram [26]. This last one is made by a nonlinear medium able to provide a self-reconfigurable index of refraction.

Two-wave-mixing in the LCLV provides both a good sensitivity and a narrow frequency bandwidth, and, can, therefore, be efficiently used to realize an AHI system [27]. The experimental setup is shown in Fig. 3.9. By means of a piezo-electrically driven mirror the signal beam E_s is phase modulated with a sinusoidal oscillation at high frequency $\wedge$ and small amplitude Δ. The frequency of the modulation is fixed at $\Omega/2\pi = 1\,\text{kHz}$, which is much greater than the bandwidth of the two-wave-mixing in the LCLV. The signal beam is sent onto the LCLV together with a reference beam E_R. The optical power of the output beams is measured through a photodiode and a lock-in amplifier synchronized to the frequency Ω of the phase modulation. The amplitude of the phase grating that is generated by the two-wave-mixing inside the LC layer is given by

$$\varphi = 2kdn_2 E_\text{R} E_\text{s} J_0(2k\Delta). \tag{3.18}$$

Because of the large value of the LCLV nonlinear coefficient n_2, the process is very efficient. The output diffracted beams are easily detected and, in particular, the output optical power at the frequency $\wedge$ can be calculated and it is given by

$$P_m = 4P_\text{R}\text{e}^{-\alpha D} K J_m(\varphi) J_{m+1}(\varphi) J_1(2k\Delta), \tag{3.19}$$

where m is the order of diffraction, $D = 1$ mm the thickness of the photoconductor and $\alpha = 0.3\,\text{cm}^{-1}$ its absorption coefficient. The ratio between the intensity of the reference and signal beam is K and J_m is the Bessel function of first kind and order m.

An important characteristic of the AHI is that when the amplitude of the phase modulation is small the detection is always linear in Δ, at variance with standard

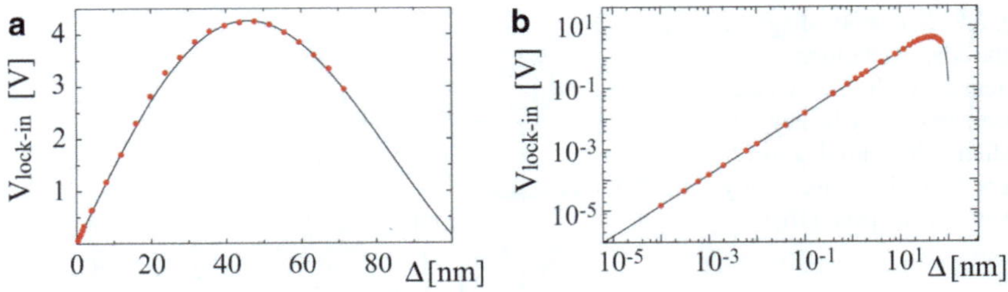

Fig. 3.10 (color online). (**a**) Signal $V_{\mathrm{lock-in}}$ detected on the $m = -1$ order as a function of the mirror displacement Δ; (**b**) the same data are plotted in logarithmic scale (from [16])

interferometers where the detection is linear only for precise values of the optical path difference between the reference and signal beam. The sensitivity of the system is obtained by considering only the photon shot-noise. The signal to noise ratio in this case is given by [27]

$$\mathrm{SNR} = \sqrt{\frac{2\eta P_{\mathrm{R}}}{\hbar\omega\Delta f}}\mathrm{e}^{-(\alpha D)/2}\frac{KJ_m(\varphi)J_{m+1}(\varphi)}{\sqrt{K^2 J_m^2(\varphi)J_{m+1}^2(\varphi)}}2k\Delta, \tag{3.20}$$

where $\eta = 0.4$ is the quantum efficiency of the photodiode and $\Delta f = 1\,\mathrm{Hz}$ is the bandwidth of the electronic detection system. The maximum signal to noise ratio is obtained for $m = -1$ and the minimum detectable phase is of the order of $7\,\mathrm{nrad/Hz}^{1/2}$. In Fig. 3.10a, the detected signal $V_{\mathrm{lock-in}}$ is plotted as a function of the mirror displacement Δ, for the $m = -1$ order, for which the theoretical curves predict the maximum sensitivity [27]. The intensity of the optical beam was $3\,\mathrm{mW/cm}^2$ and $K = 5$. In Fig. 3.10b the same data are plotted in logarithmic scale. We see that the detection is linear for small displacements and that mirror displacements as small as 0.1 pm are detected.

Finally, in order to test the ability of the system to work with complex wavefronts, we have performed an experiment in which we take as signal the optical field distribution at the exit of a multimode fiber [16, 28]. The setup is sketched in Fig. 3.11a. A high frequency modulation is created by sending on the fiber an acoustic wave at $\Omega/2\pi = 5\,\mathrm{KHz}$ and low frequency perturbations are induced by local, and small, disturbances, induced, for example, by touching the fiber. At the exit of the fiber, the optical field distribution is a speckles pattern with a slow dynamics, as shown in the inset. In the case of a classical interferometer the phase modulations of the signal are completely hidden by the noise. This is shown in Fig. 3.11b. In the case of AHI, the two-wave mixing in the LCLV provides a narrow frequency bandwidth that filters out noise low frequency noise flu, and the acoustic wave modulating the signal can be clearly distinguished. This is evident in Fig. 3.11c, where it is plotted the time evolution of the photodiode signal measured after the LCLV in the adaptive interferometer setup.

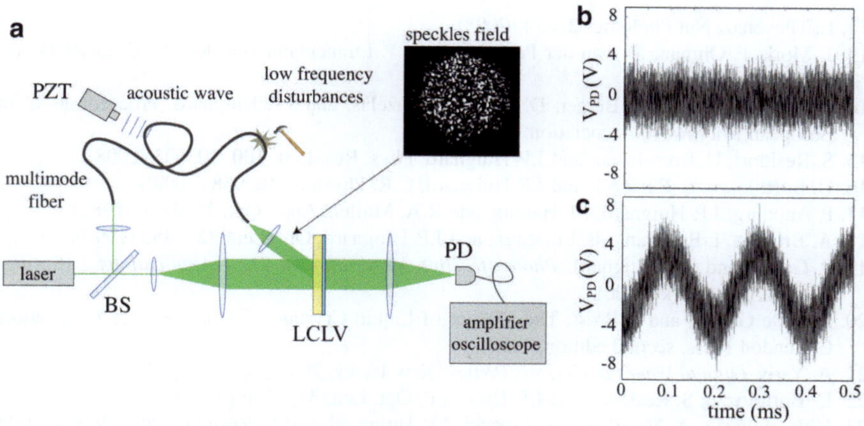

Fig. 3.11 (color online). (**a**) Schematic setup for the detection of phase modulations of a speckles field. In the inset is shown the speckles field at the output of the multimode fiber. Signal V_{PD} detected as a function of time in the case of (**b**) a Michelson interferometer and (**c**) after the LCLV in the AHI system

3.6 Conclusions

In conclusion, we have shown that by performing two-wave mixing experiments in LCLV, and by using the dispersive properties associated to the two-beam coupling process, we can obtain fast and slow-light phenomena with very large and tunable group delay. The corresponding group index due to slow-light has been exploited to enhance the sensitivity of interferometric detection systems, whereas the narrow frequency bandwidth of the gain is exploited for efficiently realizing adaptive holographic systems.

References

1. R.W. Boyd and D.J. Gauthier, In *Progress in Optics*, edited by E. Wolf (Elsevier Sceince, Amsterdam, 2002), Vol. 43, pp. 497–530.
2. Z. Shi, R.W. Boyd, D.J. Gauthier, and C.C. Dudley, Opt. Lett. **32**, 915 (2007).
3. S.E. Harris, Phys. Today **50**, 36 (1997).
4. A. Kasapi, M. Jain, G.Y. Yin, and S.E. Harris, Phys. Rev. Lett. **74**, 2447 (1995).
5. L.V. Hau, S.E. Harris, Z. Dutton, and C.H. Behroozi, Nature **397**, 594 (1999).
6. M.S. Bigelow, N.N. Lepeshkin, and R.W. Boyd, Phys. Rev. Lett. **90**, 113903 (2003).
7. E. Podivilov, B. Sturman, A. Shumelyuk, and S. Odoulov, Phys. Rev. Lett. **91**, 083902 (2003).
8. A. Shumelyuk, K. Shcherbin, S. Odoulov, B. Sturman, E. Podivilov, and K. Buse, Phys. Rev. Lett. **93**, 243604 (2004).
9. G. Zhang, R. Dong, F. Bo, and J. Xu, Appl. Opt. **43**, 1167 (2004).
10. G. Zhang, F. Bo, R. Dong, and J. Xu, Phys. Rev. Lett. **93**, 133903 (2004).
11. Y. Okawachi, M.S. Bigelow, J.E. Sharping, Z. Zhu, A. Schweinsberg, D.J. Gauthier, R.W. Boyd, and A.L. Gaeta, Phys. Rev. Lett. **94**, 153902 (2005).

12. L. Thévenaz, Nat Photonics **2**, 474 (2008).
13. J. Mork, F. Ohman, M. van der Poel, Y. Chen, P. Lunnemann, and K. Yvind, Laser Photon. Rev. 1–15 (2008).
14. S. Tonda-Goldstein, P. Berger, D. Dolfi, J. Chazelas, and J.P. Huignard, Proceedings of the European Microwave Association, 2008.
15. S. Residori, U. Bortolozzo, and J.P. Huignard, Phys. Rev. Lett. **100**, 203603 (2008).
16. U. Bortolozzo, S. Residori, and J.P. Huignard, C.R. Physique **10**, 938 (2009).
17. P. Aubourg, J.P. Huignard, M. Hareng, and R.A. Mullen, Appl. Opt. **21**, 3706 (1982).
18. A. Brignon, I. Bongrand, B. Loiseaux, and J.P. Huignard, Opt. Lett. **22**, 1855 (1997).
19. P. Gunter and J.P. Huignard, *Photorefractive Materials and Their Applications* 1 (Springer Science, New York, 2006).
20. P.G. De Gennes and J. Prost, The Physics of Liquid Crystals, (Oxford Science Publications, Clarendon Press, second edition, 1993).
21. A. Yariv, *Optical Waves in Crystals* (Wiley, New Jersey, 2003), pp. 354–358.
22. U. Bortolozzo, S. Residori, and J.P. Huignard, Opt. Lett. **31**, 2166 (2006).
23. U. Bortolozzo, A. Montina, F.T. Arecchi, J.P. Huignard, and S. Residori, Phys. Rev. Lett. **99**, 023901 (2007).
24. U. Bortolozzo, S. Residori, and J.P. Huignard, Phys. Rev. A **79**, 053835 (2009).
25. U. Bortolozzo, S. Residori, and J.P. Huignard, Laser Photon. Rev. **1** (2009).
26. A.A. Kamshilin, R.V. Romashko, and Y.N. Kulchin, J. Appl. Phys. **105**, 031101–11 (2009).
27. U. Bortolozzo, S. Residori, and J.P. Huignard, Opt. Lett. **34**, 2006–2008 (2009).
28. U. Bortolozzo, S. Residori, and J.P. Huignard, J. Holography Speckle **5**, 1 (2009).

Chapter 4
Diversity of Optical Signal Processing Led by Optical Signal Form Conversion

Tsuyoshi Konishi

Abstract This chapter reviews opportunities of optical signal form conversion as typified by time–space conversion in optical signal processing. Several examples of typical ultrafast optical signal processing using optical signal form conversion are described and their applications are introduced in respect to photonic networks, ultrafast measurement, and so on.

4.1 Introduction

Optical technology is expected to play an important role in the current and next generation industrial or scientific challenges. To make a full use of the ability of optical technology in the new stage, it is necessary to promote a fundamental review of the nature of light with taking advantage of the features offered by current existing technology. Here, let me remember our original purpose and consider the meaning of time–space conversion again. Basically, a light wave can be represented by spatiotemporal parameters based on a wave equation and we can use its spatiotemporal parameters for optical signal processing (Fig. 4.1a).

From the theoretical viewpoint, a wave equation can be modified into its simplified versions in time and space domains [1]. They are, respectively, given by

$$\left(\frac{\partial}{\partial z} - \frac{ik}{2}\cdot\frac{\partial^2}{\partial t^2}\right)\varphi = 0, \tag{4.1}$$

$$\left(\frac{\partial}{\partial z} + \frac{i}{2k_0}\cdot\Delta_\perp\right)\varphi = 0. \tag{4.2}$$

T. Konishi (✉)
Graduate School of Engineering, Osaka University, Osaka, Japan
e-mail: konishi@mls.eng.osaka-u.ac.jp

T. Fournel and B. Javidi (eds.), *Information Optics and Photonics: Algorithms, Systems, and Applications*, DOI 10.1007/978-1-4419-7380-1_4,

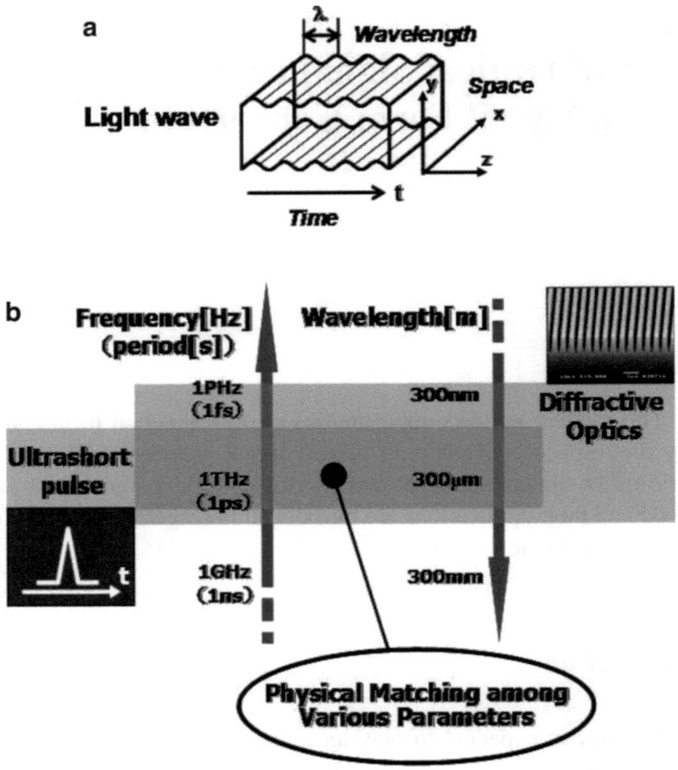

Fig. 4.1 Fundamental review of the nature of light; (**a**) light wave represented by spatiotemporal parameters and (**b**) physical matching among various physical parameters

As you can see from these two equations, temporal and spatial behaviors of light have some similarities each other. It suggests that there is a possibility of exchange between temporal and spatial phenomena.

From the phenomenalistic viewpoint, wavelength is the shortest length that can be represented by a light wave and it is at most submicrometers. In spatial domain, nanotechnology enables us to easily fabricate various dimensional structures in the range from several nanometers to several micrometers. Since this range includes a subwavelength dimension, it enables us to design artificial materials such as a photonic crystal and control a light wave more leeway. This suggests that spatial dimensions of an ultrashort pulse and an optical device come to be overlapped and ultrafast photonics and nanotechnology could go well together there. Thus, physical advancements in optical technologies encourage various kinds of physical parameters to have close interactions between them (Fig. 4.1b).

In this chapter, we review opportunities of optical signal processing assisted by optical signal form conversion for the next generation of photonics [2–5].

4.2 Optical Signal Form Conversion and Photonic Analog-to-Digital (A/D) Conversion

From the viewpoint of optical signal processing, the manner of encoding based on intensity modulation is not always suitable for processing with optical technology. In this sense, it can be said that the true necessity is not optical replacement. One promising approach appears to be employing an appropriate interface for optical signal form conversion among various physical parameters of optical signals to select an appropriate physical parameter for optical signal processing. Here, we show you representative example concerning photonic A/D conversion [6]. Figures 4.2 and 4.3 show a schematic diagram of the proposed photonic A/D convertor and its experimental results of output digital codes, respectively.

The proposed system realizes optical quantization and optical coding after the optical sampling. In optical quantization, the power of the input analog signal is converted into the extent of the center-wavelength shift by use of self-frequency

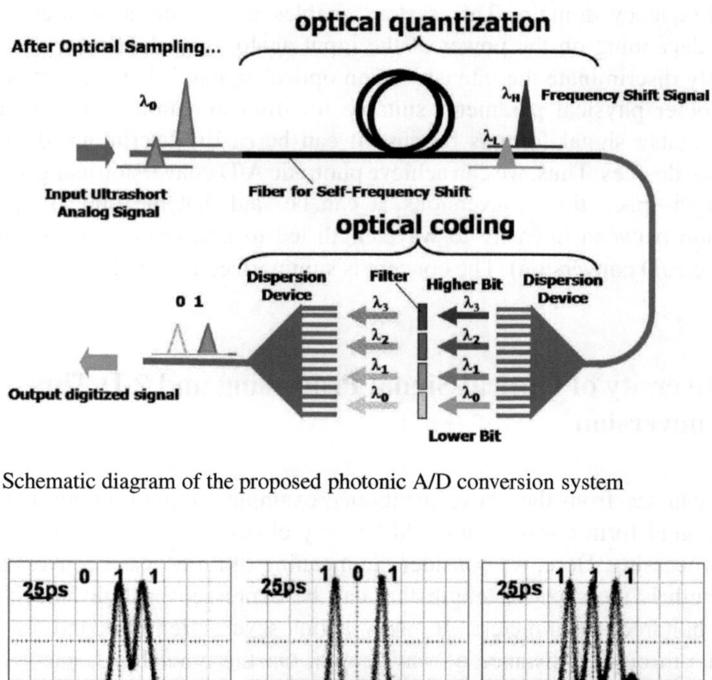

Fig. 4.2 Schematic diagram of the proposed photonic A/D conversion system

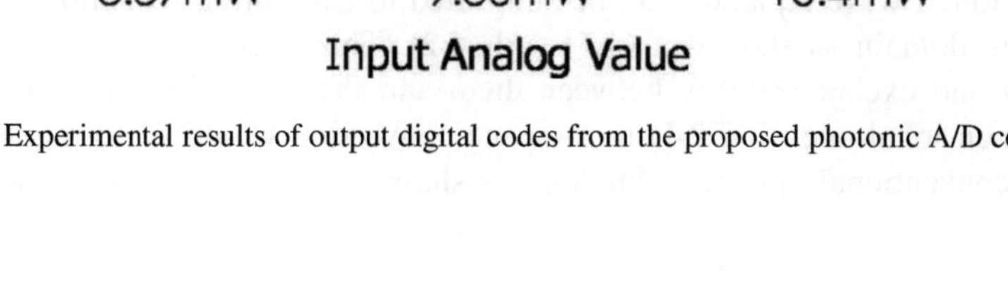

Fig. 4.3 Experimental results of output digital codes from the proposed photonic A/D conversion system

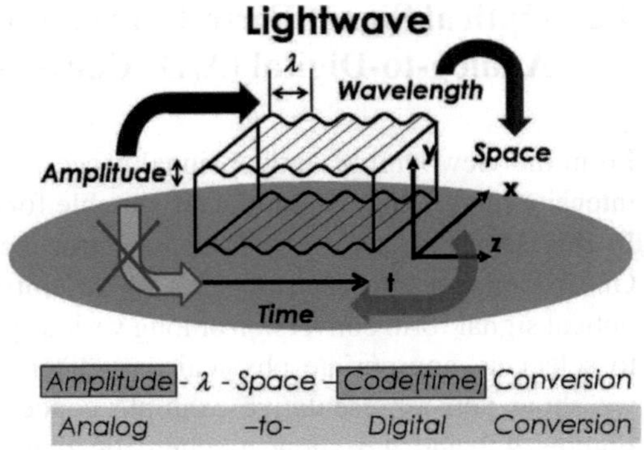

Fig. 4.4 Relationships among physical parameters in the proposed photonic A/D conversion

shifting in a fiber. Using the difference of the center wavelength, we can achieve optical quantization of an input analog signal. In optical coding, we use a pulse-shaping technique; and if we use an adequate filter and modulate the spectra of the quantized signal, we can synthesize an arbitrary shaped pulse by filtering in the frequency domain. This system enables us to output arbitrary digitized signals, depending on the power of the input analog signals. Since it is not easy to directly discriminate the intensity of an optical signal, it is necessary to convert it to another physical parameter suitable for discrimination. Wavelength is one of the suitable signal formats because it can be easily discriminated by passive dispersion devices. Thus, we can achieve photonic A/D conversion using amplitude–wavelength–space–time conversions. It can be said that the optical signal form conversion between intensity to wavelength led to new optical signal processing (photonic A/D conversion). The concept is summarized in Fig. 4.4.

4.3 Diversity of Optical Signal Processing and 2-D Time–Space Conversion

As you can see from the above-mentioned example, an appropriate interface for optical signal form conversion could be very effective for exploring new optical signal processing. Here, we consider the meaning of time–space conversion again. Wavelength is the shortest length that can be expressed by light and it is about several hundreds nanometers. It takes about several femtosecond for light to propagate through a distance of wavelength. On the other hand, nanotechnology enables to easily fabricate submicrometers structures.

It suggests that ultrafast photonics and nanotechnology could go well together. In addition, a wave equation can be simplified to two similar equations in spatial and time domain as shown in (4.1) and (4.2). This suggests the spatiotemporal analogy and exchangeability between them and the derivatives in equations are two-dimensional ones and 2-D time–space conversion would be to maximize the use of conventional optical technology as shown in Fig. 4.5. There is now much

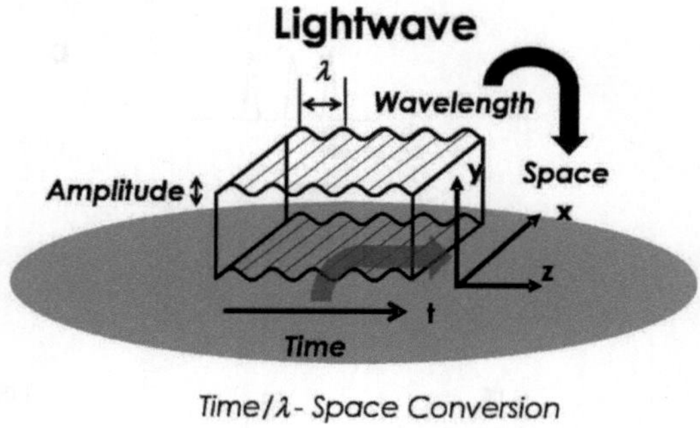

Fig. 4.5 Relationships among physical parameters in 2-D time–space conversion

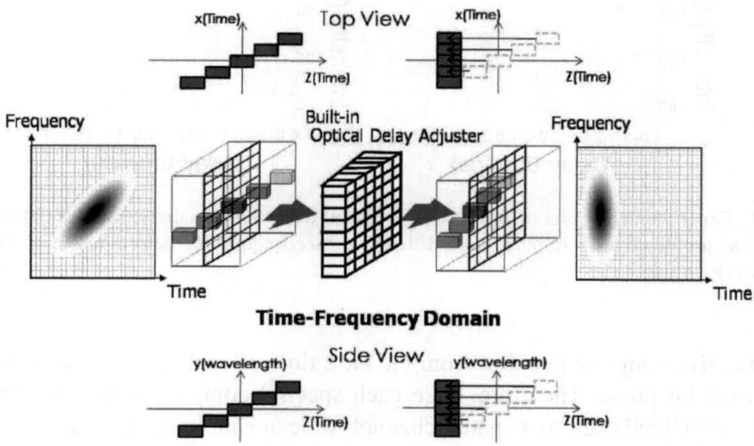

Fig. 4.6 Conceptual diagram of the proposed optical distortion equalizer

research proceeding in these topics. For example, we have proposed new approach for ultrafast optical time-resolved spectroscopy [7,8], 2-D pulse shaping [9,10], and optical distortion equalizer [11].

Here, we show you a representative example concerning an optical distortion equalizer [11]. The optical distortion equalizer is a representative example, which indicates difference from conventional time–space conversion technique [12–20]. From the viewpoint of time–frequency domain processing, all distortions are distilled to the variation in arrival times of each multiplexed spectral component in a pulse and it is equivalent to a time–frequency map. Figure 4.6 shows the conceptual diagram of operation in the proposed optical distortion equalizer. It is composed of three parts: time–frequency demultiplexing (TF-DEMUX) for retrieval of the time–frequency distribution of a distorted bit pulse in a processable and reversible way, time–frequency adjustment (TF-ADJ) for distortion cancellation by putting all multiplexed spectral components back into a base time in a time–frequency domain, and time–frequency multiplexing (TF-MUX) for returning TF-DEMUXed signals to a pulse form. A distortion can be characterized by both values of a frequency

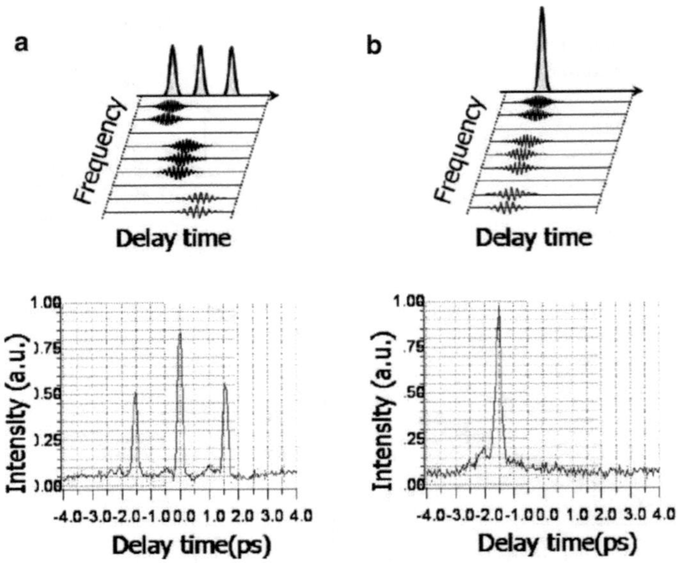

Fig. 4.7 Experimental results of the proposed optical distortion equalizer; (**a**) and (**b**) illustrate a temporal profile of a triple pulse with different center frequencies before and after TF-ADJ, respectively

and a relative temporal position from the base time of each spectral component in a distorted bit pulse. Therefore, once each spectral component in a distorted bit pulse is TF-DEMUXed to a spatial channel, we can simply achieve adjustment by setting a built-in optical phase shifter for each spatial channel so that each fixed time difference can be canceled. Since signals after TF-ADJ are still TF-DEMUXed and spatially separated, they must finally be returned to the corresponding pulse form by combining all optical delay lines for TF-ADJ into one.

We performed preliminary experiments to verify the operation of the proposed optical distortion equalizer against two types of distortions: a chromatic dispersion and a timing jitter. We regard a triple pulse with different center frequencies as a representative example of a distorted multiplexed signal with a chromatic dispersion and a timing jitter. Figures 4.7a,b show an illustrative diagram and a temporal profile of a triple pulse before and after TF-ADJ, respectively. From Fig. 4.7, we can see that a chromatic dispersion and a timing jitter are successfully compensated without any active controls.

4.4 Conclusion

This chapter reviewed opportunities of optical signal form conversion as typified by time–space conversion in optical signal processing for the current and next generation of photonics. The current physical advancements surrounding optical technology are summed up that ultrashort pulse and diffractive optics overlap each other in term of spatial dimensions. It suggests that ultrafast photonics and nanotechnology could go well together and it comes to be easier to deal with a temporal waveform in spatial domain, as well as spectral one. And conventional optical elements would be reborn again. One of the indispensable matchmaker between them would be time–space conversion. As can be seen from the example of time–space conversion, optical signal form conversion is expected to be very effective for making a full use of conventional optical technology. There is now much research proceeding in this field. For example, we have proposed new approach for optical label recognition [21, 22], and digital-to-analog (D/A) [23] conversions, and optical limiter [24].

Acknowledgements The author thank Prof. Y. Ichioka, Prof. K. Itoh, Dr. Y. Oshita, Dr. K. Iwamoto, Dr. K. Tanimura, Dr. H. Furukawa, Dr. T. Nishitani, and Dr. H. Goto for many valuable discussions and experiments during the course of this work.

References

1. S.A. Akhmanov, et al., Optics of Femtosecond Laser Pulses (AIP, New York, 1992), Chap. 1.
2. T. Konishi and Y. Ichioka, J. Opt. Soc. Am. A 16, 1076 (1999).
3. T. Konishi and Y. Ichioka, Optical Information Processing: A tribute to Adolf Lohman (SPIE, Bellingham, WA, 2002), Chap. 14.
4. T. Konishi, OFC/NFOEC2005, OTuG1 (invited paper, 2005).
5. T. Konishi, Lasers, Optics and Electro-Optics Research Trends (Nova, New York, 2007) Chap. 1.
6. T. Konishi, et al., J. Opt. Soc. Am. B 19, 2817 (2002).
7. T. Konishi and Y. Ichioka, Opt. Rev. 6, 507 (1999).
8. K. Tanimura, et al., Opt. Rev. 10, 77 (2003).
9. T. Konishi, et al., IEEE Photon. Technol. Lett. 16, 620 (2004).
10. Y. Oshita, et al., IEEE Photon. Technol. Lett., 16, 623 (2004).
11. T. Konishi, et al., IEEE J. Lightwave Technol. 24, 2693 (2006).
12. C. Froehly, B. Colombeau, and M. Vampouille, Prog. Opt. 20, 65 (1983).
13. Y.T. Mazurenko, Opt. Spectrosc. 57, 343 (1984).
14. R.N. Thurston, J.P. Heritage, A.M. Weiner, and W.J. Tomlonson, IEEE J. Quantum Electron. 22, 682 (1986).
15. A.M. Weiner, et al., IEEE J. Quantum Electron. 28, 908 (1992).
16. P.C. Sun, et al., Opt. Lett. 20, 1728 (1995).
17. M.M. Wefers and K.A. Nelson, J. Opt. Soc. Am. B 12, 1343 (1995).
18. M.M. Wefers, K.A. Nelson, and A.M. Weiner, Opt. Lett. 21, 746 (1996).
19. H.M. Ozaktas and M.C. Nuss, Opt. Comm. 131, 114 (1996).
20. P.C. Sun, Y.T. Mazurenko, Y. Fainman, J. Opt. Soc. Am. A 14, 1159 (1997)
21. T. Konishi, et.al., Opt. Lett. 18, 1445 (2001).
22. H. Furukawa, et al., Electron. Lett. 40, 1440 (2004).
23. T. Nishitani, T. Konishi, and K. Itoh, Opt. Exp. 13, 10310 (2005).
24. H. Goto, T. Konishi, and K. Itoh, J. Opt. A Pure Appl. Opt. 10, 095306 (2008).

Chapter 5
Dynamic Wavefront Sensing and Correction with Low-Cost Twisted Nematic Spatial Light Modulators

V. Durán, V. Climent, J. Lancis, E. Tajahuerce, S. Bará, J. Arines, J. Ares, P. Andrés, and Z. Jaroszewicz

Abstract Off-the-shelf spatial light modulators (SLMs), like twisted nematic liquid crystal displays (TNLCDs) used in projection systems, show some interesting features such as high spatial resolution, easy handling, wide availability, and low cost. We describe a compact adaptive optical system using just one TNLCD to measure and compensate optical aberrations. The current system operates at a frame rate of the order of 10 Hz with a four-level codification scheme. Wavefront estimation is performed through conventional Hartmann–Shack sensing architecture. The system has proved to work properly with a maximum rms aberration of 0.76 μm and wavefront gradient of 50 rad/mm at a wavelength of 514 nm. These values correspond to typical aberrations found in human eyes. The key point of our approach is careful characterization and optimization of the TNLCD for phase-only modulation. For this purpose, we exploit the so-called retarder–rotator approach for twisted nematic liquid crystal cells. The optimization process has been successfully applied to SLMs working either in transmissive or in reflective mode, even when light depolarization effects are observed.

5.1 Introduction

Clinical applications for early diagnosis and patient monitoring, as well as basic studies on physiological optics and vision science, have benefited in the last years from the development of reliable wavefront sensors that are able to measure the human eye aberrations and optical components and devices that are able to compensate them. This rapidly growing field, arising from the convergence of three distinct lines of research represented by the pioneering works of Hartmann (noninterferometric wavefront measurements) [1], Babcock (proposal of using addressable active elements for the compensation of wavefront distortions) [2] and Smirnov

V. Durán (✉)
GROC.UJI, Physics Department-INIT, Universitat Jaume I, E12080 Castelló, Spain
e-mail: vduran@sg.uji.es

T. Fournel and B. Javidi (eds.), *Information Optics and Photonics: Algorithms, Systems, and Applications*, DOI 10.1007/978-1-4419-7380-1_5,

(measurement of high-order human eye aberrations) [3] enable unprecedented advances, such as the observation and classification in vivo of the retinal cone mosaic [4].

Wavefront sensing and compensation is today a promising technology for biomedical optics. There is correspondingly a trend toward the development of reliable, low-cost, easy-to-use devices for their transfer to clinical practice. Wave aberrations can be measured using different approaches, nowadays the most efficiently used approach is based on the measurement of the local aberration slopes. This can be done with several kinds of aberrometers like the widespread Hartmann–Shack wavefront sensor [5, 6], the laser ray tracer [7], or the spatially resolved re-fractometer [8], among others. Eye aberration compensation, in turn, has been successfully demonstrated using deformable mirrors [9–13], spatial light modulators [14–18], and/or static phase plates [19, 20].

General-purpose spatial light modulators based on liquid crystal displays, like those used in video projection devices, offer some interesting features for this task. They can be configured to act basically as segmented (pixelated) piston correctors, offering a relatively high spatial resolution (of the order of 0.5–0.7 megapixels for inch-sized displays), adaptiveness, and easy addressability and control. There are also low-cost devices widely available, and it can be anticipated that their performance will improve in the near future, driven by the needs of the consumer electronics industry. With the deformable mirrors they share their ability to compensate dynamic eye aberrations (although with a lower temporal bandwidth), and with the phase plates they share their high spatial resolution. However, given the relatively low optical quality of their end surfaces and the small dynamic range of the phase retardation that can be introduced at each pixel (generally less than 2π rad at visible wavelengths), the use of conventional SLMs in eye optics and visual science was not particularly extended. The trend in the last years was rather to move towards special purpose SLMs [14–16] with higher dynamic range of phase retardation at each pixel (at the price of a drastic loss of spatial resolution) or to some high-end devices [17, 18].

5.2 Characterization of a Twisted Nematic Liquid Crystal Display

TNLCDs are optoelectronic pixelated devices that can act as programmable SLMs. Every pixel contains a thin liquid crystal layer with twisted molecular alignment and is coated with transparent electrodes. This makes it possible to apply an electric field across the cell to tilt the molecules. In order to optimize the modulation response of a TNLCD, a reliable model to describe the molecular behavior in the field-on state is required. A numerical method for computing the change in molecular orientation was first proposed by Berreman [21]. Later, Lu and Saleh presented a simplified mathematical model based on the Jones matrix formulation [22]. They assume that molecules twist linearly along the cell gap and have a constant tilt for each value of the applied voltage. In this case, an analytical solution for the Jones matrix of an activated liquid crystal cell can be obtained. However, the assumptions of Lu and Saleh approach do not hold for the current flat commercial displays. From an experimental point of view, a TNLCD can be calibrated by measuring its Jones matrices for each applied voltage with the aid of a polarimetric system [23].

5.2.1 Equivalent Retarder–Rotator Approach

An alternative method for the calibration of a TNLCD is based on the so-called equivalent retarder–rotator approach [24, 25]. It is well-known that, in the absence of polarization-dependent losses, any polarization device is optically equivalent to a system consisting of a linear retarder followed by a rotator [26]. The characteristic parameters of the equivalent system fully determine the optical behavior of the device. In the case of an off-state TNLCD, they can be used for analyzing liquid crystal (LC) modes as well as for univocally measuring the physical parameters of the LC cells [24,27]. This model can be extended to deal with an addressed TNLCD and enables to interpret the action of a LC cell as two successive rotations on the Poincaré sphere. The determination of the characteristic parameters for each value of the applied voltage may be performed by several polarimetric techniques [25,27].

Discarding depolarization and diattenuation effects, the polarization properties of a TNLCD can be described by its Jones matrix $\mathbf{J}_{\mathrm{TNLCD}}$, which can be written as

$$\mathbf{J}_{\mathrm{TNLCD}} = \exp(-i\,\beta)\,\mathbf{R}(-\phi)\,\mathbf{U}(\phi,\,\beta). \tag{5.1}$$

In (5.1), $\mathbf{R}$ is a 2×2 rotation matrix and $\mathbf{U}$ is a unitary matrix that depends on the angles ϕ and β, which are, respectively, the LC twist angle and the birefringence angle. The birefringence angle is defined as $\beta = \pi d \Delta n / \lambda$, where d is the cell thickness, Δn is the material birefringence, and λ is the light wavelength. The above factorization of $\mathbf{J}_{\mathrm{TNLCD}}$ is valid not only in the off-state but also when a voltage is applied to the LC cells [28]. In the former case, the matrix $\mathbf{J}_{\mathrm{TNLCD}}$ admits an analytical expression [29],

$$\mathbf{J}_{\mathrm{TNLCD}} = \exp(-i\,\beta)\,\mathbf{R}(-\phi)\begin{pmatrix} X - iY & Z \\ -Z & X + iY \end{pmatrix}, \tag{5.2}$$

where X, Y, and Z are real quantities given by $X = \cos\gamma$, $Y = (\beta/\gamma)\sin\gamma$, and $Z = (\phi/\gamma)\sin\gamma$, with $\gamma = (\phi^2 + \beta^2)^{1/2}$.

According to the retarder–rotator approach, a cell of a TNLCD is optically equivalent to the association of a linear retarder and a rotator. The corresponding Jones matrix can be written as

$$\mathbf{M}_{\mathrm{TNLCD}} = \exp(-i\beta)\,\mathbf{R}\left(\phi_{\mathrm{eq}}\right)\mathbf{WP}\left(2\delta_{\mathrm{eq}}, \frac{\phi + \phi_{\mathrm{eq}}}{2}\right), \tag{5.3}$$

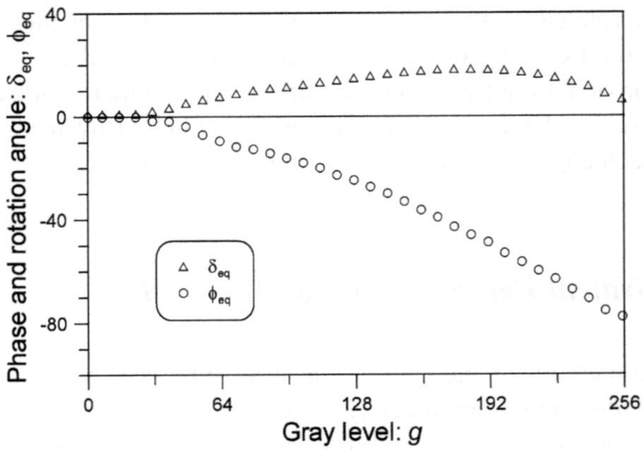

Fig. 5.1 Representation of equivalent parameters δ_{eq} and ϕ_{eq} of an off-the-shelf TNLCD as a function of the addressed gray level, g

where the rotation angle ϕ_{eq} and the phase angle δ_{eq} are two independent parameters that characterize the equivalent system. When no voltage is applied to the LC cells, both equivalent angles are related to the cell design parameters ϕ and β through the equations [24]

$$\delta_{eq} = \arcsin\left(\frac{\beta}{\gamma}\sin\gamma\right) \tag{5.4}$$

and

$$\phi_{eq} = -\phi + \arctan\left(\frac{\phi}{\gamma}\frac{\sin\gamma}{\cos\gamma}\right). \tag{5.5}$$

Calibration curves of a commercial TNLCD (Sony LCX016AL with 832×624 32–μm-sized pixels) at $\lambda = 514$ nm are represented in Fig. 5.1 [25]. The application of a voltage to the cells is performed by sending a gray-level image to the device. Thus, every gray-level g corresponds to a value of the applied voltage. The above display has an 8-bit controller that gives 256 different gray-levels in the on-state. Figure 5.1 shows that both equivalent parameters are close to zero for low values of the gray level. This means that the maximum voltage applied to the cells corresponds to $g = 0$, since in that case the LC medium becomes isotropic. A comparison of $\delta_{eq}(g)$ and $\phi_{eq}(g)$ reveals that the rotation effect is clearly the predominant one.

In the last years, TNLCDs based on liquid crystal on silicon (LCoS) technology have become widely used, since such devices are characterized by a high spatial resolution (10–μm pixel-pitch) and an excellent fill factor (>90%). LCoS displays operate in reflection and are optically equivalent to a voltage-dependent linear retarder (as the rotation effect is cancelled by the double pass of the light through the medium). However, these displays can present nonnegligible depolarization effects [30]. To deal with this fact, it is convenient to analyze light polarization devices in terms of Mueller matrices. In this case, a polarization state is not described by the complex amplitudes of the electric field but by the corresponding Stokes vector. In the Mueller–Stokes formalism, the optical behavior of a polarization element is characterized by three effects: diattenuation, retardance, and depolarization, which can be conveniently separated by the so-called polar decomposition of the Mueller matrix introduced by Lu and Chipman [31]. In an LCoS display, the diattenuation effect is negligible, but the behavior of the degree of polarization as a function of the input polarization state shows a coupling of linearly polarized light into unpolarized light than can be of about 10% [32].

The polar decomposition of Mueller matrices has been used to calibrate (and optimize) an LCoS display. First, experimental Mueller matrices of the sample display are measured by means of a suitable polarimetric technique. Then, depolarization and retardance parameters of the display are extracted with the aid of the Lu–Chipman decomposition. Once the retardance of the display has been determined, its phase response can be predicted by means of the Jones matrix calculus. On its turn, light intensity after the PSA can be calculated from experimental Mueller matrices. Therefore, combining both Jones and Mueller formalisms, the optical response of an LCoS display is fully determined [32, 33].

5.3 Optimization of the Phase Response of a Twisted Nematic Liquid Crystal Display

TNLCDs are commonly used as intensity SLMs due to its high resolution, low cost, and simple drive signal requirements. However, in many nondisplay applications it is desirable to achieve a spatial control of the phase of an input wavefront without intensity variations. In a liquid crystal cell, the electrically controlled birefringence yields a phase retardation effect. However, the twisted nematic structure not only rotates the plane of polarization (working in a wave-guiding operation mode) but also changes the ellipticity of an input polarized light beam [34]. As a result, a coupled amplitude and phase modulation is obtained when an addressed TNLCD is sandwiched between two crossed polarizers.

Different methods have been proposed to optimize the phase response of a TNLCD [23, 35–40]. As stated above, this kind of SLMs primary modulates the state of polarization (SOP) of an input light beam as the addressed voltage to cells is changed. It can be proved that a TNLCD acting as a phase-only SLM must provide a SOP modulation that is necessarily subjected to specific restrictions [38]. As an example, a phase-only response is achieved if the liquid crystal device is able to generate a set of equi-azimuth polarization states (EAPSs), i.e., a set of states that is distributed along a meridian of the Poincaré sphere. As we demonstrated in [39], a TNLCD followed by a linear retarder can work as an EAPS generator. Figure 5.2 shows the experimental results obtained with the aforementioned Sony TNLCD inserted between a linear polarizer and a quarter wave plate. When the axis of the polarizer and the slow axis of the quarter wave plate are oriented at angles $\theta_1 = -28°$ and $\theta_2 = 16°$, respectively, the system generates a set of quasi EAPSs with a mean azimuth α_0 of $-2°$. In order to achieve a phase-only modulation, a liner

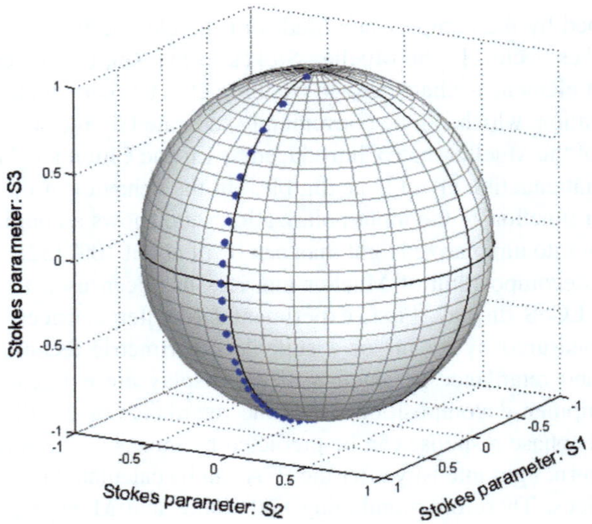

Fig. 5.2 Representation on the Poincaré sphere of the polarization states at the output of an EAPS generator implemented with a commercial TNLCD

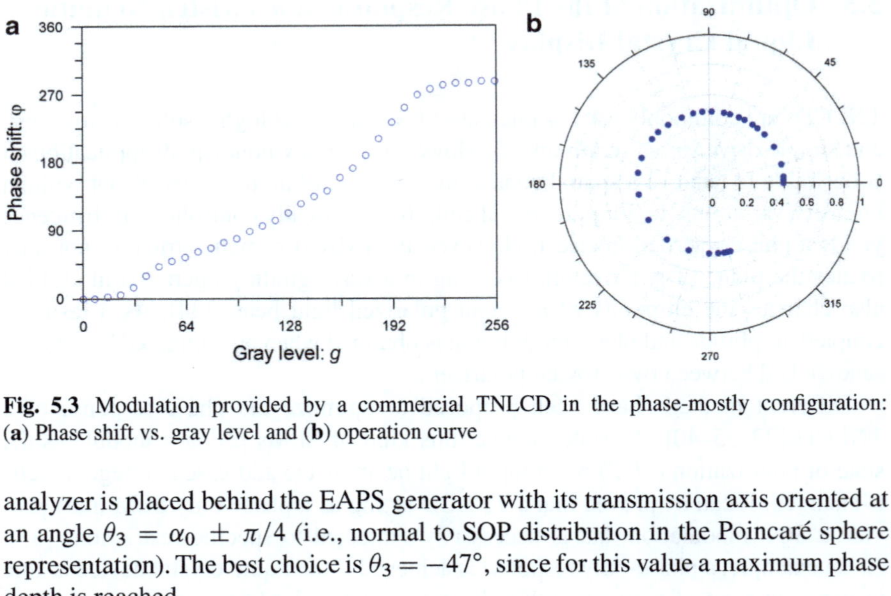

Fig. 5.3 Modulation provided by a commercial TNLCD in the phase-mostly configuration: (**a**) Phase shift vs. gray level and (**b**) operation curve

analyzer is placed behind the EAPS generator with its transmission axis oriented at an angle $\theta_3 = \alpha_0 \pm \pi/4$ (i.e., normal to SOP distribution in the Poincaré sphere representation). The best choice is $\theta_3 = -47°$, since for this value a maximum phase depth is reached.

Figure 5.3a shows a plot of the phase-shift $\Delta\varphi$ vs. the gray level g in the optimal configuration. The phase response of our SLM was measured by means of a method based on the fractional-Talbot effect [41]. The corresponding operation curve is depicted in Fig. 5.3b. The radius and polar angle of each point represent, respectively, the transmitted intensity and the phase shift for a given addressed gray level. The experimental results are close to a pure phase modulation regime. The maximum phase-modulation depth is slightly greater than $3\pi/2$ at 514 nm and the coupled intensity modulation is less than 2.5%. The values of the transmitted intensity oscillate around 0.5. This is the intensity level expected when a set of EAPS is generated [39].

5.4 Use of a Twisted Nematic Liquid Crystal Display for an Efficient Compensation of Aberrations

Off-the-shelf TNLCDs have been seldom used for the compensation of eye aberrations, mainly due to the relatively low dynamic range of the phase retardation that can be introduced at each pixel. Here we describe an efficient four-level phase encoding scheme which allows us to use a conventional TNLCD for the compensation of optical aberrations, as those typically found in human eyes [42].

5.4.1 Aberration Encoding Scheme

We used the above calibrated Sony panel as SLM, which was sandwiched between two quarter wave plates and two polarizers in order to achieve a phase-only response. The aberration to be compensated at each pixel was codified by rounding it to the nearest level using a four-level $[(0, \pi/2, \pi, 3\pi/2)$ rad] encoding scheme. In this way the SLM acts as a diffractive element. In the case of an N-level phase encoding scheme, the diffraction efficiency in the useful signal present in the first order of diffraction is equal to [43]:

$$\eta_1 = \operatorname{sinc}^2\left(\frac{1}{N}\right), \tag{5.6}$$

where $\operatorname{sinc}(x) = \sin(\pi x)/\pi x$. Then, for $N = 4$, we have $\eta_1 = 81\%$. The remaining 19% of the incident light is lost in diffraction orders other than those of interest. The diffraction efficiency of real SLMs will be somewhat lower, because the fill factor of the elementary cell is smaller than 1. Therefore, an N-step phase grating programmed in the OX direction will act as a two-dimensional grating with additional binary amplitude modulation in both directions and hence its diffraction efficiency in the $(1, 0)$ diffraction order will be equal to [44]:

$$\eta_{1,0} = \alpha_X^2 \alpha_Y^2 \operatorname{sinc}^2\left(\frac{\alpha_X}{N}\right), \tag{5.7}$$

where α_X, α_Y are the opening ratios in the OX and OY direction, respectively, i.e., the pixel's widths divided by the SLM's pitch. In our device, this limits the first-order useful signal to 26.7% of the incident energy.

As stated above, diffraction efficiency can be increased by using high values of N. However, in general this number will be limited by two factors: the maximum phase change φ_{MAX} admitted by the device $\varphi_{MAX} \geq 2\pi(N-1)/N$, and the maximum wavefront slope to be compensated, which is limited by the width of the minimum number of pixels necessary for covering the phase change of 2π rad, and is equal to $2\pi/Nd$ where d is the pixel array pitch. In our case, the number of phase steps per period was chosen as $N = 4$ [42].

5.4.2 Experimental Results

In order to evaluate the compensation performance of our TNLCD, an aberrated eye was built, consisting of a 380–*mm* focal-length lens and a CCD, in front of which different kinds of refractive phase plates were located in order to introduce different amounts and types of aberration. The compensating SLM was positioned very close to the aberrated plate without using any intermediate optical element. A more precise setup could include a relay optical system to project the SLM plane onto the plate plane, but we did not find it necessary for a proof-of-concept experiment, since for eye aberrations of typical magnitude a small axial separation between the compensating element and the eye pupil does not give rise to a noticeable loss of compensation [45]. The whole setup was illuminated @ 514 nm by a TEM_{00} expanded beam from an Ar laser, provided with a set of neutral density filters to keep irradiance at a suitable level.

The aberrated plates used in this study were manufactured at Universidade de Santiago de Compostela by gray-level photosculpture, a process already used to fabricate phase plates for the compensation of eye aberrations [19]. Figure 5.4 shows the actual interferograms of the aberrated plates (first column), the interferograms computed from the wavefront sensor measurements (second column) and

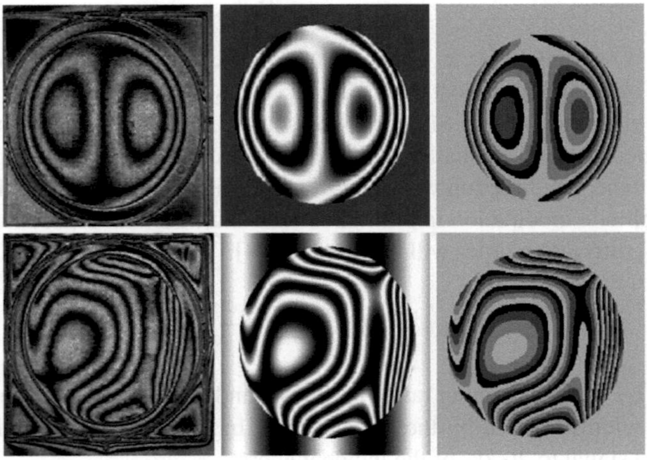

Fig. 5.4 *First column:* actual interferograms of the aberrated plates; *second column:* interferograms computed from the wavefront sensor measurements; and *third column:* the corresponding four-level phase patterns sent to the SLM

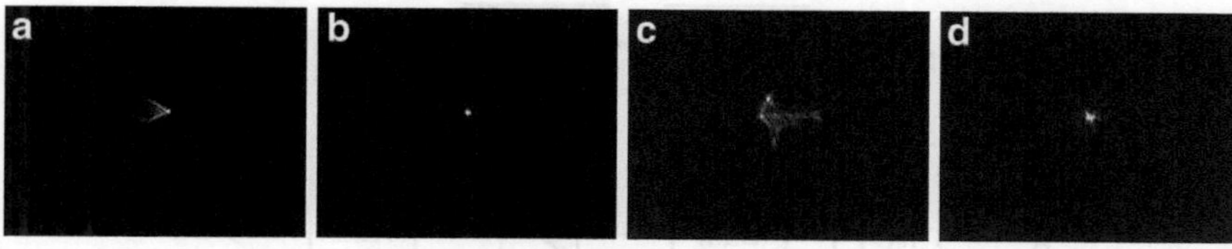

Fig. 5.5 Compensation of two types of aberration: (**a**) aberrated PSF produced by the coma term $Z_3{}^1$ and (**b**) PSF after compensation using the SLM. (**c**) An aberration pattern typical of a human eye and (**d**) PSF after compensation using the SLM

the corresponding four-level phase patterns sent to the SLM (third column). The diameter of the optical zone of the plates is 6.4 mm. The plates correspond to the Zernike coma term $Z_3{}^1$ (double-index OSA scheme [46]) in the first row and a typical pattern of a moderately highly aberrated eye in the second row. The magnitude of the aberration generated by each plate is 2.39 μm peak-to-valley (0.44 μmrms) for $Z_3{}^1$ and 4.50 μm p–v (0.76 μm rms) for the aberrated eye (all measurements were made at 587 nm wavelength).

Once measured, each phase plate was located in front of the lens-CCD setup with the SLM switched-off and an image of the corresponding aberrated point-spread function (PSF) was recorded at the CCD plane. Then, the four-level image of the compensating phase was codified in our TNLCD, as computed from the Zernike coefficients of the plate given by the wavefront sensor (taking into account that the photoresist refractive index must be corrected since we are working under 514 nm illumination), and the compensated PSF was recorded.

Experimental results are shown in Fig. 5.5. The PSF of the aberrated eye produced by each plate is shown in Fig. 5.5a, c. In their turn, Fig. 5.5b, d show the compensated PSF. It collapses in both cases to a size noticeably smaller than that of the original aberrated one.

5.5 Measurement and Compensation of Optical Aberrations Using a Single Spatial Light Modulator

TNLCDs are used not only to compensate different kinds of optical aberrations but also to generate diffractive microlens arrays to implement Hartmann–Shack wavefront sensors [47,48]. These two possibilities can be combined to build a compact and versatile adaptive optics module using a single SLM acting both as the microlens array of a Hartmann–Shack wavefront sensor and as the aberration compensating element [49].

5.5.1 Basic Layout of the Adaptive Setup

Figure 5.6 shows the basic block diagram of the adaptive system developed in this work. The core of the system is a TNLCD located in the main path of the aberrated

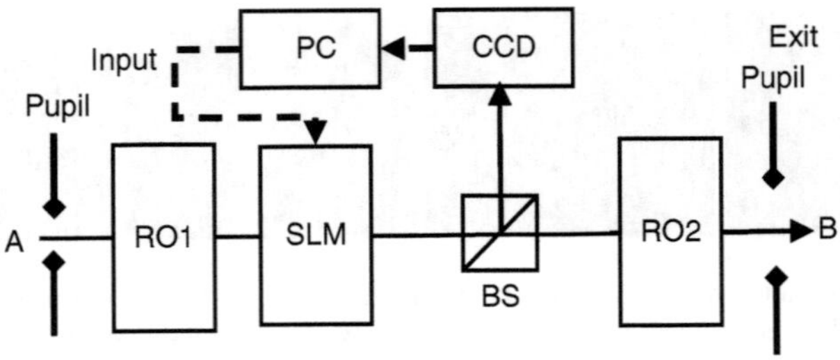

Fig. 5.6 Block diagram of the single SLM adaptive setup (see text for details)

beam. During the measurement step an array of diffractive microlenses is encoded in the TNLCD, and an image of its focal spots is taken at the CCD with the aid of a beam splitter (BS). The focal spots are detected and labeled using appropriate image processing routines running in a conventional computer (PC), their centroids are calculated and their displacements with respect to a set of reference positions are measured. The beam aberration is determined from these raw data using a suitable (e.g., least-squares) estimator [50]. This aberration, with opposite sign, is subsequently encoded into the TNLCD replacing the microlens array, starting in this way the compensation step. In case of time-varying aberrations, this loop is repeated as many times as necessary. RO1 and RO2 are two (optional) relay optics modules, useful to project the TNLCD onto the input pupil of the system, and to project the TNLCD onto the exit pupil; besides, they enable to match the respective pupil sizes, making an optimum use of the available spatial resolution of the TNLCD. During the measurement step the exit beam at B is distorted due to its interaction with the microlens array and this implementation of the adaptive unit provides no useful beam to the optical systems located behind it. However, the measurement time can be kept within very small values using an efficient design of the image acquisition and processing routines. In fact, the microlens array should be displayed only during image acquisition by the CCD. During the image processing time, the TNLCD can be displaying the previous correction phase. In this way, the blind time can be reduced to the integration time of the CCD (in our case only 30 ms). Once the aberration of the incoming beam is compensated, no further measurement has to be made until the quality of the output beam is degraded above a user-defined threshold.

5.5.2 Experimental Results

In our experiments we used the TNLCD Sony Model in a phase-only configuration that includes two polarizers and two quarter wave plates. As light source we use a green LED (Luxeon Star LXHL-MM1D^{TM}) with an emission spectrum centered in 530 nm with 35 nm bandwidth. Since the maximum depth modulation obtained with this device at 530 nm is slightly higher than $3\pi/2$ rad, we used a four-level phase

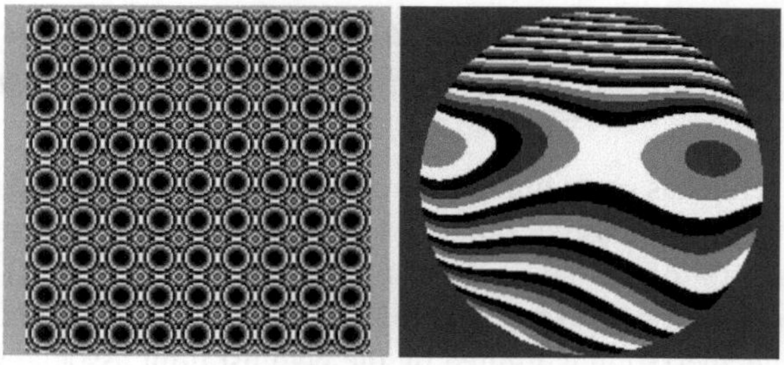

Fig. 5.7 Grayscale representation of the four-level TNLCD patterns: (*left*) for generating a 9 × 9 diffractive microlens array; (*right*) for compensating the aberration produced by an artificial eye

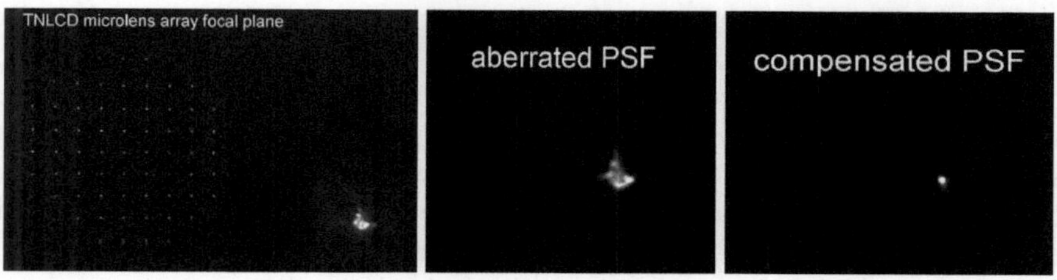

Fig. 5.8 CCD images obtained at three consecutive time intervals in a typical demonstration run

encoding scheme to display the wrapped phases corresponding to the microlens array and to the aberration compensating element. Full details of this procedure are described in Sect. 4.4.

Figure 5.7 shows typical four-level patterns sent to the TNLCD in the measurement and compensating steps. Figure 5.7 (left) corresponds to the pattern for generating a 9×9 diffractive microlens array with subpupil size 560 μm and focal length 4.2 cm (at 530 nm) used in our experiments. Figure 5.7 (right) corresponds to the pattern for compensating the aberration generated by an artificial eye with a phase plate representative of an eye aberration of average magnitude including high-order Zernike terms up to seventh radial order (35 nontrivial Zernike aberrations), with relatively strong astigmatic components and with partially corrected defocus.

Figure 5.8 shows the CCD images obtained at three consecutive slots of time in a typical demonstration run: Fig. 5.8 (left), taken during the measurement step, shows at the left side the microlens array foci, and at the lower right the image of the LED source given by the system, strongly distorted due to the passing of the aberrated beam through the microlens array; Fig. 5.8 (center) shows an enlarged view of the original aberrated LED image, with the TNLCD in off-state. Finally, Fig. 5.8 (right) shows the corrected image, after the TNLCD was fed with the aberration (with opposite sign) computed after measurement. The collapsing of the aberrated PSF after compensation can be easily noticed.

5.5.3 Discussion of Experimental Results: Limitations and Advantages of Twisted Nematic Liquid Crystal Displays

The proposed setup has some limitations common to most pixelated TNLCD-based SLMs: (1) it requires the use of polarized light, so that for completely unpolarized input beams a 50% of the power is lost at the first polarizer; (2) the light shall also be relatively narrowband for avoiding chromatic dispersion; and (3) the spatial frequency of the aberration is limited by the Nyquist limit associated to the fundamental spatial frequency of the TNLCD pixel array. There are two additional power loss factors associated with some low-cost devices as that used in our experiments: the losses due to the relatively small fill factor of the TNLCD cells and those due to the power going to diffraction orders other than that of interest.

No other fundamental factors limit the performance of the proposed device, whose temporal bandwidth is determined in our case by our particular image acquisition and processing routines. The main advantages of this setup are its great versatility, low cost, easy addressability and control, and last but not least, its seamless integration with the PC-based hardware and software commonly used in optoelectronics applications. Whether or not the light efficiency of our device is enough to apply it to near real-time compensation of human eye aberrations is still to be determined. It is, however, quite well suited to be used as an active compensation unit in optical setups where the aberrations are not very fast changing in time.

References

1. Hartmann, J. Objektivuntersuchungen. Zeitschrift für Instrumentenkunde **XXIV**, 1–21 (January), 3 and 34–47 (February), 7 and 98–117 (April) (1904)
2. Babcock, H.W. The Possibility of Compensating Astronomical Seeing. Publ. Astron. Soc. Pac. **65**, 229–236 (1953)
3. Smirnov, M.S. Measurement of the wave aberration of the human eye. Biofizika **6**, 687–703 (1961)
4. Roorda, A., Williams, D.R. The arrangement of the three cone classes in the living human eye. Nature **397**, 520–522 (1999)
5. Shack, R.V., Platt, B.C. Production and use of a lenticular Hartmann screen. J. Opt. Soc. Am. **61**, 656 (1971)
6. Liang, J.Z., Grimm, B., Goelz, S., Bille, J.F. Objective measurement of wave aberrations of the human eye with the use of a Hartmann-Shack wave-front sensor. J. Opt. Soc. Am. A **11**, 1949–1957 (1994)
7. Navarro, R., Losada, M. A. Phase transfer and point-spread function of the human eye determined by a new asymmetric double-pass method. J. Opt. Soc. Am. A **12**, 2385–2392 (1995)
8. Webb, R.H., Penney, C.M., Thompson, K.P. Measurement of ocular local wave-front distortion with a spatially resolved refractometer. Appl. Opt. **31**, 3678–3686 (1992)
9. Dreher, A.W., Bille, J.F., Weinreb, R.N. Active optical depth resolution improvement of the laser tomographic scanner. Appl. Opt. **28**, 804–808 (1989)
10. Liang, J.Z., Williams, D.R., Miller, D.T. Supernormal vision and high resolution retinal imaging through adaptive optics. J. Opt. Soc. Am. A **14**, 2884–2892 (1997)
11. Hofer, H., Chen, L., Yoon, G.Y., Singer, B., Yamauchi, Y., Williams, D.R. Improvement in retinal image quality with dynamic correction of the eye's aberrations. Opt. Express **8**, 631–643 (2001)

12. Fernández, E.J., Iglesias, I., Artal, P. Closed-loop adaptive optics in the human eye. Opt. Lett. **26**, 746–748 (2001)
13. Diaz-Santana, L., Torti, C., Munro, I., Gasson, P., Dainty, C. Benefit of higher closed-loop bandwidths in ocular adaptive optics. Opt. Express **11**, 2597–2605 (2003)
14. Love, G.D. Wave-front correction and production of Zernike modes with a liquid-crystal spatial light modulator. Appl. Opt. **36**, 1517–1524 (1997)
15. Thibos, L.N., Bradley, A. Use of liquid-crystal adaptive optics to alter the refractive state of the eye. Optom. Vision Sci. **74**, 581–587 (1997)
16. Vargas-Martin, F., Prieto, P.M., Artal, P. Correction of the aberrations in the human eye with a liquid crystal spatial light modulator: limits to performance. J. Opt. Soc. Am. A **15**, 2552–2562 (1998)
17. Shirai, T. Liquid-crystal adaptive optics based on feedback interferometry for high-resolution retinal imaging. Appl. Opt. **41**, 4013–4023 (2002)
18. Prieto, P.M., Fernández, E.J., Manzanera S., Artal, P. Adaptive optics with a programmable phase modulator: applications in the human eye. Opt. Express **12**, 4059–4071 (2004)
19. Navarro, R., Moreno-Barriuso, E., Bará, S., Mancebo, T. Phase-plates for wave-aberration compensation in the human eye. Opt. Lett. **25**, 236–238 (2000)
20. Burns, S.A., Marcos, S., Elsner, A.E., Bará, S. Contrast improvement of confocal retinal imaging by use of phase-correcting plates. Opt. Lett. **27**, 400–402 (2002)
21. Berreman, D.W. Dynamics of liquid-crystal twist cells. Appl. Phys. Lett. **25**, 12–15 (1974)
22. Lu, K.H., Saleh, B.E.A. Theory and design of the liquid-crystal TV as an optical spatial phase modulator. Opt. Eng. **29**, 240–246 (1990)
23. Moreno, I., Velasquez, P., Fernandez-Pousa, C.R., Sanchez-Lopez, M.M. Jones matrix method for predicting and optimizing the optical modulation properties of a liquid-crystal display. J. Appl. Phys. **94**, 3697–3702 (2003)
24. Duran, V., Lancis, J., Tajahuerce, E., Jaroszewicz, Z. Cell parameter determination of a twisted-nematic liquid crystal display by single-wavelength polarimetry. J. Appl. Phys. **97**, 043101–043106 (2005)
25. Duran, V., Lancis, J., Tajahuerce, E., Jaroszewicz, Z. Equivalent retarder-rotator approach to on-state twisted nematic liquid crystal displays. J. Appl. Phys. **99**, 113101–113106 (2006)
26. Hurwitz, H., Jones, R.C. A new calculus for the treatment of optical systems II. Proof of three general equivalence theorems. J. Opt. Soc. Am. **31**, 493–499 (1941)
27. Tang, S.T., Kwok, H.S. Characteristic parameters of liquid crystal cells and their measurements. J. Display Technol. **2**, 26–31 (2006)
28. Fernandez-Pousa, C.R., Moreno, I., Bennis, N., Gomez-Reino, C. Generalized formulation and symmetry properties of reciprocal nonabsorbing polarization devices: application to liquid-crystal displays. J. Opt. Soc. Am. A **17**, 2074–2080 (2000)
29. DelaTocnaye, J.L.D., Dupont, L. Complex amplitude modulation by use of liquid-crystal spatial light modulators. Appl. Opt. **36**, 1730–1741 (1997)
30. Wolfe, J.E., Chipman, R.A. Polarimetric characterization of liquid-crystal-on-silicon panels. Appl. Opt. **45**, 1688–1703 (2006)
31. Lu, S.Y., Chipman, R.A. Interpretation of Mueller matrices based on polar decomposition. J. Opt. Soc. Am. A **13**, 1106–1113 (1996)
32. Clemente, P., Duran, V., Martínez-León, Ll., Climent, V., Tajahuerce, E., Lancis, J. Use of polar decomposition of Mueller matrices for optimizing the phase response of a liquid-crystal-on-silicon display. Opt. Express **16**, 1965–1974 (2008)
33. Moreno, I., Lizana, A., Campos, J., Márquez, A., Iemmi, C. Yzuel, M.J. Combined Mueller and Jones matrix method for the evaluation of the complex modulation in a liquid-crystal-on-silicon display. Opt. Lett. **33**, 627–629 (2008)
34. Gooch, C.H., Tarry, H.A. Optical-properties of twisted nematic liquid-crystal structures with twist angles less than 90 degrees. J. Phys. D **8**, 1575–1584 (1975)
35. Konforti, N., Maron, E., Wu, S.T. Phase-only modulation with twisted nematic liquid-crystal spatial light modulators. Opt. Lett. **13**, 251–253 (1988)
36. Pezzaniti, J.L., Chipman, R.A. Phase-only modulation of a twisted nematic liquid-crystal TV by use of the eigenpolarization states. Opt. Lett. **18**, 1567–1569 (1993)

37. Davis, J.A., Moreno, I., Tsai, T. Polarization eigenstates for twisted-nematic liquid-crystal displays. Appl. Opt. **37**, 937–945 (1998)
38. Durán, V., Clemente, P., Martínez-León, Ll., Climent, V., Lancis, J. Poincaré-sphere representation of phase-mostly twisted nematic liquid crystal spatial light modulators. J. Opt. A: Pure Appl. Opt. **11**, 085403 (2009)
39. Duran, V., Lancis, J., Tajahuerce, E., Fernandez-Alonso, M. Phase-only modulation with a twisted nematic liquid crystal display by means of equi-azimuth polarization states. Opt. Express **14**, 5607–5616 (2006)
40. Duran, V., Lancis, J., Tajahuerce, E., Climent, V. Poincaré sphere method for optimizing the phase modulation response of a twisted nematic liquid crystal display. J. Displ. Technol. **3**, 9–14 (2007)
41. Serrano-Heredia, A., Lu, G.W., Purwosumarto, P., Yu, F.T.S. Measurement of the phase modulation in liquid crystal television based on the fractional-Talbot effect. Opt. Eng. **35**, 2680–2684 (1996)
42. Durán, V., Climent, V., Tajahuerce, E., Jaroszewicz, Z., Arines, J., Bará, S. Efficient compensation of Zernike modes and eye aberration patterns using low-cost spatial light modulators. J. Biomed. Opt. **12**, 014037 (2007)
43. Dammann, H. Blazed synthetic phase only holograms. Optik **31**, 95–104 (1970)
44. Jaroszewicz, Z., Koodziejczyk, A., Kowalik, A., Restrepo, R. Determination of phase step errors of kinoform gratings from their diffraction efficiencies. Opt. Eng. **40**, 692–697 (2001)
45. Bará, S., Mancebo, T., Moreno-Barriuso, E. Positioning tolerances for phase plates compensating aberrations of the human eye. Appl. Opt. **39**, 3413–3420 (2000)
46. Thibos, L.N., Applegate, R.A., Schwiegerling, J.T., Webb, R., VSIA Standards Taskforce Members: Standards for reporting the optical aberrations of eyes in Vision Science and Its Applications 2000, V. Lakshminarayanan, ed., Vol. 35 of OSA Trends in Optics and Photonics Series (Optical Society of America, Washington, D.C., 2000)
47. Seifert, L., Liesener, J., Tiziani, H.J. The adaptive Shack-Hartmann sensor. Opt. Commun. **216**, 313–319 (2003)
48. Zhao, L., Bai, N., Li, X., Ong, L.S., Fang, Z.P., Asundi, A.K. Efficient implementation of a spatial light modulator as a diffractive optical microlens array in a digital Shack-Hartmann wavefront sensor. Appl. Opt. **45**, 90–94 (2006)
49. Arines, J., Durán, V., Jaroszewicz, Z., Ares, J., Tajahuerce, E., Prado, P., Lancis, J., Bará, S., Climent, V. Measurement and compensation of optical aberrations using a single spatial light modulator. Opt. Express **15**, 15287–15292 (2007)
50. Southwell, W.H. Wave-front estimation from wave-front slope measurements. J. Opt. Soc. Am. **70**, 998–1006 (1980)

Chapter 6
Nanoinjection Detectors and Imagers for Sensitive and Efficient Infrared Detection

Omer G. Memis and Hooman Mohseni

Abstract A new class of short-wave infrared photon detectors, called Nanoinjection Photon Detectors, has been recently developed to address the main problem of using nanoscale features for light detection. Even though nanoscale sensors offer high sensitivity, their interaction with visible or infrared light is severely limited by their miniscule volume compared to the wavelength of light. Our nanoinjection detectors have highly sensitive nanometer sized electron injectors on a large and thick light-absorbing layers. These layers capture incoming light efficiently, and the nanoinjectors sense and amplify the signal. Due to the pairing of thick absorbers and nanoinjectors, high sensitivity and high efficiency can be achieved simultaneously.

6.1 Introduction

The nanoinjection detectors were developed toward operation in short-wave infrared domain from 0.95 to 1.65 μm. They were designed to exploit the properties of the type-II band alignment in InP/GaAsSb/InGaAs structure to achieve higher internal amplification, lower noise levels, and high-speed operation.

The layer structure and geometry of nanoinjection detectors were designed and optimized with a custom-built three-dimensional nonlinear finite element method simulation. Measurements on nanoinjection detectors clearly reveal their capabilities. Amplification factors reaching 10,000 have been recorded, together with low dark current densities at room temperature. Noise suppression behavior is observed at amplification factors exceeding 4,000, which lowers the detector noise to values below the theoretical shot noise limit. The devices, when properly surface treated, show bandwidths exceeding 3 GHz with an impressive time uncertainty (jitter) of 15 ps. Tests over arrays of nanoinjection detectors exhibit high uniformity with dark

H. Mohseni (✉)
Bio-Inspired Sensors and Optoelectronics Lab (BISOL), EECS Department, Northwestern University, Evanston, IL 60208, USA
e-mail: hmohsen@northwestern.edu

T. Fournel and B. Javidi (eds.), *Information Optics and Photonics: Algorithms, Systems, and Applications*, DOI 10.1007/978-1-4419-7380-1_6,

current standard deviation less than 3%. These properties make the nanoinjection photon detectors extremely suitable for demanding imaging applications, which require high efficiency, high sensitivity, and high uniformity, in addition to many applications such as nondestructive material inspection, high-speed quantum cryptography or medical optical imaging.

Arrays of nanoinjection detectors were designed, processed, and hybridized to focal plane array infrared cameras with 320×240 pixel arrays. The resonant tunneling nanoinjection imagers show responsivity values in excess of 1.3 A/W. Measured quantum efficiency was 72% at 1 V and more than 90% at 2.5 V. Daytime and nighttime images were acquired using the resonant-tunneling nanoinjection infrared camera.

6.2 Nanoinjection Single Photon Imagers

The short-wave infrared (SWIR) detectors play an important role in many modern world applications spanning diverse fields. From the optical-to-electrical converters for telecommunication wavelengths in the internet backbone to the noninvasive medical imaging methods, such as optical coherence tomography systems; from the military night-vision systems to nondestructive material inspection for homeland security, the short-wave infrared detection systems present an irreplaceable element in many essential technologies. In addition to these existing applications, emerging applications like quantum cryptography and computing, or quantum ghost imaging demand high-performance SWIR detectors.

The nanoinjection detector is a new type of SWIR photon detector. It is developed to address a common shortcoming of traditional photon detectors: the trade-off between sensitivity and detection efficiency. When detecting visible or infrared light, the energy of a photon is severely limited, and it is on the order of atto-Joules. One reliable way of sensing this small amount of energy is to use a small volume or low capacitance in the detector. However, this brings in a trade-off, where smaller volumes result in low interaction with the incoming light, and larger volumes would provide better efficiency but lower sensitivity.

The nanoinjection detectors solve this dilemma by integrating small sensing volumes with large absorption areas. They have highly sensitive nanometer-sized pillars, "nanoinjectors," on large, thick absorption layers. The large, thick layers capture incoming light efficiently, and channel the photo generated carriers towards the nanoinjectors. The injectors sense and amplify these signals. Due to the pairing of thick absorbers and nanoinjectors, high sensitivity and high efficiency can be simultaneously attained, which can satisfy a growing need for identifying and counting photons in many modern applications.

As highlighted in Fig. 6.1, the detector features nanoinjectors on a large InGaAs absorption layer. The device is based on InP/GaAsSb/InGaAs material system and therefore has type-II band alignment (Fig. 6.2), with the GaAsSb layer acting as a barrier for electrons and a trap for holes. As detailed in reference [1],

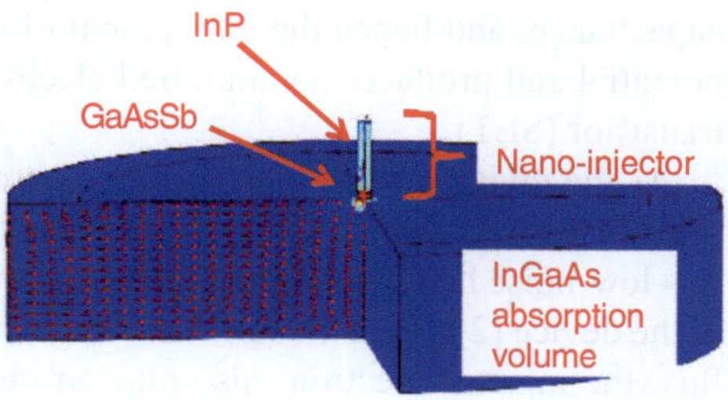

Fig. 6.1 The device geometry of a nanoinjection photon detector

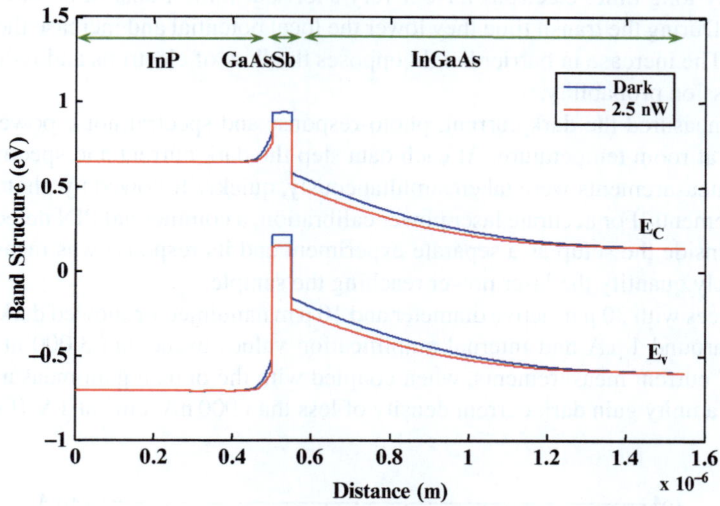

Fig. 6.2 The band structure of a nanoinjection photon detector at the central axis

the most tested active layer structure consists of 1,000 nm $In_{0.53}Ga_{0.47}As$ (n-doped), 50 nm $GaAs_{0.51}Sb_{0.49}$ (p-doped) and 500 nm InP (n-doped), from bottom to up.

The band structure across the device has InP/GaAsSb/InGaAs heterojunctions, which result in a type-II band alignment (Fig. 6.2). The conduction band incorporates a GaAsSb barrier to limit the flow of the electrons from InP side to InGaAs side. The valance band, on the other hand, exhibits a well structure, with the barriers being the interfaces of InP and InGaAs layers.

Due to the doping level and work function of each layer, the device generates an internal electric field in InGaAs region, which gets stronger when the device is biased correctly. Upon absorption, photons generate electron–hole pairs in the large absorption region. The electrons and holes are separated by the internal electric field of the device. Holes are attracted to the nanoinjector, which presents a potential trap for holes due to the type-II band alignment. With the small volume of the trap in the nanoinjector, the impact of the hole on the local voltage increases by orders of magnitude. Equivalently, the small volume of the trap represents an ultra-low capacitance, and hence the entrapment of a single hole leads to a large change of potential and produces an amplified electron injection, similar to a single electron transistor (SET).

As the multiplication mechanism is purely applied to one carrier, the amplification noise can be very small in nanoinjection detectors. One possible explanation for this low-noise behavior of the device is the negative potential feedback mechanism in the device [2]. Even though the voltage of the barrier is mainly controlled by hole flux, the injected electrons also play an important role in the regulation of barrier layer voltage. Compared to holes, which are trapped in the GaAsSb barrier for a relatively long time, electrons have a very short but finite transit time through the barrier. During the transit time they lower the local potential and increase the barrier height. The increase in barrier height opposes the flow of electrons and reduces the transmission probability.

We measured the dark current, photo-response and spectral noise power of the devices at room temperature. At each data step the dark current and spectral noise power measurements were taken simultaneously, quickly followed by photocurrent measurements. For accurate laser power calibration, a commercial PIN detector was placed inside the setup as a separate experiment and its response was measured to accurately quantify the laser power reaching the sample.

Devices with 30 μm active diameter and 10 μm nanoinjector showed dark current values around 1 μA and internal amplification values exceeding 3,000 at 0.7 V[1]. The DC current measurements, when coupled with the optical gain measurements, yielded a unity gain dark current density of less than 900 nA/cm^2 at 1 V (Fig. 6.3).

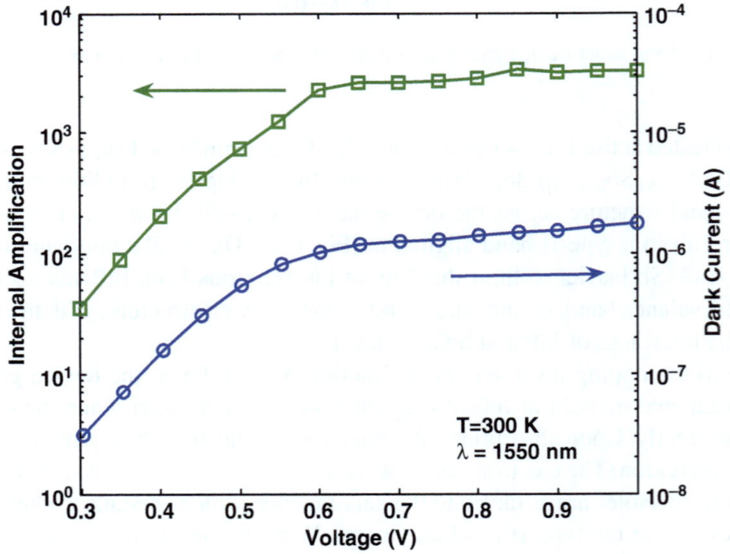

Fig. 6.3 The current–voltage and optical gain–voltage plots for a nanoinjection detector at room temperature

Similar to reference [3], the spectral noise power after amplification in unpassivated devices was measured with a spectrum analyzer around 1.5 kHz, which is beyond the $1/f$ noise knee but lower than the measured bandwidth of the device of about 4 kHz. The measured spectral power was compared to predicted spectral noise density due to Poissonian shot noise with amplification ($2qM^2 I_{\mathrm{int}} \Delta f$), where we have observed noise suppression similar to the Fano effect [4]. This phenomenonis shown to result from temporal correlation mechanisms influencing particle flow, such as Coulomb blockade [5] or Pauli exclusion principle [6]. The strength of shot noise suppression (or enhancement) is quantified by the Fano factor [7] γ as

$$\gamma = \frac{I_{\mathrm{n}}^2}{I_{\mathrm{shot}}^2} = \frac{I_{\mathrm{n}}^2}{2qI_{\mathrm{DC}}\Delta f}, \tag{6.1}$$

where I_{n} is the measured standard deviation of current or current noise; I_{shot} is the Poissonian shot noise; q is electron charge; I_{DC} is the average value of current; and Δf is the bandwidth.

For devices with internal amplification, the denominator of the right hand side needs to be modified into $2qM^2 I_{\mathrm{DC}} \Delta f$. This is because of the amplification that applies to both the signal and the noise, and hence the noise power needs to be scaled by M^2 to conserve signal-to-noise (SNR).

The devices showed high internal amplification values around 3,000 at 0.7 V and 5,000 at around 1 V[2]. The measured Fano factor was $F \sim 0.55$ (Fig. 6.4).

The noise equivalent power (NEP) of the devices was measured as $4.5\,\mathrm{fW/Hz^{0.5}}$ at room temperature without any gating, using the relation

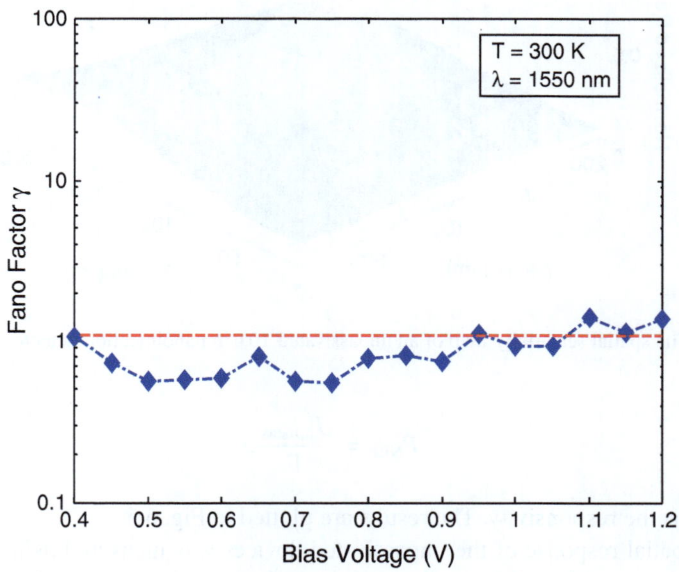

Fig. 6.4 The variation of noise vs. voltage for a 10 μm diameter detector

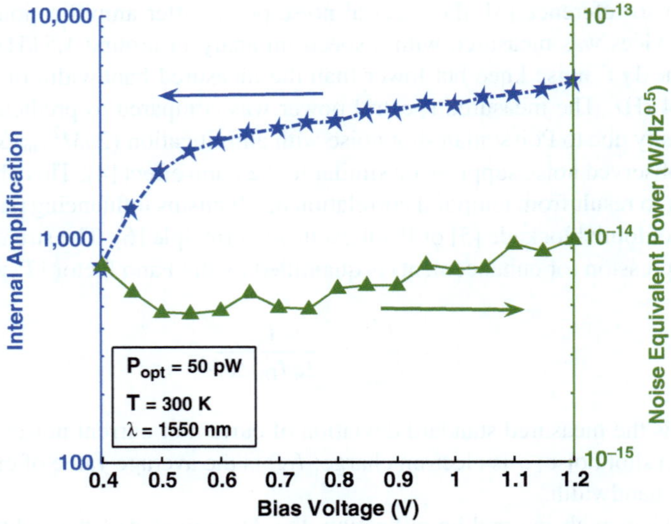

Fig. 6.5 The optical gain and noise equivalent power vs. bias voltage at room temperature

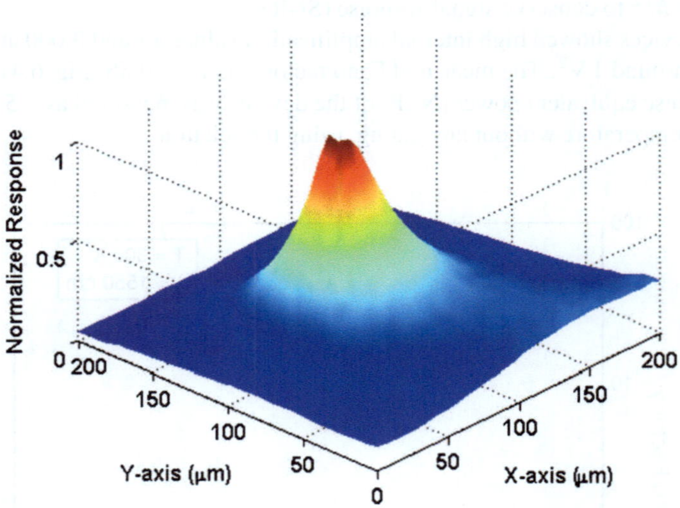

Fig. 6.6 The spatial sensitivity map of an unpassivated 10 μm nanoinjection detector

$$P_{\mathrm{NEP}} = \frac{I_{\mathrm{n,meas}}}{\Gamma}, \tag{6.2}$$

where Γ is the responsivity. The results are plotted in Fig. 6.5.

The spatial response of the unpassivated devices was measured using a surface scanning beam with $\sim 1.5\,\mu$m diameter, and 10 nm step resolution. Despite such a high gain, the device showed a very uniform spatial response as shown in Fig. 6.6, primarily due to the low internal electric field in our devices. The measured response decreases rapidly beyond a radius of about 8 μm. This property meant that two-dimensional arrays of nanoinjection detectors would not need pixel isolation methods, and that became one of the basis for the nanoinjector imagers.

The measured bandwidth of the unpassivated devices was around 3–4 kHz, much different compared to the lifetime in the GaAsSb trap layer (∼1 nanosecond with our doping levels), which we believe will be the ultimate constraint on bandwidth. We have attributed this difference to the existence and influence of surface traps.

When the devices were passivated, we have observed a drastically different behavior. The gain decreased significantly to values around 10 and the spatial response extended to beyond 100 μm when the devices were not confined by hard-etching the trap layer. However, the bandwidth of these devices exceeded 3 GHz. Rise time values of 200 ps were measured (Fig. 6.7).

Passivated nanoinjection detectors with bandwidths exceeding several gigahertz exhibited ultra-low jitter values of ∼14 ps at room temperature [8], which is a record breaking performance. The transient response of the detector was studied by exploring the relation between lateral charge transfer and jitter, where we found out that the jitter is primarily transit time limited.

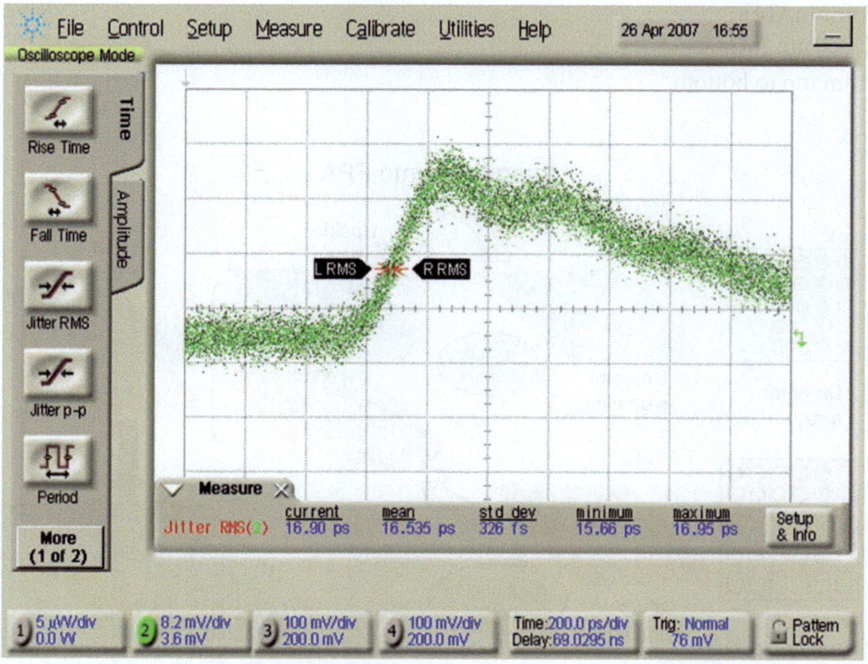

Fig. 6.7 The high-speed pulse response of unpassivated 10 μm nanoinjection detector, showing a jitter of 16.9 ps

6.3 Nanoinjection Single Photon Imagers

After individual nanoinjection detectors were designed, fabricated and evaluated, we have worked on arrays of nanoinjection detectors to form focal plane array infrared cameras. The fabrication included building an array of detectors, putting indium bumps on both the detector and the read-out array, and integrating them using indium bump bonding.

Regarding design parameters, we chose to build 320×240 pixel arrays with $30\,\mu m$ pixel pitch. We chose an off-the-shelf read-out-integrated circuit (ROIC), ISC9705 from Indigo, for hybridization. It is a 320×256 pixel ROIC with $30\,\mu m$ pixel pitch. The full well capacity is 18×10^6 electrons and the ROIC can process up to 346 frames per second at full resolution, and up to 15,600 frames with reduced region-of-interest with cooling (Fig. 6.8–6.10).

For integration, we chose $10\,\mu m$ passivated resonant tunneling injection detectors. The resonant-tunneling nanoinjection detectors are a new variant of nanoinjection technology. The active layers of this detector are as follows:

200 nm InP
5 nm InAlAs
50 nm GaAsSb
5 nm InAlAs
1,000 nm InGaAs

from top to bottom.

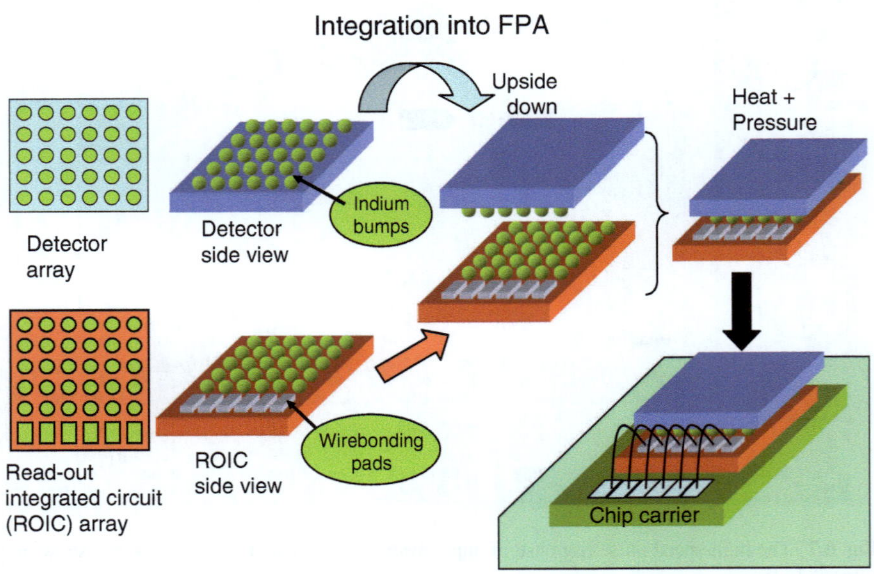

Fig. 6.8 The indium-bump hybridization process of a focal plane array and read-our integrated circuitry

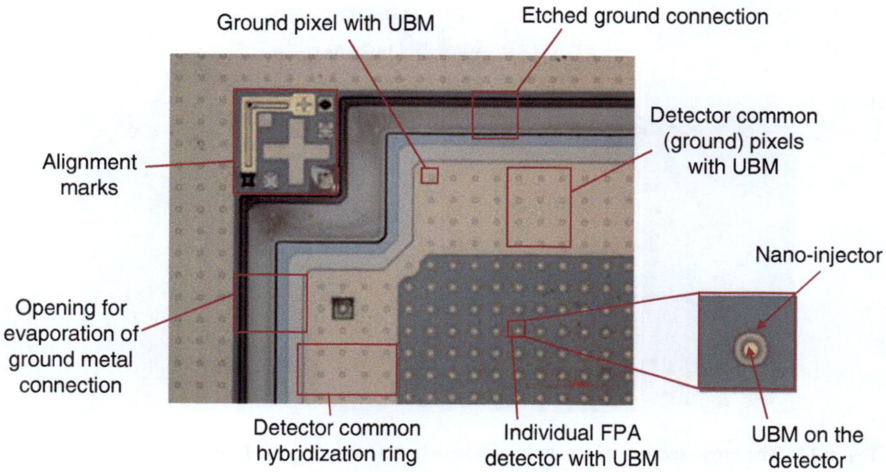

Fig. 6.9 An optical microscope image of a nanoinjection imager showing the focal plane array, the detector common hybridization ring and alignment marks

Fig. 6.10 (*Left*) An image of a processed wafer, ready for integration. (*Right*) An image of the hybridized detector array on a chip carrier

As the layers suggest, the resonant tunneling nanoinjection detectors have two thin InAlAs layers surrounding GaAsSb barrier/trap layer (Fig. 6.11). This changes the main mode of transport across the device: Instead of the thermionic emission, the devices utilize quantum tunneling through the InAlAs layers.

These detectors show low dark current and localized response under passivation. This performance is similar to a PIN-based imager (i.e., low dark current, unity gain).

To characterize the resonant tunneling nanoinjection imager, it was placed in a custom-built visible/infrared microscope setup, which doubled as a beam collimator for the laser beam. A tunable laser source at $\lambda = 1.55\,\mu\text{m}$ was used as modulated optical source with adjustable attenuation. The actual power reaching the sample was measured with a calibrated PIN detector. The detectors were tested with 1 μm

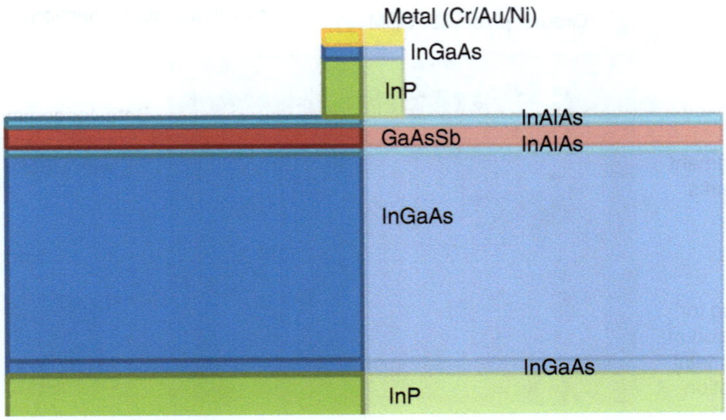

Fig. 6.11 The cross-section of a nanoinjection detector, showing the layer structure

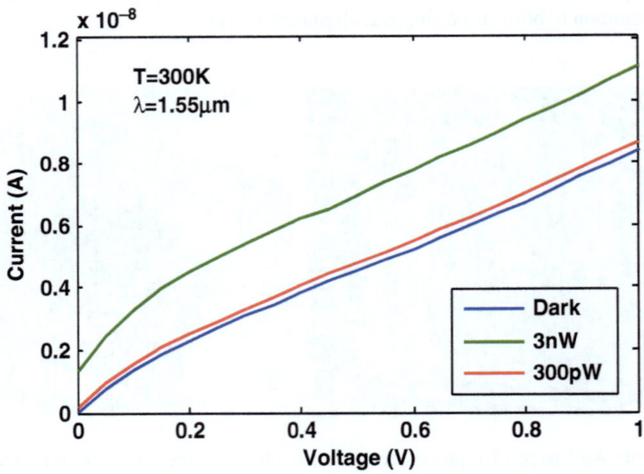

Fig. 6.12 The current voltage plot of a 10 μm resonant tunneling nanoinjection detector

radius Ni-W probes. The signal was amplified with a low-noise current preamplifier Stanford Research Systems SR-570 and recorded by an Agilent multimeter for dark and photo DC measurements. Dark current values of 2.3 nA at 200 mV bias, 4.5 nA at 500 mV and 8.4 nA at 1 V was measured at room temperature (Fig. 6.12).

Using the data from calibrated photo-current and dark current measurements, the responsivity was calculated. The responsivity vs. voltage indicated that the detector did not exhibit internal amplification, and hence the variation was due to the change in quantum efficiency.

Internal quantum efficiency was measured as 72% at 1 V and exceeded 90% at 2.5 V. Around 0 V, the devices had extremely small dark current of ∼10 pA (photovoltaic operation) with an internal quantum efficiency of 35–40% (Fig. 6.13).

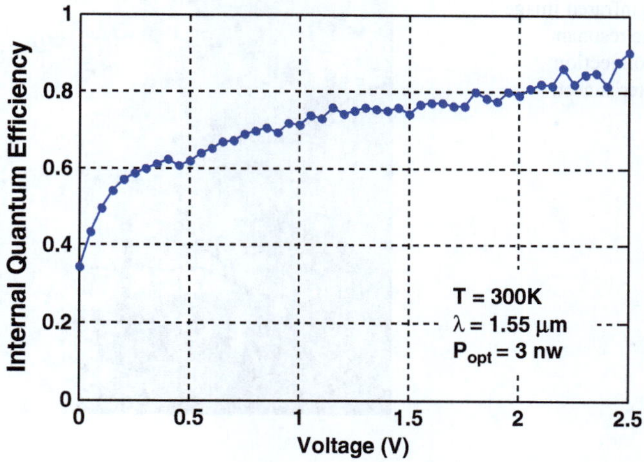

Fig. 6.13 The quantum efficiency of the resonant tunneling nanoinjection detector at different bias voltages

Fig. 6.14 An infrared image taken using the resonant tunneling nanoinjection imager just before sunset, at 8:20 pm

To find the size of the active area, the sample was placed in the microscope setup with an average optical power of 5 nW incident on the sample. The laser spot size was about 2 μm. Focused laser spot was scanned over the sample using motorized drivers with nanometer resolution. The active diameter of the devices was recorded as 30 μm.

The imager was placed in an infrared camera, and it was cooled down to a temperature of 78 K using liquid nitrogen. The devices were biased to -500 mV and the detector common was set to 6–7 V. The integration time was 15 ms and a Canon EF 50 mm f/1.8 lens was used at its maximum aperture during infrared imaging.

The infrared images taken using the resonant-tunneling nanoinjection imager are shown in Figs. 6.14–6.16.

Fig. 6.15 An infrared image taken using the resonant tunneling nanoinjection imager just after sunset, at 8:30 pm

Fig. 6.16 An infrared image taken using the resonant tunneling nanoinjection imager at night, at 9:25 pm

References

1. O. G. Memis, A. Katsnelson, S. C. Kong, H. Mohseni, M. Yan, S. Zhang, T. Hossain, N. Jin, and I. Adesida, *Appl. Phys. Lett.* 91, 171112 (2007).
2. O. G. Memis, A. Katsnelson, S. C. Kong, H. Mohseni, M. Yan, S. Zhang, T. Hossain, N. Jin, and I. Adesida, *Opt. Express* **16,** 12701 (2008).
3. F. Z. Xie, D. Kuhl, E. H. Bottcher, S. Y. Ren, and D. Bimberg, *J. Appl. Phys.* **73**, 8641 (1993).
4. P. J. Edwards, *Aust. J. Phys.* **53**, 179–192 (2000).
5. G. Kiesslich, H. Sprekeler, A. Wacker, and E. Scholl, *Semicond. Sci. Technol.* **19**, S37–S39 (2004).
6. I. A. Maione, M. Macucci, G. Iannaccone, G. Basso, B. Pellegrini, M. Lazzarino, L. Sorba, and F. Beltram, *Phys. Rev*. B 75, 125327 (2007).
7. A. Wacker E. Schöll A. Nauen F. Hohls R. J. Haug G. Kiesslich, *Phys. Status Solidi C* 0, 1293–1296 (2003).
8. O. G. Memis, A. Katsnelson, H. Mohseni, M. Yan, S. Zhang, T. Hossain, N. Jin, and I. Adesida, *IEEE Electron Device Lett.* **29**(8), 867, 2008.

Chapter 7
Biological Applications of Stimulated Parametric Emission Microscopy and Stimulated Raman Scattering Microscopy

Shin'ichiro Kajiyama, Yasuyuki Ozeki, Kiichi Fukui, and Kazuyoshi Itoh

Abstract Two types of newly developed nonlinear-optical microscopy are presented, namely, stimulated parametric emission (SPE) microscopy and stimulated Raman scattering (SRS) microscopy, together with their biological applications. These techniques allow high-resolution 3D visualization of unstained living cells. SPE microscopy is also applied to time-lapse observation of dividing cells as well as simultaneous observation of fluorescence emitted from labeled organelles and SPE signal.

7.1 Introduction

Recent advances in photonics technologies have accelerated the development of a variety of biophotonics techniques for controlling and visualizing living cells and tissues. In particular, high-resolution biomedical imaging techniques are highly demanded for exploring the origin and mechanism of life and for medical diagnosis. Currently, fluorescence microscopy is a common and powerful technique used in biological and medical laboratories because it provides us with high specificity, high contrast, and three-dimensional resolution. These advantages rely on the development of various fluorophores including fluorescent proteins [1], fluorescent dyes, and quantum dots, as well as optics/photonics techniques such as confocal detection, two-photon excitation [2], and so on. It is also possible to break the diffraction limit of spatial resolution in fluorescence microscopy. Indeed, a spatial resolution has been recently decreased down to ~20 nm [3].

Nevertheless, it is also important to explore label-free imaging techniques because labels are sometimes toxic, suffer from photobleaching, and possibly affect the property and/or dynamics of the sample. In order to visualize

K. Itoh (✉)
Department of Material and Life Science, Graduate School of Engineering, Osaka University, 2-1 Yamadaoka, Suita, Osaka 565-0871, Japan
e-mail: itoh@mls.eng.osaka-u.ac.jp

T. Fournel and B. Javidi (eds.), *Information Optics and Photonics: Algorithms, Systems, and Applications*, DOI 10.1007/978-1-4419-7380-1_7,

transparent samples, phase-contrast microscopy and differential interference-contrast microscopy are often used. However, they have poor resolution in the axial direction, and it is quite troublesome to observe the inside of thick tissues because the images are blurred by out-of-focus images. This problem is solved in some techniques such as optical coherence tomography [4] and structured-illumination technique [5].

Coherent nonlinear-optical microscopy is another interesting approach that has attracted much attention recently for label-free observation of live samples. This technique uses laser pulses, which are tightly focused in the sample and detect the nonlinear-optical interaction occurring in the vicinity of the focus. Since the signal is generated only in the focus, this technique has intrinsic three-dimensional resolution including in the axial direction. Furthermore, the signal contrast is highly dependent on the nonlinear-optical interaction involved. For example, second harmonic generation (SHG) microscopy [6] visualizes noncentrosymmetric structures, third harmonic generation (THG) microscopy [7] can detect the lipid bodies as well as the interfaces of different materials, and coherent anti-Stokes Raman scattering (CARS) microscopy [8, 9] gives molecular vibrational contrast. Considering that a variety of nonlinear-optical interactions have been explored in the field of laser spectroscopy, it is interesting and meaningful to find appropriate effects that provide us with a new contrast source in biological microscopy. In this chapter, we present two kinds of coherent nonlinear-optical microscopy techniques, namely, stimulated parametric emission (SPE) microscopy and stimulated Raman scattering (SRS) microscopy, together with their biological applications.

7.2 Stimulated Parametric Emission Microscopy

Stimulated parametric emission (SPE) microscopy [10] uses a kind of four-wave mixing (FWM) process, in which the third order nonlinear-optical effect caused by optical pulses at angular frequencies of ω_1 and ω_2 results in the generation of the FWM signal at an angular frequency of $\omega_{FWM} = 2\omega_1 - \omega_2$. In the quantum-mechanical description of nonlinear-optical effects, the SPE process is described by the energy diagram shown in Fig. 7.1, where virtual states at ω_1, $2\omega_1$, and $2\omega_1 - \omega_2$

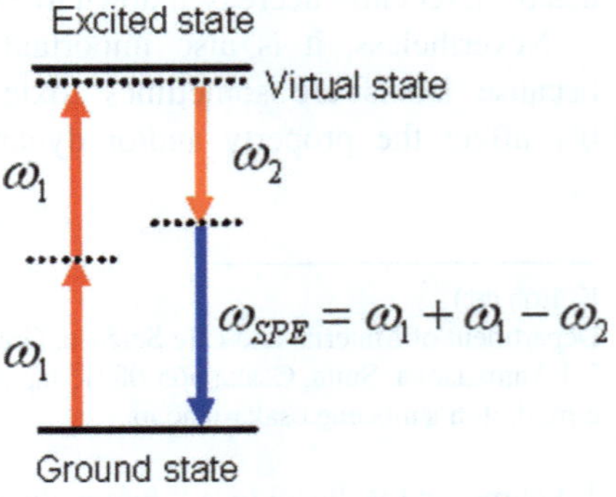

Fig. 7.1 Energy diagram of the SPE process

are involved to generate a SPE signal. SPE process is enhanced when the energy level of one of virtual states matches the energy level of the sample. Since the energy level corresponds to the absorption frequency of the sample, it is expected that the ultraviolet absorption at $2\omega_1$ can be detected as the enhanced SPE signal when near-infrared pulses are used for excitation. We call such an enhanced signal as the resonant SPE signal. Furthermore, SPE can provide a certain amount of contrast even when the sample does not have strong absorption because the signal is also dependent on the density and/or refractive index of the sample. We denote this kind of SPE signal as nonresonant SPE signal. Due to the above nature of SPE process, the microscopy provides a simple and straightforward way for three-dimensional visualization of living sample. Moreover, comparable to SHG and THG, SPE can be implemented in a single-laser configuration without using sophisticated pulse control technique, which was employed in single-laser CARS microscopy [11]. Note that, in CARS microscopy, SPE process was regarded as a source of background [8]. However, our investigations [12–14] have shown that SPE can be used to visualize 3D morphologies of living cells as well as the distributions of substances with strong absorption in the ultraviolet region.

A typical setup of SPE microscopy is shown in Fig. 7.2 [12]. A commercially available mode-locked Ti:sapphire laser is used as an excitation light source. The laser produces a 200-MHz train of optical pulses with a temporal duration of 8 fs. The wavelength of the pulse ranges from 670 to 950 nm. After appropriate compensation of group velocity dispersion of optic components, the pulses are focused in the sample by an objective lens. Then, the transmitted signal is collected by another objective lens. A set of optical filters are used to eliminate the remaining excitation pulses and extract the SPE signal in the wavelength region of 580–620 nm.

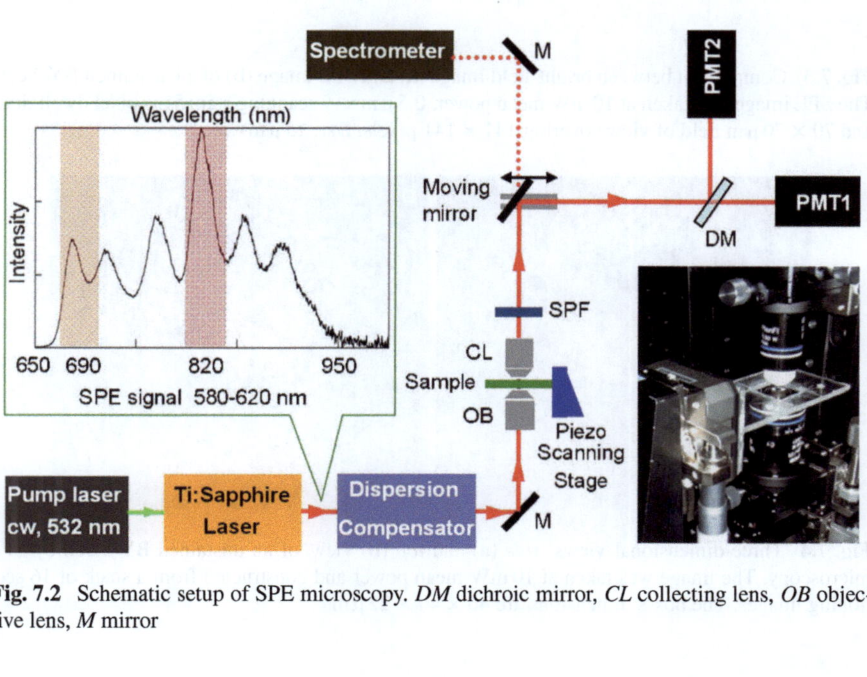

Fig. 7.2 Schematic setup of SPE microscopy. *DM* dichroic mirror, *CL* collecting lens, *OB* objective lens, *M* mirror

Then, the SPE signal is detected by photomultiplier tubes or a spectrometer. The stage-scanning mechanism based on a piezo-electric transducer is employed to collect the 3D information of the sample.

Figure 7.3 compares the ordinary bright-field image with the SPE image of tobacco BY-2 cultured cell. Since this cell is considered to have no strong absorption in the ultraviolet region, the SPE contrast originates from nonresonant SPE process. Nevertheless, the SPE image in Fig. 7.3b shows detailed inner structure of the cells with a distinct nucleus (a yellow arrow), endoplasmic reticulum (a red arrow), and a matrix of microtubules (blue arrows) which were not visualized clearly by conventional transmission optical microscopy without staining (Fig. 7.3a).

Owing to the three-dimensional resolution that originates from the confinement of the signal generation in the vicinity of a focal point, SPE microscopy is capable of optical sectioning similar to the two-photon excited fluorescence microscopy as well as other coherent nonlinear-optical microscopy techniques. Figure 7.4 showing

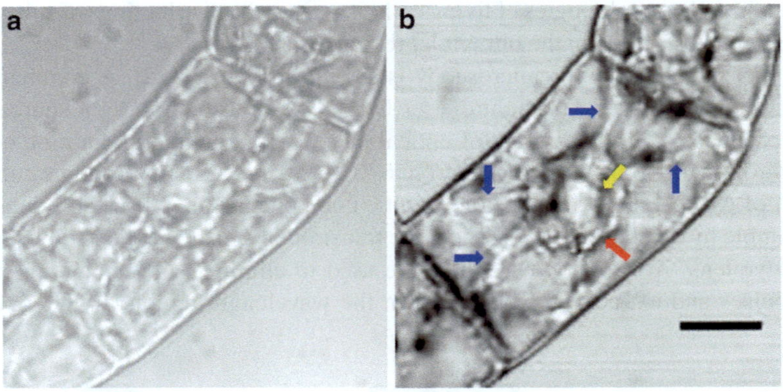

Fig. 7.3 Comparison between bright-field image (**a**) and SPE image (**b**) of an unstained BY-2 cell. The SPE image was taken at 10 mW mean power, 0.5 μm x–y scanning step, 5 ms pixel dwell time, and 70 × 70 μm field of view covering 141 × 141 pixels. Bar: 15 μm

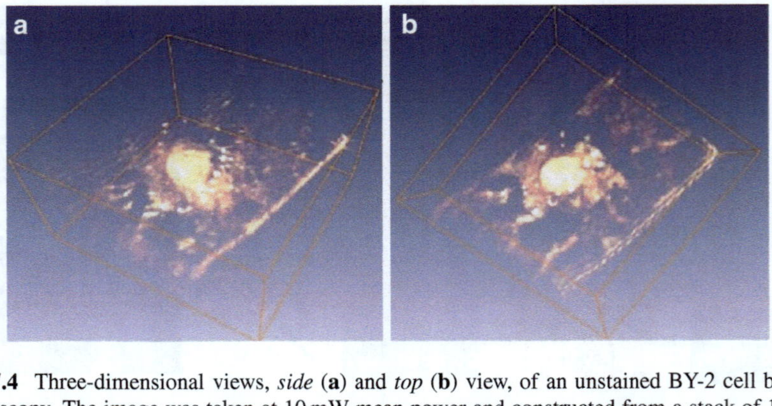

Fig. 7.4 Three-dimensional views, *side* (**a**) and *top* (**b**) view, of an unstained BY-2 cell by SPE microscopy. The image was taken at 10 mW mean power and constructed from a stack of 16 sectioning images. The box's dimensions are 40 × 40 × 22 μm

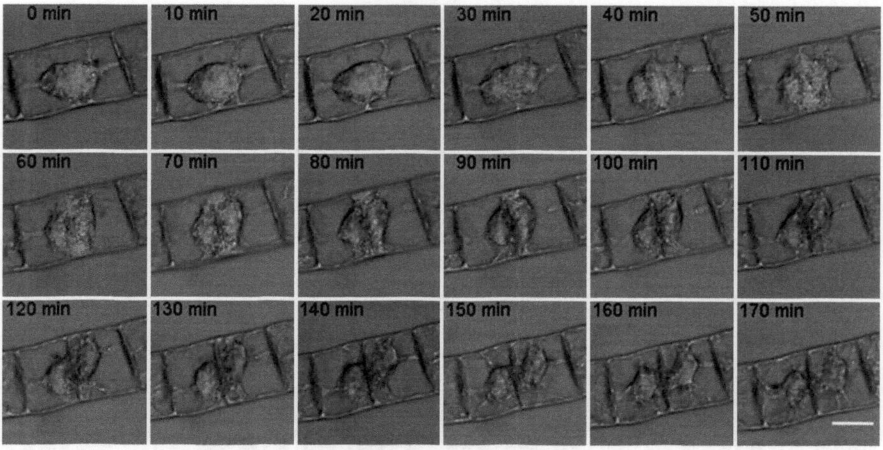

Fig. 7.5 SPE time-lapse observation of a dividing BY-2 cell at 5 mW mean power, 380 s resting time, and 10 ms pixel dwell time. Acquisition and interval times are 3.5 and 10 min, respectively. Bar: 20 μm

two different 3D views of an unstained BY-2 cell clearly indicates the ability of this microscopy for high-resolution 3D imaging of unstained living cell.

SPE microscopy can be used for long-term observation of an unstained living cell. A time-lapse observation of an unstained dividing BY-2 cell successfully proved this ability (Fig. 7.5). With the division time defined as the time elapsed from prophase to the end of cytokinesis, it took 2.5 h for a BY-2 cell to complete its division under normal light microscopy in the room condition at 20°C (data not shown). In SPE observation, compared to the division time of the cell with those of the control cases, the laser power of 5 mW did not either cause any significant effect on the cell or affect the image quality.

Note that the single broadband laser employed in this microscopy could be matched with the excitation spectra of various types of fluorophores. Thus, SPE microscopy can be combined with fluorescence microscopy by the simultaneous detection of SPE and fluorescence signals in labeled organelles. We used two kinds of cell lines, a BY-2 cell line expressing GFP-tubulin fusion protein and HeLa cells immunostained with Alexa 488 conjugated antibody against nucleolin, an abundant protein in nucleoli. In these experiments, the SPE signal was separated from the two-photon fluorescence signals by shifting the beam spectrum from the region of 670–950 nm to the region of 750–950 nm. As a result, SPE signal was restricted to the region of 660–720 nm, far from the emission region of both GFP (emission = 509 nm) and Alexa488 (emission = 519 nm). This setting enabled us to simultaneously obtain SPE and fluorescence images in both cases (GFP and Alexa 488) (Fig. 7.6a–f). In SPE images, the bunches of microtubules existing in BY-2 cells indicated by blue arrows (Fig. 7.6a) were as clearly depicted as fluorescence detection (Fig. 7.6b), and the nucleoli existing within a nucleus in

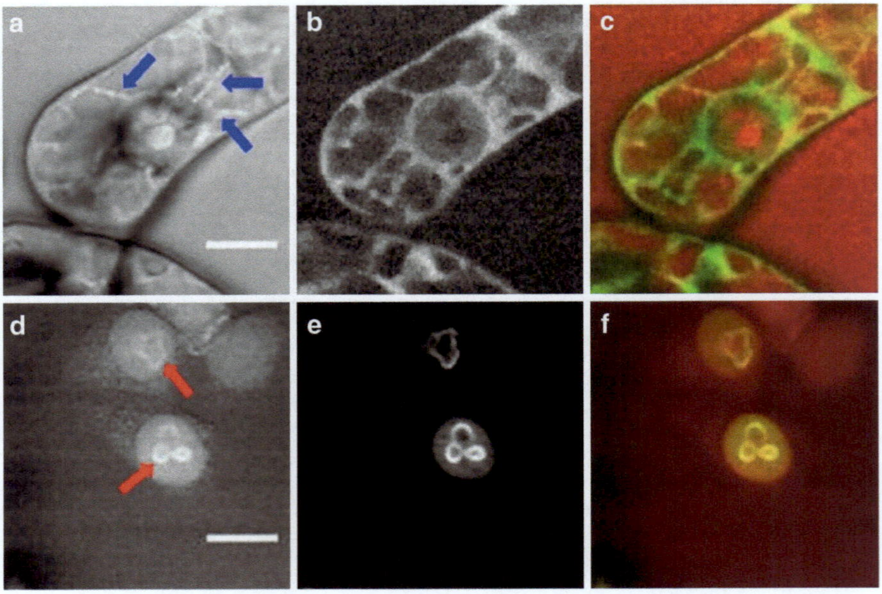

Fig. 7.6 Simulteneous SPE/fluorescence detection in SPE microscopy. (**a–c**) SPE (**a**), GFP (**b**) and merged (**c**) images of BY-2/GFP-tubulin cells. (**d–f**) SPE (**d**), Alexa488 (**e**) and merged (**f**) images of Hela/Alexa488-Nucleolin cells. SPE, *red*; GFP and Alexa, *green*. All the images were taken at 10 mW mean power, 5 ms pixel dwell time, 0.5 μm x–y scanning step, and have 60 × 60 μm field of view. *Blue* and *red arrows* indicate bunches of microtubules and nucleoli, respectively. Bars: 15 μm

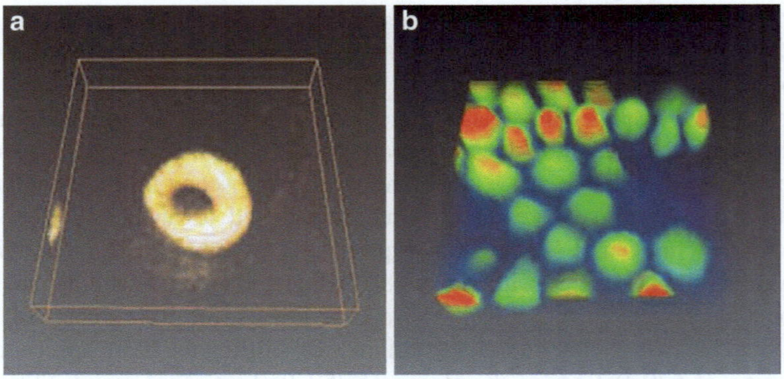

Fig. 7.7 Three-dimensional images of (**a**) a fresh mouse erythrocyte and (**b**) a purple petal tissue of *Petunia hybrida*. The image size: (**a**) 15 × 15 × 4 μm, (**b**) 80 × 80 × 14 μm

HeLa cells indicated by red arrows (Fig. 7.6d) were also demonstrated with clear correspondence to those visualized in fluorescence images (Fig. 7.6e).

Resonant SPE images can be obtained with samples that have strong ultraviolet absorption. Figure. 7.7a, b show the 3D SPE images of a fresh mouse erythrocyte

[13] and a purple petal tissue of *Petunia hybrida* [14], respectively. The SPE contrast is considered to reflect the distributions of (a) the hemoglobin and (b) the pigment, respectively.

7.3 SRS Microscopy

Among several coherent nonlinear-optical microscopy techniques, coherent anti-Stokes Raman scattering (CARS) microscopy [8, 9] is advantageous because it can provide molecular-specific image contrast based on vibrational resonances with high sensitivity. This ability is very useful to detect specific molecules in biological samples as fast as the video rate [15]. In biological applications, however, it is important to suppress so-called nonresonant signals emitted from medium or solvent because they significantly reduce the contrast of resonant CARS signals and distort the vibrational spectrum. Although various approaches have been proposed to suppress nonresonant signals [16–19], they typically suffer from signal loss and/or increased complexity of the setup. To cope with this problem, SRS has been recently exploited as a viable contrast mechanism. Three groups including ours have almost simultaneously demonstrated biological imaging by SRS microscopy [20–22].

SRS is a well-known nonlinear Raman phenomenon in which the photon energy is transferred from an anti-Stokes (AS) pulse to a Stokes (S) pulse when the frequency difference matches a Raman vibrational mode. Similar to CARS, SRS originates from third-order nonlinear susceptibility $\chi^{(3)}$. SRS reflects only vibrational resonances (i.e., Im $\chi^{(3)}$) and is unsusceptible to nonresonant nonlinearities (i.e., Re $\chi^{(3)}$), providing us with high vibrational contrast. Moreover, we pointed out in [21] that the theoretical sensitivity limit of SRS microscopy is comparable to that of CARS microscopy because SRS can be viewed as the homodyne detection of the nonlinear-optical signal with the excitation pulse. To achieve high sensitivity, it is effective to employ a high-repetition rate mode-locked laser and high-frequency lock-in detection similar to SRS spectroscopy [23]. This configuration enables us to reduce the effect of intensity noise of the laser, which significantly affects the signal-to-noise ratio of SRS signal.

Figure 7.8 shows the experimental setup [21]. A Ti:sapphire laser generated AS pulses with a pulse duration of approximately 100 fs at a wavelength of 767 nm. An optical parametric oscillator (OPO) generated S pulses with duration of approximately 200 fs at a wavelength of around 1,000 nm. The repetition rate of the pulses was 76 MHz. An acousto-optic modulator (AOM) was placed for the intensity modulation of the S pulses at a modulation frequency of 10.7 MHz. The modulation depth of AOM was >70%. These pulses were collinearly combined with a dichroic mirror and focused inside a sample by an oil-immersion objective lens with a numerical aperture (NA) of 1.4. The sample position was scanned by a piezo-electric

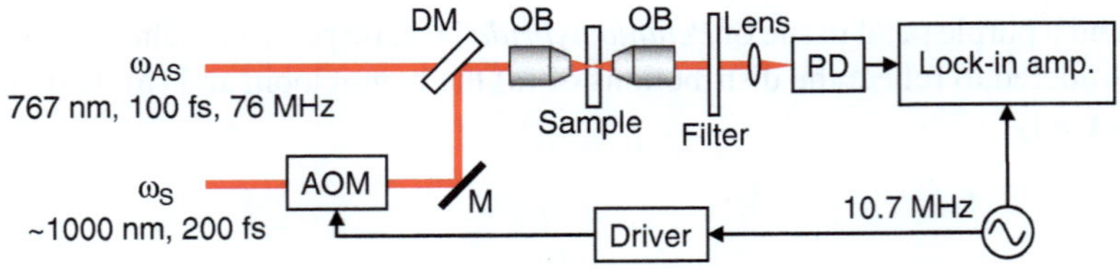

Fig. 7.8 Experimental setup of SRS microscopy. *DM* dichroic mirror, *OB* objective lens, *PD* silicon photodiode, *AOM* acousto-optic modulator

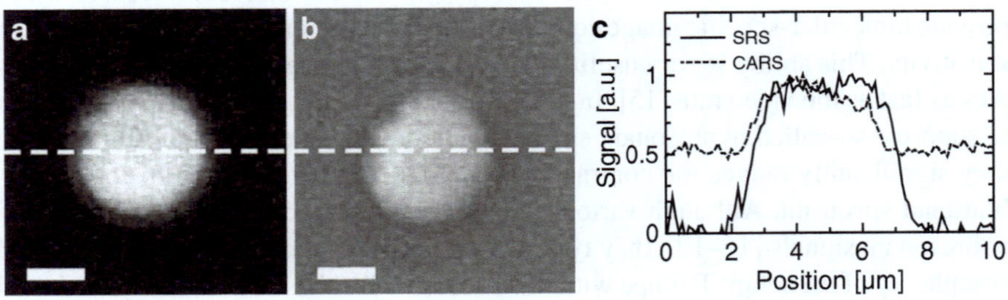

Fig. 7.9 Images of a polystyrene bead with (**a**) SRS and (**b**) CARS. Scale bar: 2 μm. Pixel dwell time: 2 ms. Raman shift: 2,967 cm^{-1}. Optical power: 0.6 mW. (**c**) Cross-sections at the broken lines in (**a**) and (**b**)

transducer stage. The output pulses from the sample were collected by another lens with the same NA. After S pulses were removed by an optical filter, AS pulses were detected by a silicon photodiode (PD), and its photocurrent was measured by a lock-in amplifier.

Figure 7.9a, b compares an SRS image of a polystyrene bead with a diameter of 4.5 μm in water with a CARS image of the same. In the SRS image, the water did not emit any signal, leading to a high contrast. In turn, the CARS image was accompanied by a large amount of background from the nonresonant water. These characteristics were clearly presented in one-dimensional cross-sections shown in Fig. 7.9c. The SNRs of SRS and CARS signals from polystyrene bead were almost the same and were > 20 dB. The spatial resolution was measured by observing a 50-nm polystyrene bead to be 0.5 and 1.0 μm in the lateral and axial directions, respectively.

SRS microscopy is readily applicable to the observation of unstained living cells. Considering that the lock-in signal is proportional to Im $\chi^{(3)}$, we constructed images in such a way that the image intensity is proportional to $(\text{Im}\,\chi^{(3)})^2$. As shown in Fig. 7.10, nuclei and cell walls of a tobacco BY-2 cultured cell were clearly visualized in 2D and 3D, based on vibrational contrast by CH stretching modes.

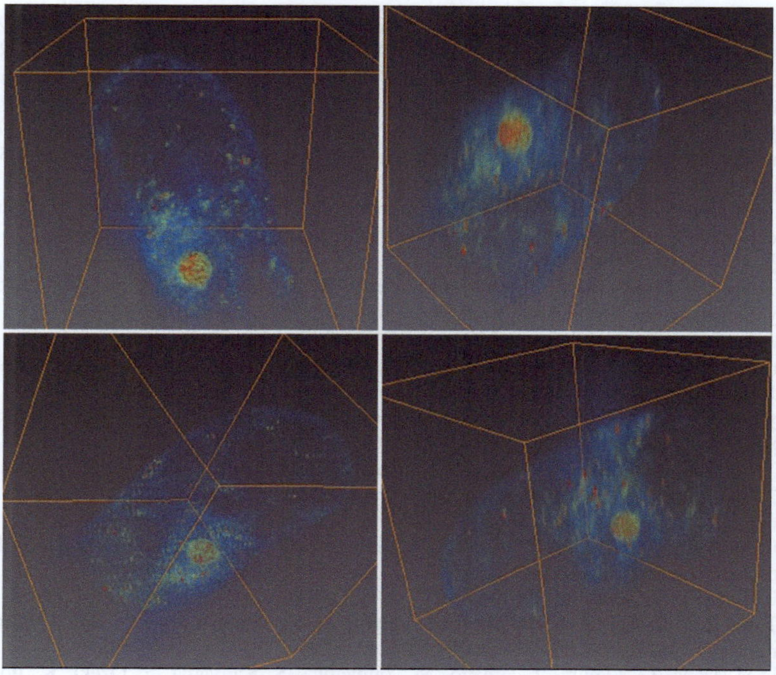

Fig. 7.10 Three-dimensional SRS images of an unstained tobacco BY-2 cell observed from four different directions. Image size: 40 × 40 × 40 μm. Pixel number: 160 × 160 × 40. Pixel dwell time: 3 ms. Raman shift: 2,967 cm^{-1}. Optical power: 1.0 mW

7.4 Conclusion

In summary, SPE and SRS microscopy allow high-resolution 2D and 3D imaging of unstained live cells. Simultaneous observation of fluorescence from labeled organelles and SPE are considered very useful for cellular studies. As for the SRS microscopy, high-contrast vibrational imaging is possible by suppressing the nonresonant background. This novel microscopy technique would be advantageous especially in high-contrast molecular mapping of weak substances in solvent, which is typical in the observation of biological samples. All these features have proved that SPE and SRS microscopy would be new and effective tools for life science and biomedical applications.

References

1. M. Chalfie, Y. Tu, G. Euskirchen, W. W. Ward, and D. C. Prasher, "Green fluorescent protein as a marker for gene expression," Science, vol. 263, no. 5148, pp. 802–805 (1994).
2. W. Denk, J. H. Strickler, W. W. Webb, "Two-photon laser scanning fluorescence microscopy," Science, vol. 248, no. 4951, pp. 73–76 (1991).

3. S. W. Hell, "Far-field optical nanoscopy," Science, vol. 316, no. 5828, pp. 1153–1158 (2007).
4. D. Huang, E. A. Swanson, C. P. Lin, J. S. Schuman, W. G. Stinson, W. Chang, M. R. Hee, T. Flotte, K. Gregory, C. A. Puliafito, and J. G. Fujimoto, "Optical coherence tomography," Science, vol. 254, no. 5035, pp. 1178–1181 (1991).
5. M. A. A. Neil, R. Juskaitis, and T. Wilson, "Method of obtaining optical sectioning by using structured light in a conventional microscope," Opt. Lett., vol. 22, no. 24, pp. 1905–1907 (1997).
6. P. J. Campagnola, M.-D. Wei, A. Lewis, and L. M. Loew, "High-resolution nonlinear optical imaging of live cells by second harmonic generation," Biophys. J., vol. 77, no. 6, pp. 3341–3349 (1999).
7. Y. Barad, H. Eisenberg, M. Horowitz, and Y. Silberberg, "Nonlinear scanning laser microscopy by third harmonic generation," Appl. Phys. Lett., vol. 70, no. 8, pp. 922–924 (1997).
8. A. Zumbusch, G. R. Holtom, and X. S. Xie, "Three-dimensional vibrational imaging by coherent anti-Stokes Raman scattering," Phys. Rev. Lett., vol. 82, no. 20, pp. 4142–4145 (1999).
9. M. Hashimoto, T. Araki, and S. Kawata, "Molecular vibration imaging in the fingerprint region by use of coherent anti-Stokes Raman scattering microscopy with a collinear configuration," Opt. Lett., vol. 25, no. 24, pp. 1768–1770 (2000).
10. K. Isobe, S. Kataoka, R. Murase, W. Watanabe, T. Higashi, S. Kawakami, S. Matsunaga, K. Fukui, and K. Itoh, "Stimulated parametric emission microscopy," Opt. Express, vol. 14, no. 2, pp. 786–793 (2006).
11. N. Dudovich, D. Oron, and Y. Silberberg, "Single-pulse coherently controlled nonlinear Raman spectroscopy and microscopy," Nature, vol. 418, no. 6897, pp. 512–514 (2002).
12. H. M. Dang, T. Kawasumi, G. Omura, T. Umano, S. Kajiyama, Y. Ozeki, K. Itoh, and K. Fukui, "Three-dimensional unstained live-cell imaging using stimulated parametric emission microscopy," Jpn. J. Appl. Phys., vol. 48, p. 097003 (2009).
13. H. M. Dang, G. Omura, T. Umano, M. Yamagiwa, S. Kajiyama, Y. Ozeki, K. Itoh, and K. Fukui, "Label-free imaging by stimulated parametric emission microscopy reveals a difference in hemoglobin distribution between live and fixed erythrocytes," J. Biomed. Opt., vol. 14, p. 040506 (2009).
14. M. Yamagiwa, Y. Ozeki, G. Omura, T. Suzuki, S. Kajiyama, K. Fukui, and K. Itoh, "Nonlinear phase imaging using two-beam interferometry in stimulated parametric emission microscopy," Jpn. J. Appl. Phys., vol. 48, p. 062501 (2009).
15. C. L. Evans, E. O. Potma, M. Puoris'haag, D. Côte', C. P. Lin, and X. S. Xie, "Chemical imaging of tissue in vivo with video-rate coherent anti-Stokes Raman scattering microscopy," Proc. Natl. Acad. Sci. U.S.A., vol. 102, no. 46, pp. 16807–16812 (2005).
16. A. Volkmer, J.-X. Cheng, and X. S. Xie, "Vibrational imaging with high sensitivity via epi-detected coherent anti-Stokes Raman scattering microscopy," Phys. Rev. Lett., vol. 87, no. 2, pp. 23901–23904 (2001).
17. J. -X. Cheng, L. D. Book, and X. S. Xie, "Polarization coherent anti-Stokes Raman scattering microscopy," Opt. Lett. vol. 26, no. 17, pp. 1341–1343 (2001).
18. E. O. Potma, C. L. Evans, and X. S. Xie, "Heterodyne coherent anti-Stokes Raman scattering (CARS) imaging," Opt. Lett., vol. 31, no. 2, pp. 241–243 (2006).
19. F. Ganikhanov, C. L. Evans, B. G. Saar, and X. S. Xie, "High-sensitivity vibrational imaging with frequency modulation coherent anti-Stokes Raman scattering (FM CARS) microscopy," Opt. Lett., vol. 31, no. 12, pp. 1872–1874 (2006).
20. C. W. Freudiger, W. Min, B. G. Saar, S. Lu, G. R. Holtom, C. He, J. C. Tsai, J. X. Kang, and X. S. Xie, "Label-free biomedical imaging with high sensitivity by stimulated Raman scattering microscopy," Science, vol. 322, no. 5909, pp. 1857–1861 (2008).
21. Y. Ozeki, F. Dake, S. Kajiyama, K. Fukui, and K. Itoh, "Analysis and experimental assessment of the sensitivity of stimulated Raman scattering microscopy," Opt. Express, vol. 17, no. 5, pp. 3651–3658 (2009).
22. P. Nandakumar, A. Kovalev, and A Volkmer, "Vibrational imaging based on stimulated Raman scattering microscopy," New J. Phys., vol. 11, p. 033026 (2009).
23. B. F. Levine, C. V. Shank, and J. P. Heritage, "Surface vibrational spectroscopy using stimulated Raman scattering," IEEE J. Quantum Electron, vol. QE-15, no. 12, pp. 1418–1432 (1979).

Part II
3D Passive/Active Imaging and Visualization

Chapter 8
Novel Approaches in 3D Sensing, Imaging, and Visualization

Robert Schulein, M. DaneshPanah, M. Cho, and B. Javidi

Abstract Three-dimensional (3D) imaging systems are being researched extensively for purposes of sensing and visualization in fields as diverse as defense, medical, art, and entertainment. When compared to traditional 2D imaging techniques, 3D imaging offers advantages in ranging, robustness to scene occlusion, and target recognition performance. Amongst the myriad 3D imaging techniques, 3D multiperspective imaging technologies have received recent attention due to the technologies' relatively low cost, scalability, and passive sensing capabilities. Multiperspective 3D imagers collect 3D scene information by recording 2D intensity information from multiple perspectives, thus retaining both ray intensity and angle information. Three novel developments in 3D sensing, imaging, and visualization systems are presented: 3D imaging with axially distributed sensing, 3D optical profilometry, and occluded 3D object tracking.

8.1 Introduction

New technologies that can be used for purposes of remote sensing, visualization, and information exploitation are always in demand. In recent years, there has been an increased interest among researchers to develop 3D imaging technologies for these purposes. Because 3D imaging technologies have dramatically increased information gathering capabilities when compared to traditional 2D imaging technologies, they offer advantages in ranging, robustness to scene occlusion, and target recognition performance. Because of the numerous advantages, 3D imaging technologies have emerged as standards in fields as diverse as defense, medical, art, and entertainment.

B. Javidi (✉)
Department of Electrical and Computer Engineering, University of Connecticut, U-2157, Storrs, CT 06269-2157, USA
e-mail: bahram@engr.uconn.edu

T. Fournel and B. Javidi (eds.), *Information Optics and Photonics: Algorithms, Systems, and Applications*, DOI 10.1007/978-1-4419-7380-1_8,

Amongst the numerous 3D imaging technologies like holography, laser radar, stereoscopy, and tomography, the passive multiperspective 3D imaging technique has gained recent attention for its unique properties [1–7]. Multiperspective 3D imaging systems obtain 3D information by recording incoherent images from multiple viewing locations. Because different perspectives of a scene are recorded, angular information of optical rays is retained in addition to intensity [7], allowing for full 3D reconstruction of the captured scene.

Because multiperspective 3D imaging systems use 2D imagers, systems can be built using a single camera with a lenslet array or an array of inexpensive imagers [6]. Additionally, 3D multiperspective imaging systems can be deployed for short- or long-range applications, making the method more scalable than many of the competing 3D imaging technologies.

The state-of-the-art 3D multiperspective imaging systems make use of either microlens arrays or arrays of imagers to capture the various perspective images, each being called an elemental image. The depth cues of objects in a captured scene can be regenerated optically for 3D displays or computationally for quantitative analysis. The latter is of interest to this investigation and is further explored. For 3D computational reconstruction in the spatial domain, one approach is to back-propagate the elemental images through simulated optics and superimpose them at a desired reconstruction plane. For a traditional 3D multiperspective imaging system with sensors aligned on a lateral grid, plane-by-plane computational reconstruction is achieved according to [5]:

$$I(x,y,z_0) = \sum_{k=0}^{K-1}\sum_{l=0}^{L-1} O_{kl}\left[-x+\left(1+\frac{1}{M}\right)S_x k, -y+\left(1+\frac{1}{M}\right)S_y l\right]/R^2(x,y),$$

$$R^2(x,y) = (z_0+g)^2 + \left[(Mx-S_x k)^2 + (My-S_y l)^2\right]\times\left(1+\frac{1}{M}\right)^2 \qquad (8.1)$$

where z_0 is the distance between the simulated optics and the desired reconstruction plane, g denotes the distance of the image plane from each simulated lens, $M = z_0/g$ is the magnification factor, and S_x and S_y are the respective x and y lateral sensor separations. An object will appear in focus in a reconstruction image at plane z_0 equal to the distance of the object from the pickup optics. That same object will appear out of focus at other reconstruction planes. By computing many reconstruction plane images, depth information of a scene can be extracted using the 3D multiperspective imaging technology. The reconstruction process is illustrated in Fig. 8.1.

There has been significant research into novel 3D imaging techniques based on the 3D multiperspective imaging idea [6, 7]. Presented here are three novel 3D multiperspective imaging approaches. A technique called 3D imaging with axially distributed sensing is presented in Sect. 8.2; axially distributed sensing breaks from the traditional lateral grid to record perspective images by translating a camera and taking pictures along its optical axis. Profilometry and optical slicing, techniques for extracting volumetric surface information from 3D multiperspective datasets,

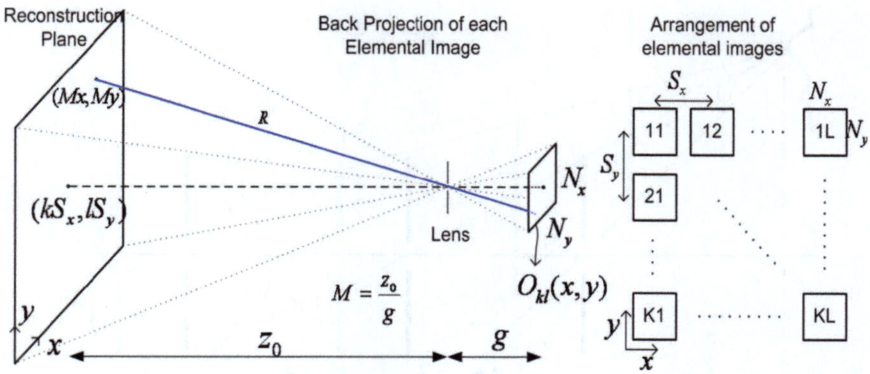

Fig. 8.1 Computational plane-by-plane 3D reconstruction process for traditional planar 3D multiperspective imaging architectures

are presented in Sect. 8.3. A method for tracking 3D targets through occlusion using 3D multiperspective imaging and a Summation of Absolute Difference algorithm is presented in Sect. 8.4. Conclusions are given in Sect. 8.5.

8.2 Three-Dimensional Imaging with Axially Distributed Sensing

Traditional 3D multiperspective imaging systems use a planar pickup grid that is perpendicular to a common imager optical axis to record perspective images. While these traditional planar approaches are good for 3D acquisition and provide advanced capabilities such as ranging and occlusion removal, collection of lateral parallax results in large sensor sizes. Thus, there has been recent interest in developing new sensing approaches that break from the traditional lateral grid. Many other sensor architectures have been recently proposed with different purposes [4, 8, 9]. This section highlights a novel approach for acquiring and reconstructing 3D scenes called axially distributed sensing [9]. In the axially distributed sensing method, a single imager is translated along its optical axis, collecting intermittent perspective images. A computational reconstruction approach based on elemental image back propagation is developed along with experimental results.

As shown in the diagram of the axially distributed sensing architecture in Fig. 8.2, a single camera takes intermittent images as it is translated along its optical axis. As with the traditional lateral grid 3D multiperspective imaging discussed in Sect. 8.1, each image is called an elemental image and contains a unique perspective of the scene. For simplicity, a pinhole approximation is adopted, and the focal length of each lens is approximated as the distance between the pinhole and sensor, g. Because all of the elemental images share a common optical axis, a cylindrical coordinate system is used with the optical axis aligned with the coordinate z axis.

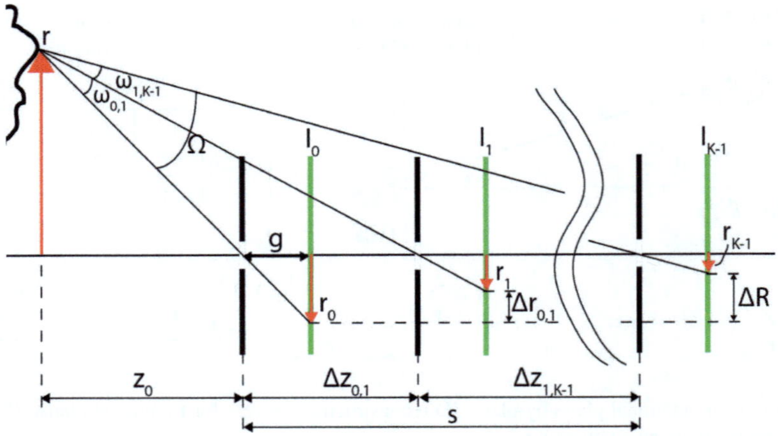

Fig. 8.2 Diagram of the axially distributed sensing 3D imaging optical pickup process. K imagers are translated along a common optical axis, each collecting a different perspective image of a scene

Object points are defined by their radial distance from the optical axis, r. For acquisition, a total of K elemental images are captured, with the image closest to the scene defined as $k = 0$ and the image furthest from the scene defined as $k = K-1$. Consider an object located a distance of z_0 from the closest imager. The distance between this object and the k–th imager is $z_k = z_0 + \Delta z_{0,k}$, where $\Delta z_{0,k}$ is the distance between the 0th and k-th imagers. The magnification of the object in each elemental image will be different because of the difference in distance. For each elemental image, the object will appear with magnification $M_k = z_k/g$ at an offset of $r_k = gr/z_k$ from the image center.

Like traditional 3D multiperspective imaging methods, plane-by-plane reconstructions can be performed for axially distributed sensing. A desired plane in object space can be reconstructed by back-projecting each elemental image through a virtual pinhole and superimposing the ensemble at the desired plane. To create plane-by-plane reconstructions, each elemental image is enlarged by a magnification factor relative to the distance between its virtual pinhole and the desired reconstruction distance in virtual object space. Each projected elemental image is superimposed at the desired reconstruction plane to form a reconstruction image via:

$$I(x, y, z_0) = \frac{1}{K} \sum_{k=0}^{K-1} I_k \left(\frac{x}{M_k}, \frac{y}{M_k} \right), \quad \text{with } M_k = \frac{z_k}{z_0} \tag{8.2}$$

where M_k denotes the relative magnification of the k–th imager with respect to the 0th imager. Note that distances to computational reconstruction planes will be given as distances from the closest, k_0, imager. This method is extendable to cases where the pinhole gap varies among imagers [8]. The closest imager is chosen as the reference image because this involved magnifying every other image. If a further imager was chosen as the reference, the closer images would have to be demagnified for reconstruction, causing the reconstruction scheme to lose some higher spatial frequencies.

The 3D information gathering capability of the axially distributed sensing architecture is not uniform across the entire field of view. For an object point located at a distance z_0 from the $k = 0$th imager and r away from the optical axis, the total information gathering performance of the system can be quantified by the angle subtended by all sensors and written as:

$$\Omega = \tan^{-1}\left(\frac{rs}{r^2 + z_0^2 + z_0 s}\right), \tag{8.3}$$

where $s = \Delta z_{0,K-1}$ is the total axial span of the sensors. Equation (8.3) holds the implicit assumption that the object point of interest (z_0, r) is observable by all K imagers.

Another important factor when analyzing axially distributed sensing performance is the minimum axial separation required between sensors to retain nonredundant information. When considering the pixelized structure of sensors, it is necessary for two imagers to image the object point of interest (z_0, r) onto different pixels locations for depth extraction to be possible. By requiring the i−th and j−th sensors to be located such that the difference between the object imaging point on the two sensors, $\Delta r_{i,j}$, be greater than or equal to one pixel, the lower bound of the sensor separation can be derived [10]. For sensor pixel size μ, the minimum axial separation between two adjacent sensors to satisfy the nonredundancy condition for an object point z_i away from the closer of the two sensors at radial distance $r > 0$ is:

$$\text{Min}\,\Delta z_{i,j} = \frac{\mu z_i^2}{\mu z_i + rg}. \tag{8.4}$$

The relationships in (8.3) and (8.4) show that the placement of the imagers in the system can be tailored to provide optimal 3D collection capacity for a given region of interest. In addition, it can be inferred from (8.4) that imagers should be placed closer together the closer they are to the region of interest. Furthermore, all imagers will image on axis object coordinates ($r = 0$) onto the center of their images. Thus, no 3D information can be collected for on-axis object coordinates. This further shows that there is a greater capacity for 3D information collection for objects further from the optical axis (large r) and closer to the imaging system (small z_0).

To demonstrate the performance of the 3D imaging with axially distributed sensing architecture, an optical laboratory setup is performed. A toy car is positioned approximately 435 mm away from the $k = 0$th imager at a radial distance of $r \approx 60$–110 mm. A transparency with black concentric circles is placed at $z_0 \approx 575$ mm. A toy police car is placed at $z_0 \approx 670$ mm and $r \approx 60$–140 mm with a toy fire truck placed at $z_0 \approx 805$ mm and $r \approx 80$–180 mm. A CMOS camera with 4368(H) $\times$ 2912(V) pixels and 8.2 μm pitch is used for an imager with a lens with focal length $f = 70$ mm. The pinhole gap, g, is approximated at 70 mm

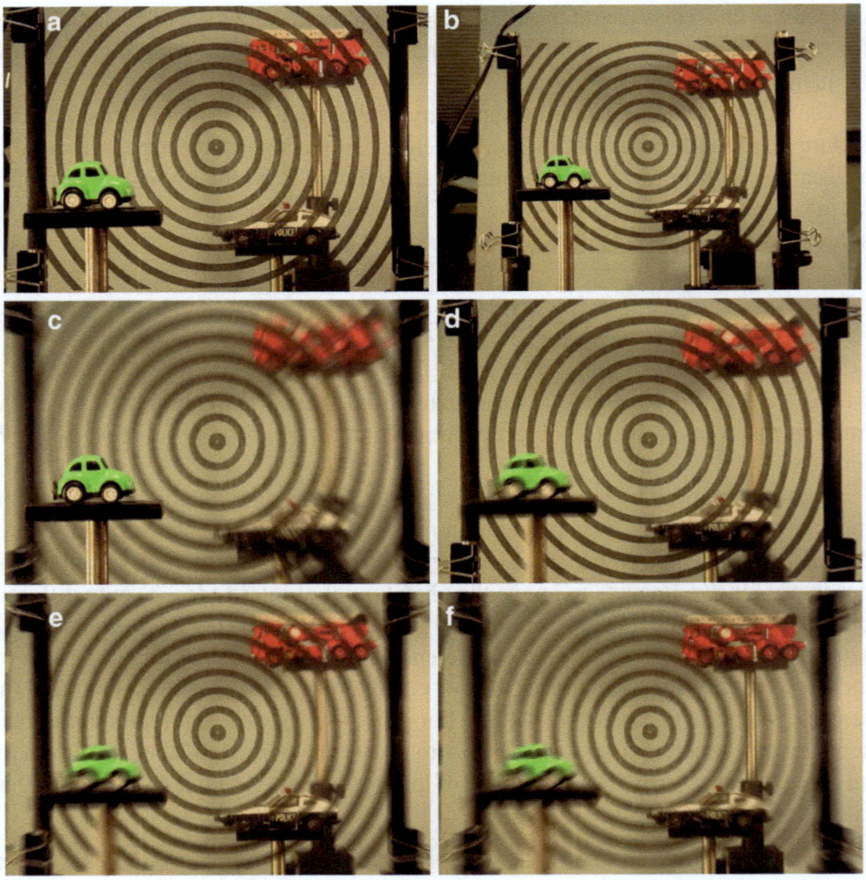

Fig. 8.3 Axially distributed sensing experiment. Elemental image (**a**) closest ($k = 0$) to the scene and (**b**) furthest ($k = 40$) from the scene. Computational reconstructions bring respective objects into focus at z_0 equal to (**c**) 435 mm, (**d**) 575 mm, (**e**) 670 mm, and (**f**) 805 mm

for computational reconstructions. The camera is translated at $\Delta z_{i,i+1} = 5$ mm for a total of $K = 41$ elemental images displaced over $s = 200$ mm. The closest, $k = 0$, elemental image is shown in Fig. 8.3a and the furthest, $k = 40$, elemental image is shown in Fig. 8.3b.

Computational reconstruction images are made with (8.2) over planes of interest. Four reconstruction images at the object locations are shown in Fig. 8.3c–f. From (8.3) and (8.4), it can be seen that the larger the object r, the more out of focus it will appear in out-of-focus reconstruction planes. This is well observed by the concentric rings that are in focus in Fig. 8.3d at reconstruction plane $z_0 = 575$ mm and out of focus in Figs. 8.3e, f at reconstruction planes $z_0 = 670$ mm and $z_0 = 805$ mm. The rings further from the optical axis (larger r) appear severely defocused compared to the rings closer to the optical axis (smaller r). From (8.3), the ensemble of imagers subtends a smaller angle Ω for smaller rings than the larger rings and as a result, the first two rings are almost in focus in the whole range of reconstructions. This illustrates the fact that the 3D information collectivity of such sensor architecture is nonuniform and decreases as the objects get closer to the optical axis.

8.3 Profilometry and Optical Slicing

The reconstruction algorithms described earlier require a priori knowledge of object distance or object features to ascertain correct computational reconstruction distances. This knowledge is not always available for an unknown scene. However, information about the third dimension (depth) of the object space is encoded as variable shifts (disparity) of image features in the perspective images. The magnitude of disparity depends on the depth of the object point as well as intrinsic characteristics of the pickup sensors. Separate threads of research exist that pursue passive depth extraction apparatus and algorithms. Stereopsis [11] and multibaseline techniques [12] utilize disparity to find the object depth based on triangulation. Another category of depth extraction algorithms, known as depth from defocus, relies on different measurements of a certain defocus metric at different camera settings and infers the object depth from the results [13]. Here, a different method is presented for profilometry and depth estimation that uses the multiperspective elemental image ensemble without need for feature extraction and matching [14].

For any given voxel in 3D space that appears in the common field of view of multiple imagers, some redundant intensity information is captured by the imagers. Voxels can be categorized into *surface* and *nonsurface* groups. Surface voxels, as the name suggests, contain an object surface that reflects light in all directions. Nonsurface voxels are all remaining voxels in space that are either in front, within, or behind the objects in the space. Assuming that reflection off of object surfaces is diffusive in all directions, one can expect that surface voxels will be imaged with the same intensity by all imagers. However, the information recorded for a nonsurface voxel originates from different parts of the object or background and will be registered as different intensities by the multiperspective ensemble. This variation in observed intensity is exploited in the search for the 3D locations of object surfaces.

The optical intensity field in 3D space is modeled as the spectral radiation pattern (SRP), defined as:

$$L(\theta, \phi, \lambda) \quad \text{(a.u. of intensity)}, \tag{8.5}$$

in which $-\pi/2 \leq \theta \leq \pi/2$ and $0 \leq \varphi \leq 2\pi$ are the respective zenith and azimuthal parameters that determine the ray angle, and λ denotes wavelength. The SRP parameter definition is depicted in Fig. 8.4b. This function can be calculated for each object space voxel from a multiperspective imaging dataset.

As mentioned earlier, if surface reflections are isotropic within the solid angle subtended by all imagers, the SRP is expected to be almost uniform for surface

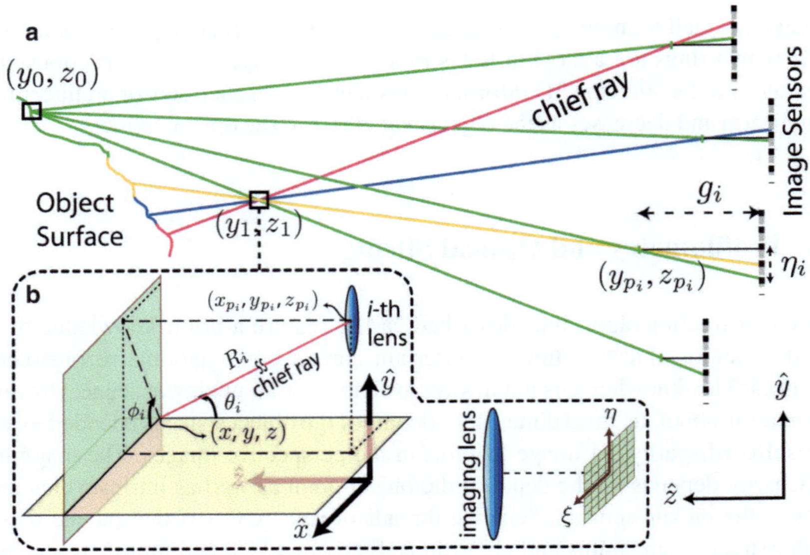

Fig. 8.4 (**a**) Ray diagram showing distinct spectral radiation patterns (SRPs) at surface and free-space voxels. (**b**) SRP parameter definition

voxels. However, voxels that do not contain object surfaces will have a SRP that is more likely to vary. The difference in SRP between surface and nonsurface voxels is illustrated in Fig. 8.4a, in which points (y_0, z_0) and (y_1, z_1) manifest two distinct SRP functions. Therefore, the depth of each transversal point (x, y) can be extracted by finding z such that SRP variation is minimized over the range of interest $Z = [z_{\min}, z_{\max}]$. The following can be used to calculate a depth map from a typical 3D multiperspective RGB dataset:

$$\hat{z}(x, y) = \underset{z \in Z}{\arg\min} \sum_{j=\{\mathrm{R,G,B}\}} \sum_{i=1}^{N} \left[L(\theta_i, \phi_i, \lambda_i) - I(\lambda_j) \right]^2_{(x,y,z)}, \tag{8.6}$$

with,

$$I(\lambda)|_{(x,y,z)} = \frac{1}{N} \sum_{i=1}^{N} L(\theta, \varphi, \lambda)|_{(x,y,z)}, \tag{8.7}$$

with N being the total number of elemental images.

To demonstrate the proposed method, results of a 3D multiperspective imaging experiment with 121 elemental images is presented. As described in Sect. 8.1, elemental images are captured on a regular grid with 5 mm spacing. The scene, as seen in Fig. 8.5 inset, includes a green car, a yellow car, and a red helicopter, located respectively in the range 28–33, 36–42, and 45–52 cm from the sensors all with $g_i = 35$ mm (Fig. 8.4). The recovered depth map of the 3D scene is illustrated in Fig. 8.5.

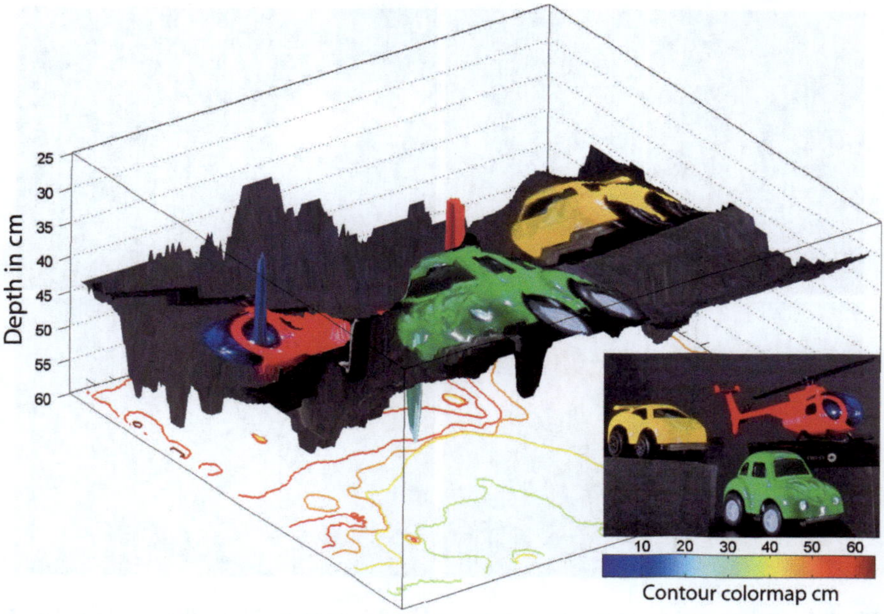

Fig. 8.5 Extracted depth information using (8.6) and (8.7) with contours of equal depth. Inset shows a 2D image of the original scene

After calculating the depth map, the scene can be sliced computationally into a set of reconstruction planes only containing in-focus voxels by only selecting voxel values over a certain range. Optical sections of the 3D profilometry image depicted in Fig. 8.5 are shown in Fig. 8.6. In contrast to the back-projection based 3D reconstruction techniques discussed in Sects. 8.1 and 8.2 where objects out of the reconstruction plane appear out of focus, the optical slicing process automatically removes voxels outside the range of interest. In this way, applications like 3D object recognition can be achieved with greater efficiency, and the chances for false detection are diminished. For comparison, a computational reconstruction of the plane sliced in Fig. 8.6a is depicted in Fig. 8.6d.

The profilometry and optical slicing method is accurate for convex objects with Lambertian (diffusive) surfaces. However, due to the redundancy in 3D intensity information conveyed in SRP and the statistical metric used for depth estimation in (8.7), the proposed method is robust to deviation from Lambertian and convex conditions to some degree. However, the accuracy of depth estimations will suffer when these two conditions are not fully met. Profilometry and optical slicing shows promise for more standardized implementation to represent 3D multiperspective datasets.

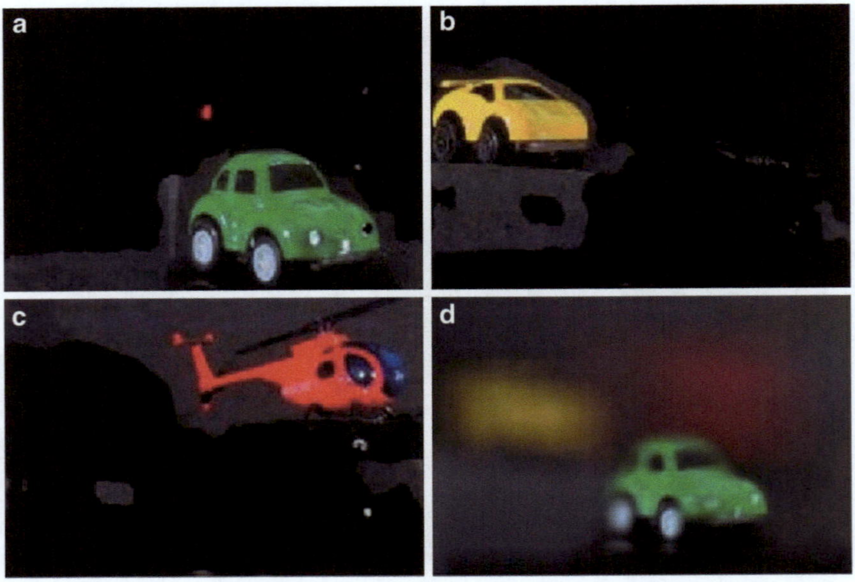

Fig. 8.6 Optical sections of the scene shown in Fig. 8.5 at (**a**) 28–33 cm, (**b**) 36–42 cm, (**c**) 45–52 cm. (**d**) Conventional reconstruction contaminated by out of focus objects

8.4 Occluded Target Tracking in 3D

Tracking objects in real-world situations is of great interest to researchers. Oftentimes, imaging conditions are not ideal, and optical degradations or heavy occlusion surrounding a target may make tracking with conventional 2D imaging difficult or impossible. The effects of occlusion and other optical degradations can be remedied by 3D multiperspective imaging systems and computational reconstructions. Because occlusion effects are diminished, a summation of absolute difference (SAD) algorithm can be implemented on 3D reconstruction images to track objects as they move through volumetric space [15].

The SAD algorithm [16] is a block-based correspondence algorithm that computes absolute differences between two images. If one of the images is a target image, the location of minimum difference between the two images corresponds to the location of the target image in the other image. It is simple because it only requires two frames and can therefore be used for tracking fast-moving objects. SAD can be applied to 3D multiperspective reconstruction images to track an object as it moves through volumetric space. SAD computes the light intensity differences for each pixel in a small target window t_x by t_y as follows:

$$\mathrm{SAD}\,(x,y) = \sum_{u=0}^{t_x-1} \sum_{v=0}^{t_y-1} |R_n(x+u, y+v) - T_{n-1}(u,v)|, \tag{8.8}$$

where $R_n(\cdot)$ is the n-th frame reconstructed 3D image, $T_{n-1}(\cdot)$ is the target pixel matrix from the $(n-1)$-th frame reconstructed 3D image, $x = 1, 2, 3, \ldots, r_x$, $y = 1, 2, 3, \ldots, r_y$, when r_x, r_y are the number of pixels for each frame image in the x and y directions.

To show the occluded object tracking abilities of the 3D multiperspective imaging system with SAD algorithm, an optical lab experiment is performed. A toy car object is placed behind occlusion and moved to several different locations in volumetric space. For each object location, a 3D multiperspective imaging dataset is collected. A central elemental image, sized 3072(H) × 2048(V) pixels, of the scene at the first object location is shown in Fig. 8.7a, and a set of elemental images for the same location is shown in Fig. 8.7b.

Fig. 8.7 Three-dimensional occluded object tracking experiment. Central elemental image (**a**) and set of elemental images (**b**) of the first frame shows an occluded car object behind tree branches. Computational reconstructions and tracking results of the object in the (**c**) first, (**d**) second, (**e**) seventh, and (**f**) thirteenth frames

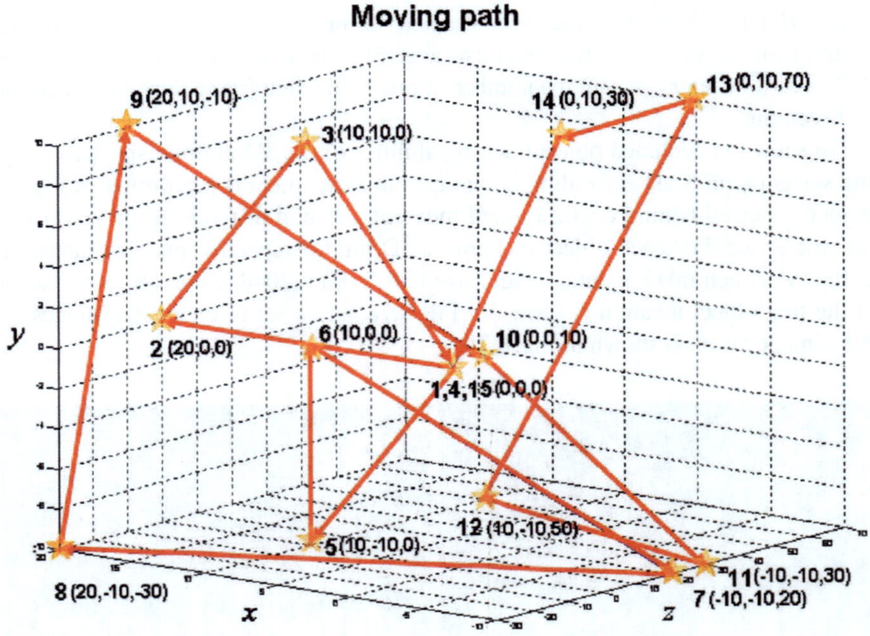

Fig. 8.8 Three-dimensional moving path of occluded object tracking experiment. The object is moved over a 3D volume with the number illustrating the step in the sequence

Clearly, 2D imaging cannot be used for object recognition because the heavy occlusion blocks most of the object in each perspective image. To verify the tracking method, the object is randomly moved in a 3D path to 15 different locations as shown in Fig. 8.8.

A range of computational reconstruction images are made with each object location 3D multiperspective dataset with (8.1). The depth information of the occluded object is assumed, and the correct reconstruction plane image is chosen for each object location and called the frame image. The minimum SAD value of each frame image is computed using (8.8). The minimum SAD value in the current frame is found and determines the location of the object. Absolute values of edges are used instead of pixel values to enhance differences between the minimum and other SAD values. The image may contain some noise pixels with values similar to the target object of interest that would normally cause recognition problems. However, the edge values of the object will be well defined and are used as criterion to segment the target.

The minimum SAD value is used to determine the correct object location in each frame. By measuring the correct object location in all of the frames, the object is tracked through 3D volumetric space. Four of the reconstruction image frames with bounding box showing the correctly located object are shown in Fig. 8.7c–f. It is found that the SAD algorithm is successfully able to track the car object as it moves through space. This is another example of the abilities that make 3D multiperspective imaging a more powerful tool than traditional 2D imaging.

8.5 Conclusions

There has been recent interest in 3D imaging and visualization systems for fields as diverse as defense, medical, art, and entertainment. Because 3D imaging technologies have dramatically increased information gathering capabilities when compared to traditional 2D imaging technologies, they offer advantages in ranging, robustness to scene occlusion, and target recognition performance. Of all the 3D imaging technologies, the passive multiperspective 3D imaging technique has gained recent interest because of its advantages in cost and scalability. Presented here are three novel 3D multiperspective imaging technologies. Axially distributed sensing uses a single imager translated along its optical axis to collect 3D images. Profilometry and optical slicing are techniques for extracting volumetric surfaces from 3D multiperspective datasets. Three-dimensional targets can be tracked in space through heavy foreground occlusion using 3D multiperspective imaging and a Summation of Absolute Difference algorithm. Three-dimensional multiperspective imaging is a compelling field that will continue to see development and new novel implementations in the coming years.

References

1. Benton SA (ed) (2001) Selected Papers on Three-Dimensional Displays. SPIE Milestones Series MS162, SPIE Press, Belingham
2. Javidi B, Okano F, Son J (eds) (2008) Three Dimensional Imaging, Visualization, and Display Technology. Springer, New York
3. Okoshi T (1976) Three-Dimensional Imaging Techniques. Academic, New York
4. Stern A, Javidi B (2006) Three-dimensional image sensing, visualization, and processing using integral imaging. Proc. IEEE 94(3):591–607
5. Tavakoli B, Daneshpanah M, Javidi B, Watson E (2007) Performance of 3D integral imaging with position uncertainty. Opt. Express 15:11889–11902
6. Martinez-Cuenca R, Saavedra G, Martinez-Corral M, Javidi B (2009) Progress in 3-D multiperspective display by integral imaging. Proc. IEEE 97(6):1067–1077
7. Levoy M (2006) Light fields and computational imaging. IEEE Comput 39(8):46–55
8. Daneshpanah M, Javidi B, Watson E A (2008) Three dimensional imaging with randomly distributed sensors. Opt. Express 16:6368–6377
9. Schulein R, Daneshpanah M, Javidi B (2009) 3D imaging with axially distributed sensing. Opt. Lett. 34:2012–2014
10. Kishk S, Javidi B (2003) Improved resolution 3D object sensing and recognition using time multiplexed computational integral imaging. Opt. Express 11:3528–3541
11. Verri A, Torre V (1986) Absolute depth estimate in stereopsis. J. Opt. Soc. Am. A 3:297–299
12. Okutomi M, Kanade T (1993) A multiple-baseline stereo. IEEE Trans. Pattern Anal. Mach. Intell. 4:353–363

13. Rajagopalan A N, Chaudhuri S, Mudenagudi U (2004) Depth estimation and image restoration using defocused stereo pairs. IEEE Trans. Pattern Anal. Mach. Intell. 11:1521–1525
14. Daneshpanah M, Javidi B (2009) Profilometry and optical slicing by passive three-dimensional imaging. Opt. Lett. 34:1105–1107
15. Cho M, Javidi B (2008) Three-dimensional tracking of occluded objects using integral imaging. Opt. Lett. 33:2737–2739
16. Vassiliadis S, Hakkennes E A, Wong J S S M, Pechanek G G (1998) The sum-absolute-difference motion estimation accelerator. Proc 24th EUROMICRO Conference, Volume 2:20559

Chapter 9
Overview of Free-viewpoint TV (FTV)

Masayuki Tanimoto

Abstract We have developed a new type of television named Free-viewpoint TV (FTV). FTV is an innovative visual medium that enables us to view a 3D scene by freely changing our viewpoints. FTV is based on the ray-space method that represents one ray in real space with one point in the ray space. We have also developed new type of ray capture and display technologies such as a 360° mirror-scan ray capturing system and a 360° ray-reproducing display. The international standardization of FTV has been made in MPEG. The first phase of FTV is Multiview Video Coding (MVC), and the second phase is 3D Video (3DV).

9.1 Introduction

We have developed a new type of television named Free-viewpoint TV (FTV) [1–7]. FTV is an innovative visual medium that enables us to view a 3D scene by freely changing our viewpoints as if we were there. It is easy to realize the free viewpoint for virtual scenes made by computer graphics. However, it is very difficult and has not yet been realized for real scenes. Therefore, FTV that has achieved such a function for real scenes will bring an epochal change in the history of visual media.

We proposed the concept of FTV and verified its feasibility with the world's first real-time system including the complete chain of operation from image capture to display. The viewpoint was controlled with a mouse, and free-viewpoint images of real moving scenes were displayed on a 2D display in real time [8, 9].

We have been developing technologies for FTV and realized FTV on a single PC and a mobile player. We also realized FTV with free listening-point audio [10]. FTV is based on the ray-space method [11–14].

We proposed FTV to Moving Picture Experts Group (MPEG) [15]. MPEG regarded FTV as the most challenging 3D medium and started the international

M. Tanimoto (✉)
Graduate School of Engineering, Nagoya University, Chikusa-ku, Furo-cho, Nagoya 464-8603, Japan
e-mail: tanimoto@nuee.nagoya-u.ac.jp

T. Fournel and B. Javidi (eds.), *Information Optics and Photonics: Algorithms, Systems, and Applications*, DOI 10.1007/978-1-4419-7380-1_9,

standardization activities of FTV. The first phase of FTV was Multiview Video Coding (MVC) [16], and the second phase is 3D Video (3DV [17]). 3DV is a standard that targets to serve a variety of 3D displays. MVC was completed in May 2009.

9.2 FTV System

At the sender side, a 3D scene is captured by multiple cameras.

9.2.1 Configuration of FTV System

Figure 9.1 shows a basic configuration of the FTV system. The captured images contain the misalignment and luminance differences of the cameras. They must be corrected to construct the ray space. The corrected images are compressed for transmission and storage by the MVC encoder.

At the receiver side, reconstructed images are obtained by the MVC decoder. The ray space is constructed by arranging the reconstructed images and interpolating them. Free-viewpoint images are generated by cutting the ray space vertically and are displayed on a 2D/3D display. Photorealistic, free-viewpoint images of the moving scene are generated in real time.

9.2.2 Video Capture

We constructed a 1D-arc capturing system shown in Fig. 9.2 for a real-time FTV system [18]. It consists of 16 cameras, 16 clients, and 1 server. Each client has one camera, and all clients are connected to the server with Gigabit Ethernet.

A "100-camera system" has been developed to capture larger space by Nagoya University (Intelligent Media Integration COE and Tanimoto Laboratory) [19]. The system consists of one host-server PC and 100 client PCs (called "nodes")

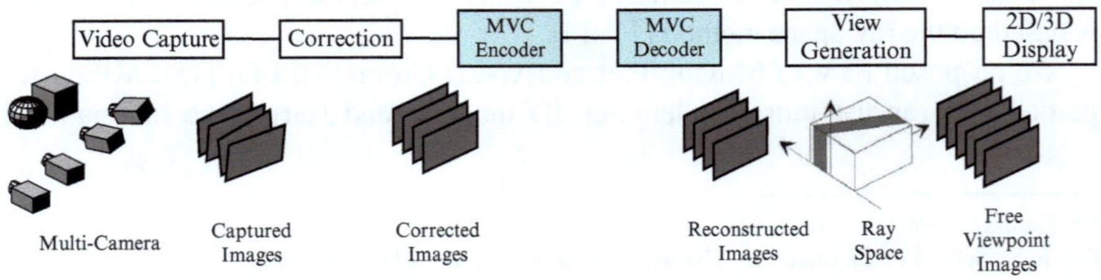

Fig. 9.1 A basic configuration of FTV system

Fig. 9.2 One-dimensional arc capturing system of FTV

Table 9.1 Specification of 100-camera system

Image resolution	$1,392(\mathrm{H}) \times 1,040(\mathrm{V})$
Frame rate	29.4118 (fps)
Color	Bayer matrix
Synchronization	Less than 1 (μs)
Sampling rate of A/D	96 (kS/s) maximum
Maximum number of nodes	No limit (128 max for one sync output)

Fig. 9.3 100-camera system (**a**) linear arrangement, (**b**) circular arrangement, and (**c**) two-dimension arrangement

that are equipped with JAI PULNiX TM-1400CL cameras. The interface between camera and PC is Camera Link. The host PC generates a synchronization signal and distributes it to all of the nodes. This system is capable of capturing not only high-resolution video with 30 fps but also analog signals of up to 96 kHz. The specification of the 100-camera system is listed in Table 9.1.

The camera setting is flexible as shown in Fig. 9.3. MPEG test sequences "Pantomime," "Champagne_tower," and "Dog" shown in Fig. 9.4 were taken in linear arrangement Fig. 9.3a.

Fig. 9.4 MPEG test sequences

9.2.3 Correction

The geometric correction [20, 21] and color correction [22] of multicamera images are performed by measuring the corresponding points of images. This measurement is made once the cameras are set.

9.2.4 MVC Encoding and Decoding

Multiview images have high temporal and interview correlations. MVC reduces these correlations [16,23,24]. The standardization of MVC is described in Sect. 9.4.

9.2.5 View Generation

Ray space is formed by placing the reconstructed images vertically and interpolating them. Free-viewpoint images are generated by making a cross-section of the ray space.

Examples of the generated free-viewpoint images are shown in Fig. 9.5. Complicated natural scenes, including sophisticated objects such as small moving fish, bubbles, and reflections of light from aquarium glass, are reproduced very well.

The quality of the generated view images depends on the ray-space interpolation. The ray-space interpolation is achieved by detecting depth information pixel by pixel from the multiview video. We proposed several interpolation schemes of the ray space [25–31].

Previously, free-viewpoint images were generated by a PC cluster. Now, they can be generated by a single PC, and FTV on a PC can be accomplished in real time [29]. FTV is also implemented on a portable player as shown in Fig. 9.6.

Fig. 9.5 An example of generated FTV images at various times and viewpoints

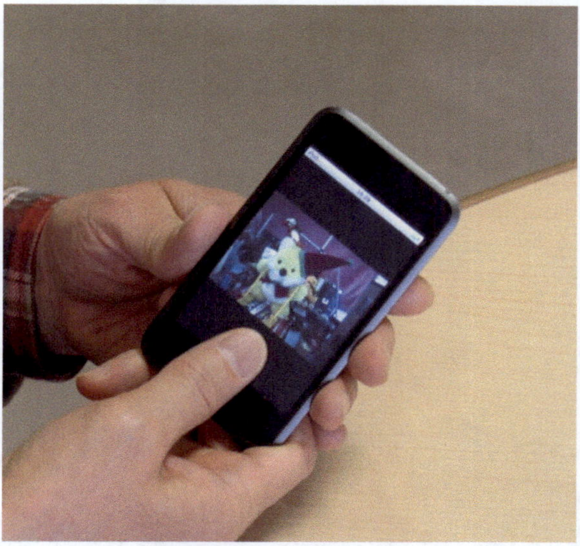

Fig. 9.6 FTV on a portable player

9.2.6 2D/3D Display

FTV needs a new user interface to display free-viewpoint images. Two types of display, 3D display and 2D/3D display with a viewpoint controller, are used for FTV as shown in Fig. 9.7.

Viewpoint control by head tracking is shown here. Many head-tracking systems have been proposed using magnetic sensors, various optical markers, infrared cameras, retroreflective light from retinas, etc. Our head-tracking system uses only

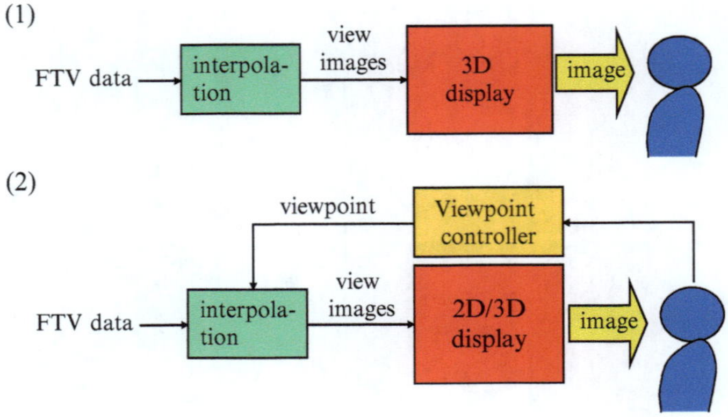

Fig. 9.7 Two types of display for FTV

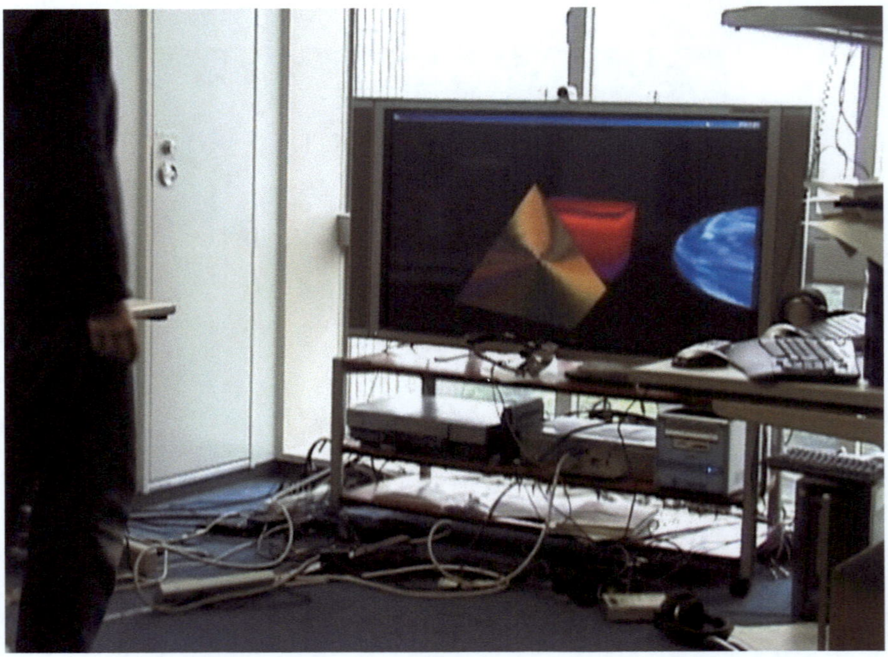

Fig. 9.8 Two-dimensional display with eye tracking (**a**) without head tracking, (**b**) with head tracking

a conventional 2D camera and detects the position of a user's head by image processing. The user does not need to attach any markers or sensors.

In the user interface using a 2D display, the location of the user's head is detected with the head-tracking system, and the corresponding view image is generated. Then, it is displayed on the 2D display as shown in Fig. 9.8.

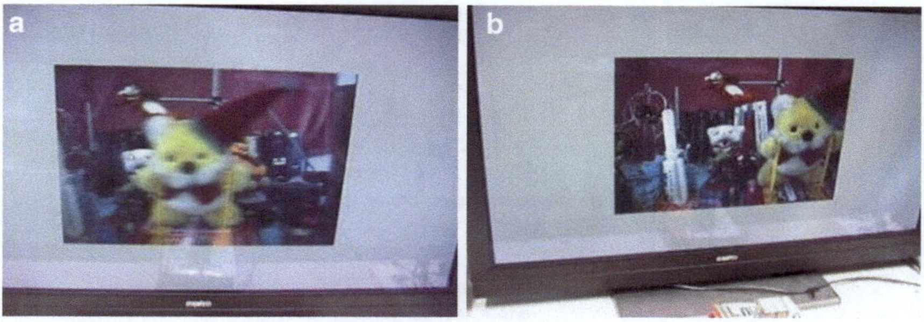

Fig. 9.9 Three-dimensional display (**a**) with head tracking, and (**b**) without head tracking

Automultiscopic displays enable a user to see stereoscopic images without special glasses. However, there are two problems: a limited viewing zone and discreteness of motion parallax. Because the width of the viewing zone for each view approximates the interpupillary distance, the view image does not change with the viewer's movement within the zone. On the other hand, when the viewer moves over the zone, the view image changes suddenly.

In the user interface using the automultiscopic display, the function of providing motion parallax is extended by using the head-tracking system. The images fed to the system change according to the movement of the head position to provide small motion parallax, and the view channel for feeding the images is switched for handling large motion. This means that binocular parallax for the eyes is provided by automultiscopic display, while motion parallax is provided by head tracking and changing the image adaptively as shown in Fig. 9.9.

9.3 Ray-Space Technology

We are developing ray-space technology such as ray capture [32–34], ray processing [35–38], and ray display [39].

9.3.1 Ray Capture

We have developed a 360° mirror-scan ray capturing system [33, 34] as shown in Fig. 9.10. This system uses two parabolic mirrors. Incident rays that are parallel to the axis of a parabolic mirror gather at the focus of the parabolic mirror. Hence, rays that come out of an object placed at the focus of the lower parabolic mirror gather at the focus of the upper parabolic mirror. Then, the real image of the object is generated at the focus of the upper parabolic mirror, and a rotating aslope mirror scans rays at the focus of the upper parabolic mirror. Finally, the image from the aslope mirror is captured by a high-speed camera. By using this system, we can capture all-around convergent views of an object as shown in Fig. 9.11.

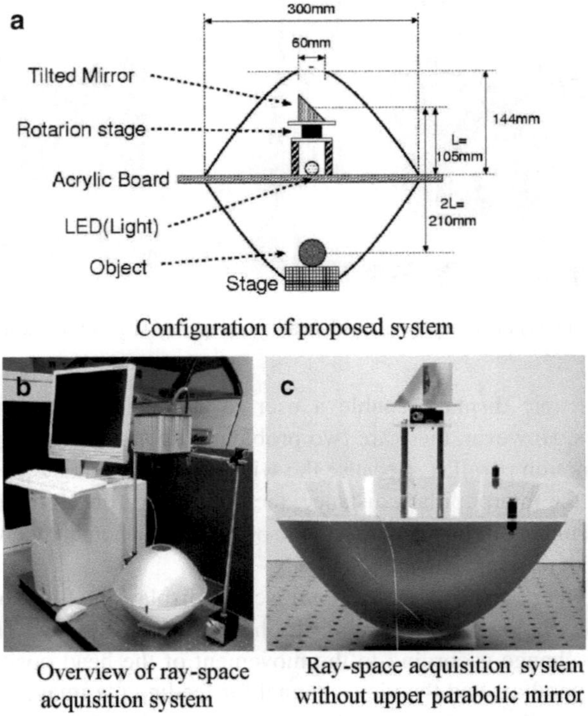

Fig. 9.10 360° mirror-scan ray capturing system (**a**) configuration of proposed system, (**b**) overview of ray-space acquisition system, and (**c**) ray-space acquisition system without upper parabolic mirror

9.3.2 Ray Display

Figure 9.12 shows the SeeLINDER [39], a 360°, ray-producing display, that allows multiple viewers to see 3D FTV images. It consists of a cylindrical parallax barrier and one-dimensional light-source arrays. LEDs are aligned vertically for the one-dimensional light-source arrays. The cylindrical parallax barrier rotates quickly, and the light-source arrays rotate slowly in the opposite direction. If the aperture width of the parallax barrier is sufficiently small, the light going through the aperture becomes a thin flux, and its direction is scanned by the movement of the parallax barrier and the light-source arrays. By synchronously changing the intensity of the light sources with the scanning, pixels whose luminance differs for each viewing direction can be displayed. We can see the 3D image naturally, and the images have the strong depth cues of natural binocular disparity. When we move around the display, the image changes corresponding to our viewing position. Therefore, we perceive the objects just as if they were floating in the cylinder.

We are going to connect these two systems directly in real time.

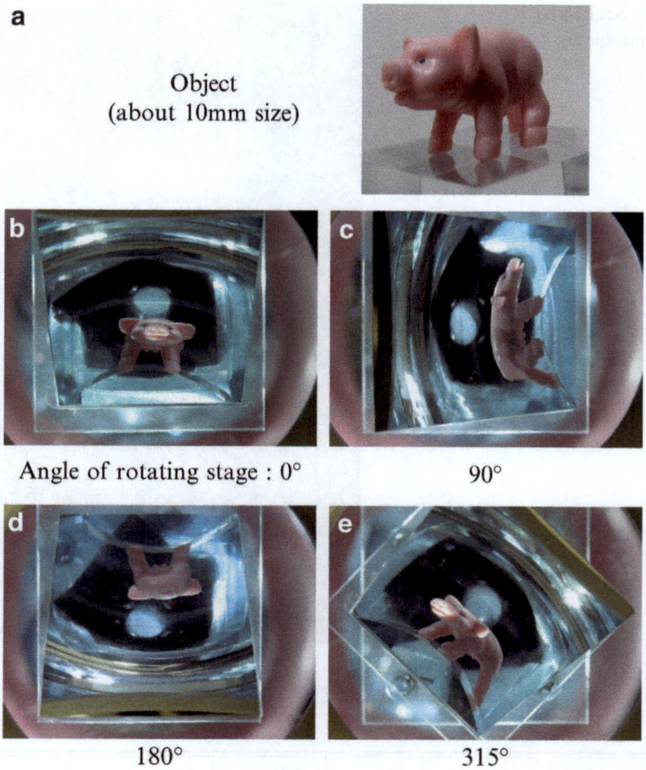

Fig. 9.11 Object and captured images of 360° ray capturing system (**a**) object, (**b**) captured view at 0-degree, (**c**) captured view at 90-degree, (**c**) captured view at 180-degree, and (**e**) captured view at 315-degree

9.4 International Standardization

We proposed FTV to MPEG in December 2001. Figure 9.13 shows the history of FTV standardization at MPEG.

In the 3D Audio Visual (3DAV) group of MPEG, many 3D topics such as omnidirectional video, FTV, stereoscopic video, and 3DTV with depth disparity information were discussed. According to the comments of the industry [40], the discussion was converged on FTV in January 2004.

Then, the standardization of the coding part of FTV started as MVC [16]. The MVC activity moved to the Joint Video Team (JVT) of MPEG and ITU for further standardization processes in July 2006. The standardization of MVC is based on H.264/MPEG4-AVC and was completed in May 2009. MVC was the first phase of FTV.

FTV cannot be constructed by coding part alone. We proposed to standardize the entire FTV [41], and MPEG started the standardization activity of FTV in April 2007 [42].

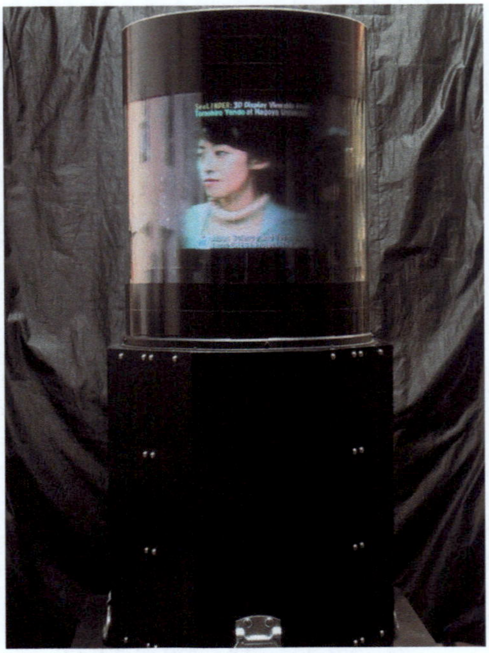

Fig. 9.12 The SeeLINDER, a 360° ray-reproducing display

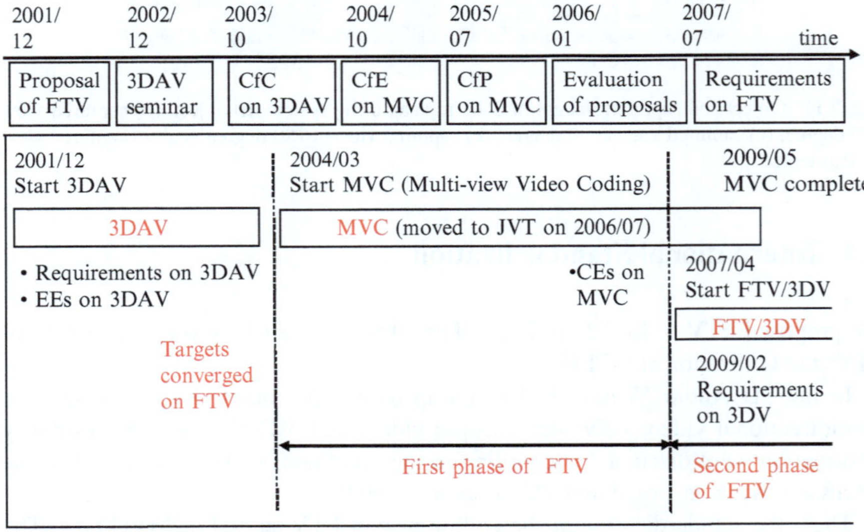

Fig. 9.13 History of FTV standardization at MPEG

The FTV reference model [43] is defined as shown in Fig. 9.14. Possible standardization items are FTV data format, decoder, and interpolation. Interpolation might be informative. Thus, FTV is a new framework that includes a coded representation for multiview video and depth information to support the generation

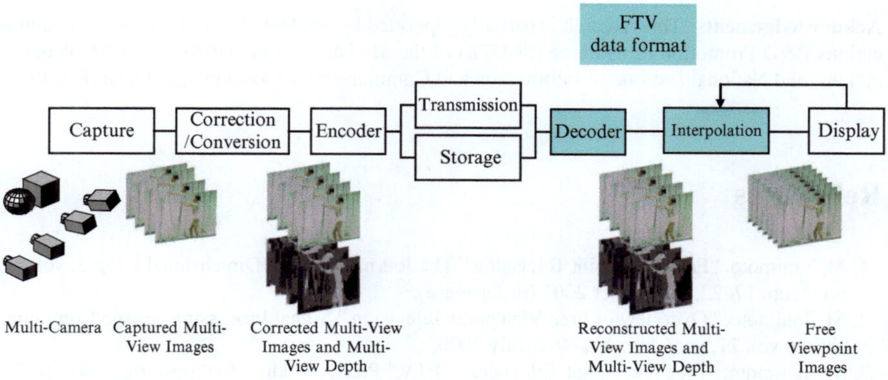

Fig. 9.14 FTV reference model and standardization items

of high-quality intermediate views at the receiver. This enables free-viewpoint functionality and view generation for 2D/3D displays. Free-viewpoint images can be generated efficiently from multiview images using multiview depth information [44].

In January 2008, MPEG-FTV targeted the standardization of 3DV. 3DV means video for 3D displays and is a standard that targets to serve a variety of 3D displays. 3DV is the second phase of FTV. The introduction to 3DV is described in the reference [17].

9.5 Conclusion

FTV is an innovative medium that allows one to view a 3D world by freely changing the viewpoint. FTV is the most advanced visual system and will bring an epochal change in the history of television.

FTV is an ultimate 3DTV that captures and transmits all visual information of a 3D scene, a natural interface between human and environment, and an innovative tool to create new types of content and art.

FTV will find many applications in the fields of broadcast, communication, amusement, entertainment, advertising, exhibition, education, medicine, art, archives, security, surveillance, and so on.

The international standardization of FTV has been conducted at MPEG. The first phase of FTV, MVC, has been recently completed. The second phase of FTV, 3DV, will be completed within the next 2 years. The introduction of FTV is not far because the interactive view generation of FTV has already been realized in real time. Therefore, we can enjoy FTV on a PC or a portable player if the standards of FTV are available, and FTV contents are delivered over internet or with packaged media.

Acknowledgements This research is partially supported by Strategic Information and Communications R&D Promotion Programme (SCOPE) of the Ministry of Internal Affairs and Communications, and National Institute of Information and Communications Technology, Japan (NICT).

References

1. M. Tanimoto, "Free Viewpoint Television," The Journal of Three Dimensional Images, vol. 15, no. 3, pp. 17–22, September 2001 (in Japanese).
2. M. Tanimoto, "Overview of Free Viewpoint Television," Signal Processing: Image Communication, vol. 21, no. 6, pp. 454–461, July 2006.
3. M. Tanimoto, "Free Viewpoint Television – FTV," Picture Coding Symposium 2004, Special Session 5, December 2004.
4. M. Tanimoto, "FTV (Free Viewpoint Television) Creating Ray-Based Image Engineering," Proceedings of ICIP2005, pp. II-25–II-28, September 2005.
5. M. Tanimoto, "Free Viewpoint Television," OSA Topical Meeting on Digital Holography and Three-Dimensional Imaging, DWD2, June 2007 (Invited Paper).
6. M. Tanimoto; "FTV (Free viewpoint TV) and Creation of Ray-Based Image Engineering," ECTI Transaction on Electrical Engineering, Electronics and Communications, vol. 6, no. 1, pp.3–14, February 2008 (Invited Paper).
7. M. Tanimoto, "FTV (Free viewpoint TV) and Ray-Space Technology," IBC2008, The cutting edge, part two (September 2008).
8. M. Sekitoh, K. Toyota, T. Fujii, T. Kimoto, M. Tanimoto, "Virtual Bird's-Eye View System based on Real Image," EVA 2000 Gifu, 8, pp. 8-1–8-7, October 2000.
9. M. Sekitoh, T. Fujii, T. Kimoto, M. Tanimoto, "Bird's Eye View System for ITS," IEEE, Intelligent Vehicle Symposium, pp. 119–123, May 2001.
10. M. P. Tehrani, K. Niwa, N. Fukushima, Y. Hirano, T. Fujii, M. Tanimoto, K. Takeda, K. Mase, A. Ishikawa, S. Sakazawa, A. Koike, "3DAV Integrated System Featuring Arbitrary Listening-point and Viewpoint Generation," Proceedings of IEEE Multimedia Signal Processing, MMSP 2008, PID-213, pp. 855–860, October 2008 (Best Paper Award).
11. T. Fujii, "A Basic Study on Integrated 3-D Visual Communication," Ph.D. Dissertation in Engineering, The University of Tokyo, 1994 (in Japanese).
12. T. Fujii, T. Kimoto, M. Tanimoto, "Ray Space Coding for 3D Visual Communication," Picture Coding Symposium 1996, pp. 447–451, March 1996.
13. M. Tanimoto, A. Nakanishi, T. Fujii, T. Kimoto, "The Hierarchical Ray-Space for Scalable 3-D Image Coding," Picture Coding Symposium 2001, pp. 81–84, April 2001.
14. T. Fujii, M. Tanimoto, "Free-Viewpoint Television Based on the Ray-Space Representation," Proceedings of SPIE ITCom 2002, pp. 175–189, August 2002.
15. M. Tanimoto, T. Fujii, "FTV – Free Viewpoint Television," ISO/IEC JTC1/SC29/WG11, M8595, July 2002.
16. "Introduction to Multi-view Video Coding," ISO/IEC JTC 1/SC 29/WG11, N7328, July 2005. http://www.chiariglione.org/mpeg/technologies/mp-mv/index.htm
17. "Introduction to 3D Video," ISO/IEC JTC1/SC29/WG11 N9784, May 2008. http://www.chiariglione.org/mpeg/technologies/mp3d/index.htm
18. P. Na Bangchang, T. Fujii, M. Tanimoto, "Experimental System of Free Viewpoint TeleVision," Proceedings of IST/SPIE Symposium on Electronic Imaging, vol. 5006–66, pp. 554–563, January 2003.
19. T. Fujii, K. Mori, K. Takeda, K. Mase, M. Tanimoto, Y. Suenaga, "Multipoint Measuring System for Video and Sound: 100-Camera and Microphone System," IEEE 2006 International Conference on Multimedia & Expo (ICME), pp. 437–440, July 2006.
20. K. Matsumoto, T. Yendo, T. Fujii, M. Tanimoto, "Multiple-Image Rectification for FTV," Proceedings of 3D Image Conference 2006, P-19, pp. 171–174, July 2006.

21. N. Fukushima, T. Yendo, T. Fujii, M. Tanimoto, "A Novel Rectification Method for Two-Dimensional Camera Array by Parallelizing Locus of Feature Points," Proceedings of IWAIT2008, B5–1, January 2008.
22. K. Yamamoto, T. Yendo, T. Fujii, M. Tanimoto, "Colour Correction for Multiple-camera System by using Correspondences" The Journal of the Institute of Image Information and Television Engineers vol. 61, no. 2 pp. 213–222, February 2007.
23. Y. He, J. Ostermann, M. Tanimoto, A. Smolic; "Introduction to the Special Section on Multi-view Video Coding," IEEE Transactions on Circuits and Systems for Video Technology, vol. 17, no. 11, pp. 1433–1435, November 2007.
24. K. Yamamoto, M. Kitahara, T. Yendo, T. Fujii, M. Tanimoto, S. Shimizu, K. Kamikura, Y. Yashima; "Multiview Video Coding Using View Interpolation and Color Correction," IEEE Transactions on Circuits and Systems for Video Technology, vol. 17, no. 11, pp. 1436–1449, November 2007 (Invited Paper).
25. A. Nakanishi, T. Fujii, T. Kimoto, M. Tanimoto, "Ray-Space Data Interpolation by Adaptive Filtering using Locus of Corresponding Points on Epipolar Plane Image," The Journal of the Institute of Image Information and Television Engineers (ITE), vol. 56, no. 8, pp. 1321–1327, August 2002.
26. M. Droese, T. Fujii, M. Tanimoto, "Ray-Space Interpolation based on Filtering in Disparity Domain," Proceedings of 3D Conference 2004, pp. 213–216, Tokyo, Japan, June 2004.
27. M. Droese, T. Fujii, M. Tanimoto, "Ray-Space Interpolation Constraining Smooth Disparities Based on Loopy Belief Propagation," Proceedings of IWSSIP 2004, pp. 247–250, Poznan, Poland, September 2004.
28. N. Fukushima, T. Yendo, T. Fujii, M. Tanimoto, "Real-Time Arbitrary View Interpolation and Rendering System Using Ray-Space," Proceedings of SPIE Three-Dimensional TV, Video, and Display IV, vol. 6016, pp. 250–261, November 2005.
29. N. Fukushima, T. Yendo, T. Fujii, M. Tanimoto, "An Effective Partial Interpolation Method for Ray-Space," Proceedings of 3D Image Conference 2006, pp. 85–88, July 2006.
30. N. Fukushima, T. Yendo, T. Fujii, M. Tanimoto, "Free Viewpoint Image Generation Using Multi-Pass Dynamic Programming," Proceedings of SPIE Stereoscopic Displays and Virtual Reality Systems XIV, vol. 6490, pp. 460–470, Feburuary 2007.
31. Y. Mori, N. Fukushima, T. Yendo, T. Fujii, M. Tanimoto,"View Generation with 3D Warping Using Depth Information for FTV," Signal Processing: Image Communication, vol. 24, no. 1–2, pp. 65–72, January 2009.
32. T. Fujii, M. Tanimoto, "Real-Time Ray-Space Acquisition System," SPIE Electronic Imaging, vol. 5291, pp. 179–187, January 2004.
33. K. Manoh, T. Yendo, T. Fujii, M. Tanimoto, "Ray-Space Acquisition System of All-Around Convergent Views using a Rotation Mirror," Proceedings of SPIE, vol. 6778, pp. 67780C-1–8, September 2007.
34. T. Fujii, T. Yendo, M. Tanimoto; "Ray-Space Transmission System with Real-Time Acquisition and Display," Proceedings of IEEE Lasers and Electro-optics Society Annual Meeting 2007, pp. 78–79, October 2007.
35. R. Takano, T. Yendo, T. Fujii, M. Tanimoto, "Scene Separation in Ray-Space," Proceedings of IMPS2005, pp. 31–32, November 2005.
36. R. Takano, T. Yendo, T. Fujii, M. Tanimoto, "Scene Separation and Synthesis Processing in Ray-Space," Proceedings of IWAIT2007, P6–23, pp. 878–883, January 2007.
37. N. Chimura, T. Yendo, T. Fujii, M. Tanimoto, "New Visual Arts by Processing Ray-Space," Proceedings of Electronic Imaging & the Visual Arts (EVA) 2007 Florence, pp. 170–175, March 2007.
38. N. Chimura, T. Yendo, T. Fujii, M. Tanimoto, "Image Generation with Special Effects by Deforming Ray-Space," Proceedings of NICOGRAPH, S1–4, November 2007.
39. T. Endo, Y. Kajiki, T. Honda, M. Sato, "Cylindrical 3-D Video Display Observable from All Directions," Proceedings of Pacific Graphics 2000, pp. 300–306, October 2000.
40. "Call for Comments on 3DAV," ISO/IEC JTC1/SC29/WG11 N6051, October 2003.

41. M. Tanimoto, T. Fujii, H. Kimata, S. Sakazawa, "Proposal on Requirements for FTV," ISO/IEC JTC1/SC29/WG11, M14417, April 2007.
42. "AHG on FTV (Free Viewpoint Television)," ISO/IEC JTC1/SC29/WG11 N8947, April 2007.
43. "Applications and Requirements on FTV," ISO/IEC JTC1/SC29/WG11, N9466, October 2007.
44. M. Tanimoto, T. Fujii, K. Suzuki, "Experiment of view synthesis using multi-view depth," ISO/IEC JTC1/SC29/WG11, M14889, October 2007.

Chapter 10
Presentation and Perception of Digital Hologram Reconstructions of Real-World Objects on Conventional Stereoscopic Displays

R. Näsänen, T.M. Lehtimäki, T. Pitkäaho, and T.J. Naughton

Abstract We describe how stereo image pairs can be computed from digital holograms of real-world objects by using a left-side and a right-side window in reconstruction. We also show how the stereo pairs can be displayed on spatially interlaced and anaglyphic of stereoscopic display devices. In addition, we studied the question of how the window size affects the perceptual quality of stereoscopically presented hologram reconstructions. With decreasing window size, the perceived amount of noise increases, the perception of details and the overall perceptual quality of images decreases. Further, we studied the question of how the visual system fuses images shown to the two eyes when the perceived depth of focus is increased by presenting a near-focused image to one eye and a far-focused image to the other eye. The fusion can be described as a spatially weighted mean, where the local contrast of each image is used as a weight. This can be a useful way of extending the depth of focus in stereoscopic presentation of hologram reconstructions.

Keywords Three-dimensional image processing · Digital holography · Numerical reconstruction · Perceptual quality · Stereoscopic display · Depth of focus

10.1 Introduction

Digital holograms [6, 8, 12] capture the 3D information of objects. A digital hologram, recorded on a digital camera, results from the interference between a reference wave front and an object wave front reflected from or transmitted through an object. For visual observation, digital holograms can be reconstructed optoelectronically or numerically. In principle, optoelectronic reconstruction allows one to view the

T.J. Naughton (✉)
RFMedia Laboratory, Oulu Southern Institute, University of Oulu, Vierimaantie 5, FI-84100 Ylivieska, Finland
and
Department of Computer Science, National University of Ireland, Maynooth, Ireland
e-mail: tomn@cs.nuim.ie

T. Fournel and B. Javidi (eds.), *Information Optics and Photonics: Algorithms, Systems, and Applications*, DOI 10.1007/978-1-4419-7380-1_10,

depicted object or scene from different angles and accommodate (focus) to different depths. Correspondingly, by using numerical reconstruction, it is possible to compute different angular views of the hologram to different depths. While writing this paper, the development of the optoelectronic holographic display technology is still in an early stage. Therefore, before this technology becomes more matured, displaying numerical hologram reconstructions on conventional displays seems to be a useful alternative.Optical displays for computer generated holograms of 2D stereo pairs have also been considered [1].

Conventional stereoscopic displays are not truly three-dimensional. Therefore, it is of interest to investigate the question of how information should be presented on these displays to optimise the perception of 3D and 2D information. Understanding the relationship between information presentation and perception will allow the design of efficient ways of mediating hologram information on conventional displays.

The purpose of this paper is (1) to describe how holographic data can be prepared for presentation on different kinds of conventional stereo displays, (2) to show how subjective image quality depends on the window size used for numerical reconstruction and (3) how human visual system fuses a near- and a far-focused image when they are presented to different eyes in order to extend the perceived depth of focus in hologram reconstructions.

10.2 Preparation of Holograms for Displaying on Conventional Displays

10.2.1 How to Encode Different Perspectives in Numerical Hologram Reconstructions

Different areas in a hologram represent different perspectives, i.e. angles of view, of the object (see, e.g. [4]). Therefore, in order to compute a reconstruction corresponding to a specific perspective, an area of the hologram is selected for computation. This is described formally below.

When the hologram centre is located at point (0, 0), two different subsets of a hologram can be obtained as follows:

$$\Lambda[H(x,y)] = \begin{cases} 0, & \text{if } x \geq 0 \\ H(x,y) & \text{otherwise} \end{cases} \tag{10.1}$$

and

$$\mathrm{P}[H(x,y)] = \begin{cases} 0, & \text{if } x < 0, \\ H(x,y) & \text{otherwise.} \end{cases} \tag{10.2}$$

From these subsets of a whole hologram, two perspectives, corresponding to a left (L)- and right (R)-hand view of an object, can be reconstructed as follows:

$$L = F_{\Lambda,z} \tag{10.3}$$

and

$$R = F_{\mathrm{P}}, z, \tag{10.4}$$

where F is the Fresnel approximation at any distance z given by:

$$F_{H,z}(x, y) = \left| \frac{-i}{\lambda z} H(x, y) \cdot \exp\left[i\pi \frac{x^2 + y^2}{\lambda z} \right] \right|^2, \tag{10.5}$$

In (10.5), the full hologram H can be replaced by Λ or P, λ is the wavelength of light and "•" denotes the convolution operation. When Λ and P are used, the equations above show how to obtain two horizontal perspectives. In this case, half of the hologram data is used for each perspective (Fig. 10.1). To get multiple perspectives, we can use various parts of the hologram data for reconstruction.

For a single view display, i.e. a conventional 2D display, one hologram reconstruction from one perspective is sufficient. This can be the perspective of a scene corresponding to the angle of hologram capture, and, in this case, the whole hologram data can be used. For a specific depth z, this can be obtained as

$$D(x, y) = F_{H,z}(x, y), \tag{10.6}$$

where D is the display in question.

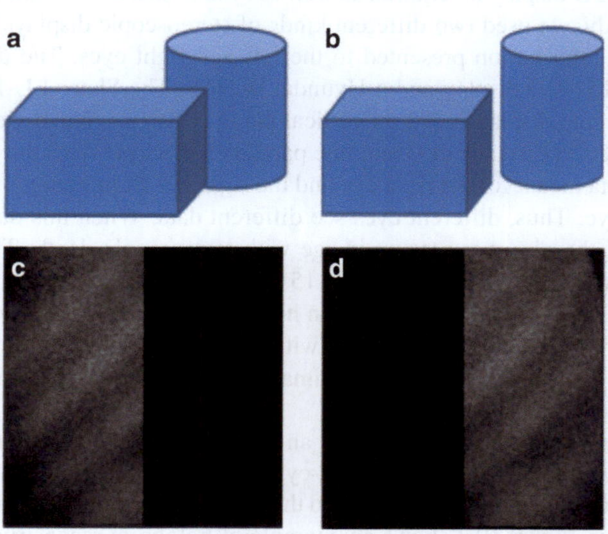

Fig. 10.1 (**a**) The scene shown from the left perspective, (**b**) the scene shown from the right perspective, (**c**) hologram data of the left perspective and (**d**) hologram data of the right perspective

For a stereoscopic display, two different perspectives of the scene are required. These can be obtained as:

$$D_0 = F_{\Lambda,z}, \tag{10.7}$$
$$D_1 = F_{\mathrm{P},z}, \tag{10.8}$$

where D_0 and D_1 are the views to be displayed to the left and right eye, respectively. For multi-view displays, multiple perspectives of the scene are usually required. Reconstructions for the needed amount of perspectives can be obtained as:

$$D_0 = F_{\Lambda,z}, \tag{10.9}$$
$$D_1 = F_{\mathrm{P},z}, \tag{10.10}$$
$$D_N = F_{\mathrm{P},z}, \tag{10.11}$$

where D_0, D_1 and D_N are spatially interlaced in left-right mode or over-under depending on the display in question. The number of reconstructions needed is dependent on the display.

10.2.2 Conventional Stereoscopic Displays

In a stereoscopic display, the views can be presented as spatially interlaced either in left-right mode or in over-under mode as appropriate for the display in question. Another possibility is to use the shutter glass method where successive frames containing D_0 and D_1 are presented to the left and right eye in repetition. In this case, a high-speed 2D display is required as well as synchronised LCD shutter glasses.

In our study, we used two different kinds of stereoscopic displays based on interlacing the information presented to the left and right eyes. The displays were a Sharp LL-151–3D display and a Hyundai W240S. The Sharp LL-151–3D is an auto-stereoscopic display based on vertical parallax barrier. When viewed in front of the display at a distance of 60 cm, the parallax barrier prevents the left eye from seeing the data meant for the right eye and the right eye from seeing the data meant for the left eye. Thus, different eyes see different data. When this data comprises a stereo pair, the observer sees an image with stereoscopic depth. The way pixel columns are organised on the Sharp LL-151–3D display is described in Fig. 10.2 a. The Hyundai W240S display is based on horizontal interlacing and polarisation. It uses a filter in front of the LCD screen with alternating polarising properties pixel row by pixel row to construct a stereo image from a stereo pair. This principle is illustrated in Fig. 10.2b.

An anaglyphic display only requires an ordinary 2D colour display and colour filter glasses, typically red-green or red-cyan. In anaglyphic presentation, one eye sees a view presented with one colour and the other eye sees an image presented with another colour. Figure 10.3 shows an example of hologram reconstructions for the left and right eyes and an anaglyphic combination of the reconstructions prepared for viewing with red-green filter glasses.

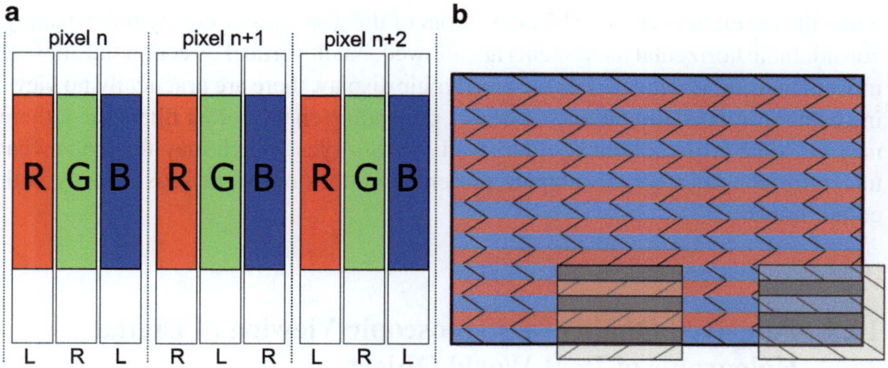

Fig. 10.2 (**a**) Interlacing in Sharp LL151–3D LCD display. Each pixel is divided into three sub-pixels (R, G and B). Sub-pixels are shown to the left (L) or right (R) eye. (**b**) How image is formed in horizontally interlaced polarised display. Small squares on the bottom represent polarised filters in front of the eyes. By using correctly polarisation-coded filters (acting as analysers), one eye sees every second row, while the other eye sees the others

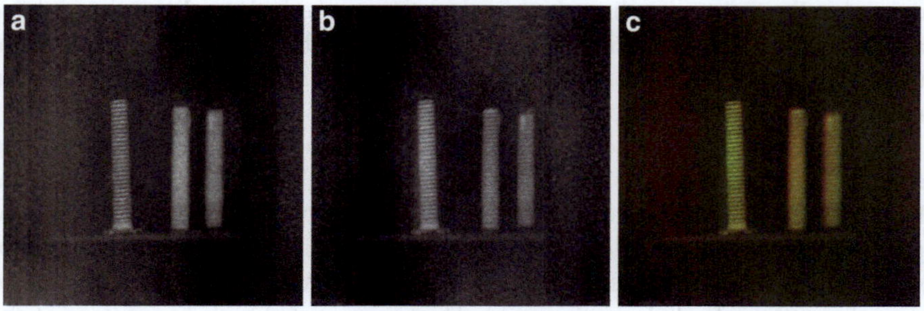

Fig. 10.3 Speckle-filtered reconstructions from two different perspectives: (**a**) intensity of the left perspective reconstruction, (**b**) intensity of the right perspective reconstruction and (**c**) an anaglyphic image made from (**a**) and (**b**)

10.2.3 Conclusions

In this section, it is shown that it is possible to reconstruct holograms as image pairs that can be presented as stereo images. There are some limitations related to the generation of stereo pairs, however. Using a small window size, it is possible to better utilise the angular range available in a hologram. However, the amount of noise in reconstructions increases with decreasing window size. There are also limitations related to stereoscopic displays. The spatially interlaced displays have limitations with respect to viewing position. Wrong viewing position leads to increased cross-talk and even a complete break-down of stereo perception. The Sharp LL-151–3D display requires that the viewer is located quite accurately at a distance of 60 cm directly in front of the display. The Hyundai W240S allows more freedom of viewing position, but the cross-talk in different parts of the display surface is lowest when the observer is at about 90 cm in front of the display. For the Hyundai display, considerable horizontal movements are allowed, but in vertical direction the allowed movement is more limited. For the anaglyphic display, there are practically no viewing position-related limitations. However, depending on the colour filters used, there may be some cross-talk and colour rivalry across eyes, which may feel somewhat uncomfortable. Further, anaglyphic presentation does not work satisfactorily with colour images.

10.3 Perceived Quality in Stereoscopic Viewing of Digital Holograms of Real-World Objects

10.3.1 Introduction

Stereoscopic depth perception is based on the fact that a three-dimensional object or view produces slightly different images in the two eyes. The local differences of corresponding points, i.e. disparities, in these images code three-dimensional shape and depth (see, e.g. [7]). The real-world situation can be mimicked to some extent by means of two 2D images that correspond to the images received by the eyes in natural viewing of 3D objects.

As described in the previous section, stereoscopic images can be created by using spatially limited areas, i.e. windows, of the left- and right-hand side of a digital hologram in numerical reconstruction. The use of small window sizes allows to produce a greater angular difference between perspectives and better perception of stereoscopic depth. However, as window size is decreased, a smaller proportion of the information of a hologram is used. This is associated with an increase of the amount of noise and lower visibility of image details. There are indications that lower the visibility of stimulus, either due to reduced contrast or due to the presence of noise, can have a detrimental effect on stereo perception (cf. [5, 10]).

The purpose of this study was to find out how the size of reconstruction window, the visibility of noise and the visibility of details are related to the perceived image quality of non-stereo and stereoscopic digital hologram reconstructions.

10.3.2 Methods

10.3.2.1 Subjects and Apparatus

Our experiment was performed using 13 subjects with normal or corrected-to-normal vision (ten females and three males aged between 23 and 53). Their stereovision was evaluated with stereo pair photographs before the experiment to ensure they could be able to perceive stereoscopic depth: the subjects were shown a set of three images – left-left image, left-right image and right-right image. Observers were asked to evaluate which of the images could be seen stereoscopically. The dominating eye of the subjects was also determined.

Stimuli were presented on a Sharp LL-151–3D LCD monitor, which is an auto-stereoscopic display. It allows users to see stereoscopic images without using special glasses. It consists of two LCD panels, one in front of another. In 3D mode, the foreground LCD panel acts as a parallax barrier so that the right and left eyes of a viewer receive different images. The screen resolution was $1,024 \times 768$ pixels (512×768 pixels for each eye). The distance between the subject and the screen was approximately 60 cm, which is required for the minimisation of cross-talk.

10.3.2.2 Stimuli

We used a set of six reconstructed stereo pairs of five digital holograms. Each set of the stereo pairs was reconstructed using six different window sizes 64, 128, 256, 512, 768 and 1,024. An example of a stereo pair set is presented in Fig. 10.4. Each hologram has its own reconstruction depth where the whole object was perceived in focus.

10.3.2.3 Procedure

We presented subjects with reconstructions from five different holograms, using six different window sizes (apertures on the hologram data) and two perspectives (left and right) from each hologram. We asked subjects to evaluate reconstructions of individual left and right perspectives, and resulting stereo pairs. Subjects were required to answer either yes or no to the questions: "Does noise disturb?" and "Are details visible?" The results were presented as the percentage of "yes" responses of all responses. They also evaluated the quality of images on scale 1–5 (1 = very poor, 5 = very good quality). The results were averaged across all subjects and all hologram reconstructions used.

10.3.3 Results

Results from the evaluations are presented in Fig. 10.5. With increasing window size, the noise in reconstructions was experienced less disturbing as indicated by the frequency of "yes" responses to the question "Does noise disturb?" At the same time, the visibility of details increased as indicated by the increasing frequency of "yes" responses to the question "Are details visible?" The overall subjective image quality was closely related to the visibility of details and inversely related to the experienced disturbing effect of noise, that is, subjective image quality increased with the window size.

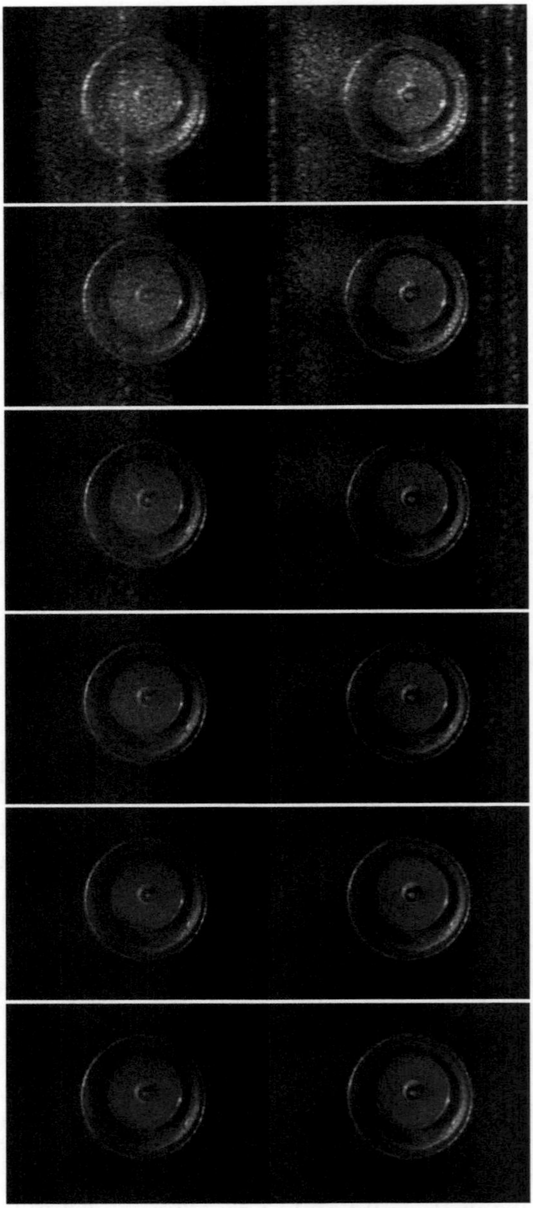

Fig. 10.4 An example of stereo pair set reconstructed with window sizes 64, 128, 256, 512 and 1,024

Despite the noise in the reconstructions and their general poor quality, all subjects perceived stereoscopic depth in the stereo pairs. Stereoscopic images were perceived typically with less noise than the non-stereo images used to compose a stereo image. If the members of a stereo pair were not equal in quality, subjects rated the quality of the stereo pair nearly equal to the quality of the better member independently of eye dominance.

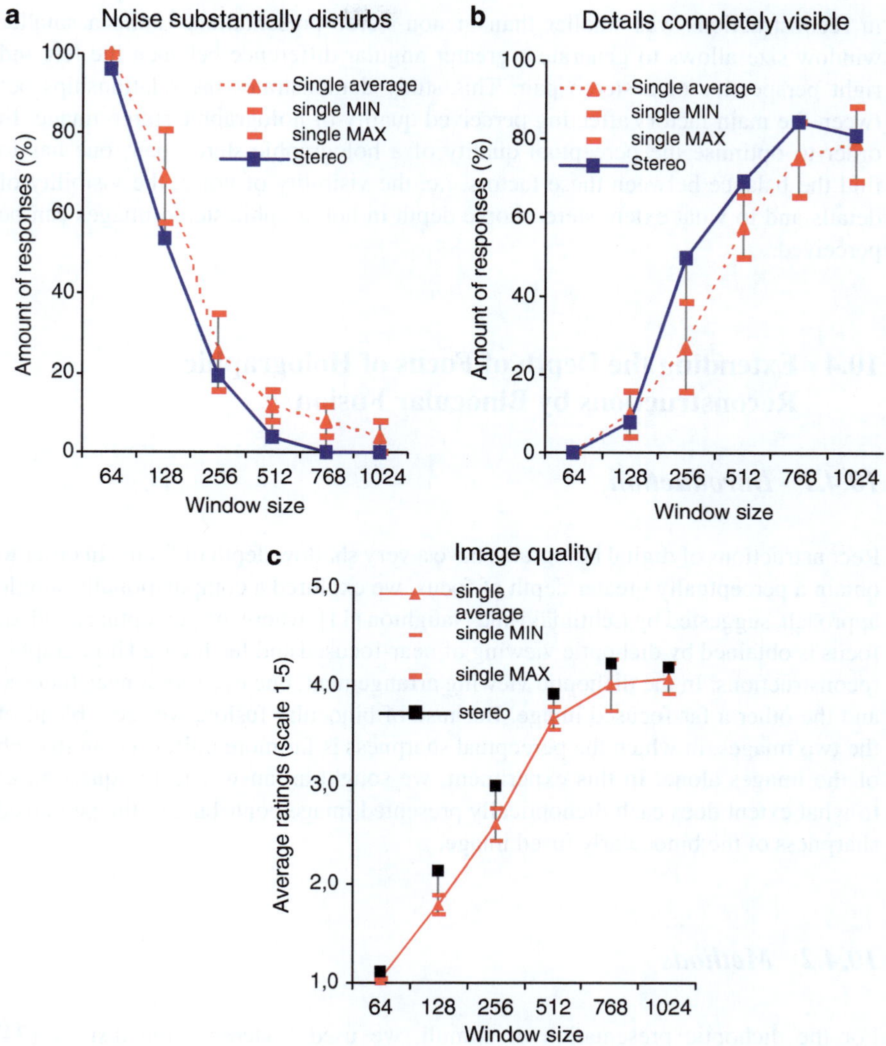

Fig. 10.5 The disturbing effect noise (**a**), visibility of details (**b**) and the evaluated perceived quality of images reconstructed with different window sizes (**c**). In (**a**) and (**b**), the vertical axis gives the percentage of "yes" responses of all responses

10.3.4 Conclusions

The results showed that with increasing window size the noise in reconstructions was experienced less disturbing. At the same time, the visibility of details increased. The overall subjective image quality was closely related to the visibility of details and inversely related to the experienced disturbing effect of noise. Further, the results showed that in stereoscopic presentation the perceived amount of noise in reconstructions was smaller than in non-stereo presentation. Using a smaller window size allows to generate a greater angular difference between the left and right perspective of a stereo pair. This study demonstrates the relationships between the main factors affecting perceived quality of holographic stereo image. In order to optimise the perceptual quality of a holographic stereo pair, one has to find the balance between these factors, i.e. the visibility of noise, the visibility of details and to what extent stereoscopic depth in holographic stereo images can be perceived.

10.4 Extending the Depth of Focus of Holographic Reconstructions by Binocular Fusion

10.4.1 Introduction

Reconstructions of digital holograms have a very shallow depth of focus. In order to obtain a perceptually greater depth of focus, we explored a computationally simple approach, suggested by Lehtimäki and Naughton [11], where the perceptual depth of focus is obtained by dichoptic viewing of near-focused and far-focused holographic reconstructions. In the dichoptic viewing arrangement, one eye sees a near-focused and the other a far-focused image. Because of binocular fusion, we see a blend of the two images, in which the perceptual sharpness is far more uniform than in each of the images alone. In this experiment, we sought an answer to the question of to what extent does each dichoptically presented image contribute to the perceived sharpness of the binocularly fused image.

10.4.2 Methods

For the dichoptic presentation of stimuli, we used a stereoscopic display (24′ Hyundai W240S), which was viewed with circular polarising glasses. On the left half of the display, the subjects saw the near-and far-focused images dichoptically, which by binocular fusion produced a perceptually increased depth of focus. On the right half of the display, they saw a computationally fused image, which consisted of locally weighted averages of the near- and far-focused images. The computational fusion was obtained by using (10.12).

$$I_{\text{fused}}(x,y) = \frac{|C_{\text{left}}(x,y)|^p I_{\text{left}}(x,y) + |C_{\text{right}}(x,y)|^p I_{\text{right}}(x,y)}{|C_{\text{left}}(x,y)|^p + |C_{\text{right}}(x,y)|^p}, \tag{10.12}$$

where $I_{\text{left}}(x,y)$ and $I_{\text{right}}(x,y)$ are the images received by the left and right eyes, respectively. Local root-mean-square (rms) contrasts of the left and right eye images,

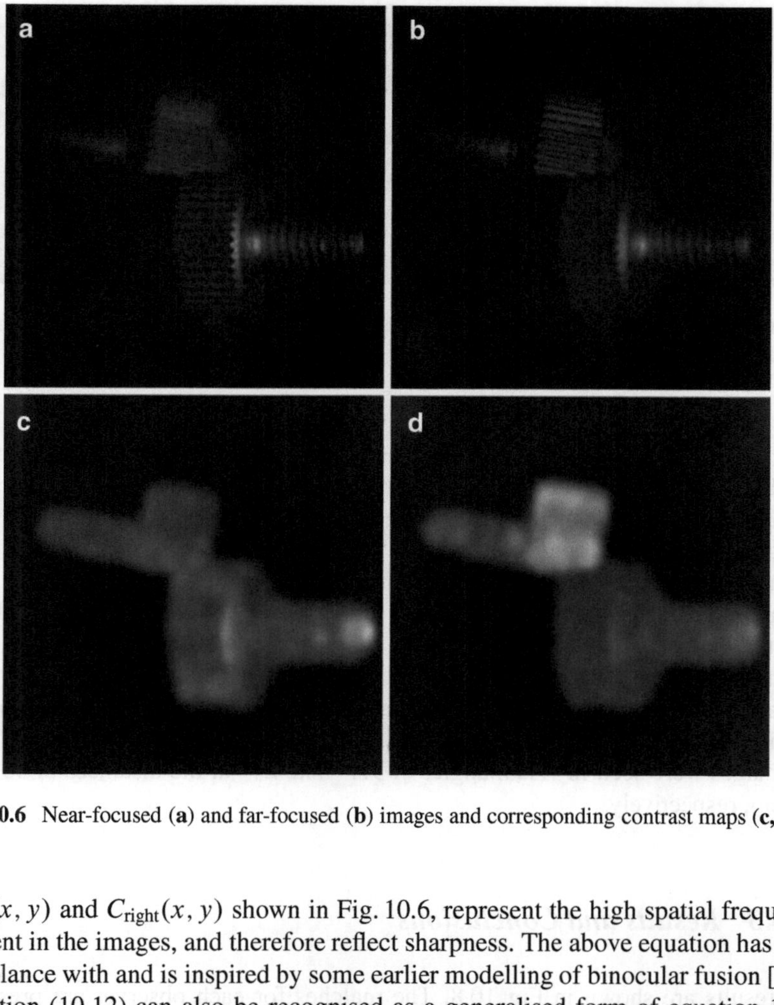

Fig. 10.6 Near-focused (**a**) and far-focused (**b**) images and corresponding contrast maps (**c, d**)

$C_{\text{left}}(x, y)$ and $C_{\text{right}}(x, y)$ shown in Fig. 10.6, represent the high spatial frequency content in the images, and therefore reflect sharpness. The above equation has a resemblance with and is inspired by some earlier modelling of binocular fusion [3,9]. Equation (10.12) can also be recognised as a generalised form of equation 10 of [2]. They suggested the equation as a simplified description of binocular summation. The difference between our and their equation is that the weights $[C_{\text{left}}(x, y)$ and $C_{\text{right}}(x, y)]$ are spatially varying contrasts instead of modulation coefficients. Therefore, our equation is applicable to complex images instead of simple low spatial frequency grating patches alone used by [2].

In (10.12), exponent p varied from -4 to $+8$ with steps of 0.5. When p is equal to zero, the equation reduces to an ordinary arithmetic mean. When p is above zero, a high value of rms contrast, i.e. sharp image, is emphasised relative to a low value of rms contrast, i.e. blurred image. When p is negative, the blurred image is emphasised relative to the sharp image. Thus, the series of values of exponent p gave a series of computationally fused images, where sharpness varied from blurred to sharp. Figure 10.7 shows examples of images fused computationally with values -4, 0, and 8 of exponent p.

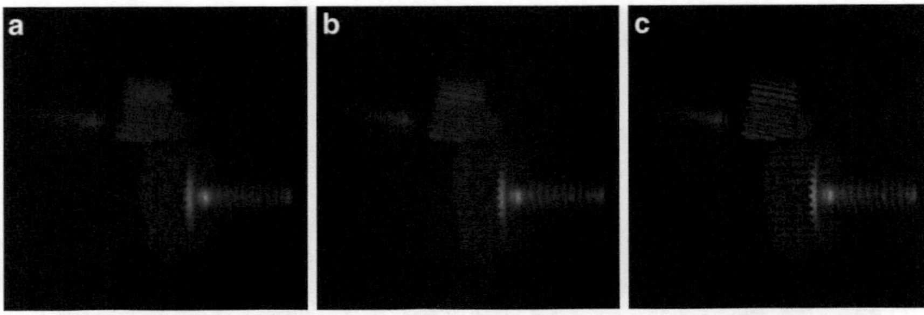

Fig. 10.7 Examples of computationally fused images. The values of p are -4, 0 and $+8$ from left to right (**a–c**)

In the experiment, the observer could vary the sharpness of the computationally fused image in real time by using a graphical slider. The task of the observers was to match the computationally fused image to the binocularly fused image with respect to perceived sharpness. Ding and Sperling [2] showed that weighted mean summation describes binocular fusion well for their stimuli. If the weighted mean summation rule applies to high spatial frequencies as well, our experiment should directly give the exponent in (10.12) that characterises the summation rule used by human observers in binocular fusion of complex objects.

Each measurement consisted of 20 matches. Measurements were made for two viewing distances, 150 and 300 cm. The image size was $24.6 \times 24.6\,\text{cm}^2$. Thus, the images were seen in visual angles of 9.4° and 4.7° at the distances of 150 and 300 cm, respectively.

10.4.3 Results and Conclusions

The results are shown in Fig. 10.8. The matches for each observer are presented as cumulative distributions of exponent p of the computational fusing function of (10.12). All values of exponent p are clearly above zero. This suggests that in the perception of the binocularly fused image, the in-focus areas of each image, i.e. the sharp parts, had a relatively greater contribution than the out-of-focus areas, i.e. the blurred parts. However, the dominance of in-focus areas was far below complete, since all matches were lower than the highest value of exponent $p = 8$. The extent to which sharp image is emphasised relative to the blurred image varied across individuals. Thus, there appears to be observer-dependent differences in binocular fusion.

In Fig. 10.9, we have studied the binocular fusion for the overall contrast. The task of the observer was to match the contrast of a binocularly presented image to a stimulus where one eye saw a similar image and the other eye saw a zero-contrast image. The median-matched binocular contrast computed across observers (0.3305) was practically identical to the contrast seen by one eye (0.33). In (10.12),

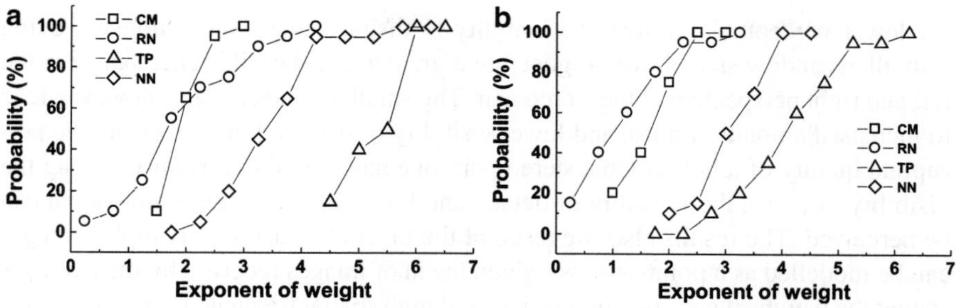

Fig. 10.8 Cumulative distributions of perceived sharpness matches between binocularly fused and computationally fused images presented as cumulative distributions of the exponent of weight. In panel (**a**), the viewing distance was 150 cm and in panel (**b**) it was 300 cm

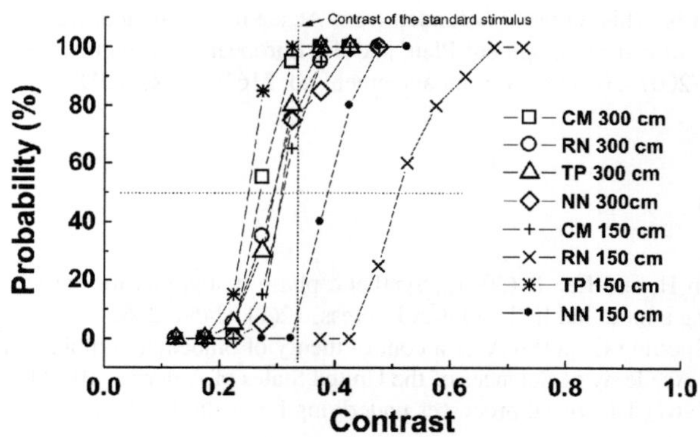

Fig. 10.9 Cumulative distributions of perceived contrast matches. The contrast of a binocularly presented image was matched to stimulus where a similar image was seen by one eye and zero-contrast image by the other eye

the zero-contrast image has a zero weight, and, therefore, the fused image is the same as the non-zero image. According to (10.12), the image shown to both eyes should also be perceived as similar to the image presented to one eye only. Thus, the experimental result is in agreement with (10.12). However, there is variability in the results across observers and conditions, which suggest that there are individual differences as well as some dependence on viewing condition such as viewing distance.

10.5 General Conclusions

Stereoscopic image pairs can be computed from digital holograms by choosing windows from the left and right side of the hologram. The stereo pairs can be displayed on stereoscopic displays in order to increase perceived depth in hologram reconstructions. There are some limitations of spatially interlaced stereoscopic displays related to the viewing position. The limitations of the anaglyphic method are lower perceptual comfort and inability to display good colour images. Using a smaller window size allows to generate a greater angular difference between the left and right perspective of the stereo pair. The smaller window sizes, however, lead to increased amount of noise and lower visibility of details. For optimising the perceptual quality of a holographic stereo pair, one has to find the balance among the visibility of noise, the visibility of details and how well the stereoscopic depth can be perceived. The results also suggest that the binocular fusion of complex images can be modelled as a point-wise weighted mean of images received by the two eyes so that the weight at each position is the local high spatial frequency contrast of each image raised to a power greater than zero. Dichoptic or stereoscopic presentation of near- and far-focused images may be a useful and simple way to extend the depth of focus of holographic reconstructions.

Acknowledgments This research is funded from Academy of Finland, Science Foundation Ireland under the National Development Plan, and the European Community's Seventh Framework Programme FP7/2007–2013 under grant agreement no. 216105 ("Real 3D").

References

1. Choi K, Kim H, and Lee B (2004). Synthetic phase holograms for auto-stereoscopic image displays using a modified IFTA. Optics Express, 12(11):2454–2462
2. Ding J and Sperling G (2006) A gain-control theory of binocular combination. Proceedings of the National Academy of Sciences of the United States of America, 103:1141–1146
3. Engel GR (1967) The visual processes underlying binocular brightness summation. Research, 7:753–767
4. Frauel Y, Naughton TJ, Matoba O, Tajahuerce E, and Javidi B (2006) Three-Dimensional Imaging and Processing Using Computational Holographic Imaging. Proceedings of the IEEE, 94:636–653
5. Glennester A and Parker AJ (1997) Computing stereo channels from masking data. Vision Research, 37:2143–2152
6. Goodman JW and Lawrence RW (1967) Digital image formation from electronically detected holograms. Applied Physics Letters, 11:77–79
7. Howard IP and Rogers BJ (2002) Seeing in depth, vol. 2. Porteous, Toronto
8. Kronod MA, Merzlyakov NS, and Yaroslavskii LP (1972) Reconstruction of a hologram with a computer. Soviet Physics – Technical Physics, 17:333–334
9. Legge GE (1984) Binocular contrast summation – II. Quadratic summation. Vision Research, 24:385–394
10. Legge GE and Gu YC (1989) Stereopsis and contrast. Vision Research, 29:989–1004
11. Lehtimäki T and Naughton TJ (2007) Stereoscopic viewing of digital holograms of real-world objects. Capture, transmission and display of 3D video, article no. 39, Kos, Greece, 7–9 May
12. Schnars U and Jueptner W (2005) Hologram Recording, Numerical Reconstruction, and Related Techniques, Springer, Berlin

Chapter 11
Parallel Phase-Shifting Digital Holography Based on the Fractional Talbot Effect

L. Martínez-León, M. Araiza-E, B. Javidi, P. Andrés, V. Climent, J. Lancis, and E. Tajahuerce

Abstract Parallel phase-shifting digital holography techniques allow dynamic measurements with just one shot of the image sensor. Here, we briefly review some of these techniques and report a method for recording on-axis single-shot digital holograms based on the self-imaging phenomenon. A simple binary two-dimensional periodic amplitude grating is used to codify the reference beam in a Mach–Zehnder interferometer, generating a periodic three-step phase distribution with uniform irradiance over the sensor plane by fractional Talbot effect. An image sensor records only one shot of the interference between the light field scattered by the object and the codified parallel reference beam. Images of the object are digitally reconstructed from the digital hologram through the numerical evaluation of the Fresnel diffraction integral. This scheme provides an efficient way to perform dynamic phase-shifting interferometric techniques. Unlike other parallel phase-shifting techniques, neither complex pixelated polarization devices nor special phase diffractive elements are required. Experimental results confirm the feasibility and flexibility of our method.

11.1 Introduction

The complex amplitude distribution of a diffracted light beam can be measured and reconstructed through digital holography techniques [1]. From chemical recording and wet developing, conventional holography evolved toward digital sampling and numerical processing, thanks to the advances in solid-state sensors and the improvement in computation resources [2–5]. Nowadays, digital holography is effectively applied in many applications, including three-dimensional (3D) imaging and microscopy [6–13], encryption [14–16], and pattern recognition [17–19].

E. Tajahuerce (✉)
GROC·UJI, Departament de Física, Universitat Jaume I, 12071 Castelló, Spain
and
INIT (Institute of New Imaging Technologies), Universitat Jaume I, 12071 Castelló, Spain
e-mail: enrique.tajahuerce@uji.es

T. Fournel and B. Javidi (eds.), *Information Optics and Photonics: Algorithms, Systems, and Applications*, DOI 10.1007/978-1-4419-7380-1_11,

The basic digital holography procedure begins with the electronic recording of the pattern resulting from the interference between the light scattered by a sample and a known reference beam. Next, this interference pattern is numerically processed, to reconstruct an image of the diffracting object by simulating the reference and using a Fresnel propagation algorithm. As in conventional holography, a problem arises because the three patterns generated in the reconstruction, the real and virtual images and the zero order, overlap. Nevertheless, several approaches have been proposed to prevent the noisy terms or the superposition of them. One of the most direct methods is recording the digital hologram with off-axis geometries, introducing a tilt between the object and reference beams [5, 8, 14]. Thus, the real and virtual images and the zero order appear physically separated in the reconstruction window. However, since common digital cameras have a pixel size of a few micrometers, the maximum spatial frequency of the interference pattern is limited, and consequently the angle between the object and the reference beams must remain small. On the other hand, the spatial separation of the different orders in the reconstruction window restricts the size of the object to be reconstructed. Therefore, an acceptable trade-off between image resolution and the size of the reconstructed area must be contemplated when recording the hologram.

Phase-shifting techniques [6, 7, 11] are a frequent choice to overcome the problems posed by off-axis digital holography, as they guarantee that the spatial resolution of digital sensors is efficiently exploited. Several interference patterns with various phase retardations for the reference are acquired in this scheme. A suitable combination of these patterns provides the complex amplitude distribution generated only by the object beam at the image sensor. In other words, the term associated to the virtual image can be conveniently isolated. Four different interference patterns are usually captured, although recently it has been proved that it is possible to get good results with just two [20]. In this way, this sort of time multiplexed in-line digital holography cannot be used when the object wave front experiences faster changes than the recording time needed for grabbing each sequence of holograms.

Parallel phase-shifting techniques allow dynamic measurements by recording several phase-shifted interference patterns with just one shot of the image sensor [21–27]. Usually, spatial-multiplexing techniques are employed and, accordingly, acquisition times are only limited by the sensor capabilities. Phase plates and polarizing elements are used to generate the phase-shifted interferograms. Among the possible approaches, in some of them the area of the sensor is divided into a few separated but continuous regions, each one receiving a reference wave with a phase shifted at different values [21]. The spatial multiplexing is possible thanks to diffractive optical elements. In other words, the various reference waves are periodically distributed, as a pattern repeated all over the sensor, with a unit cell that occupies several pixels of the detector [22–27]. In the latter case, pixelated phase-shifting arrays are required, though these polarization arrays are difficult to fabricate and hard to integrate with the CCD sensor. Besides, sometimes it is necessary to use an imaging system between the phase-shifting array and the image sensor, likely to bring in aberrations. Despite these drawbacks, these methods are indeed able of recording digital holograms using three or even just two phase-shifting steps [24–27].

Some one-shot schemes perform computing operations on the recorded digital hologram to suppress the zero order or also the unwanted real or virtual unfocussed image of the object in the reconstruction window [28–31]. That implementation allows to deal with vibrations and to work with moving objects. Some others prove that it is possible to use only one interferogram, even working with on-axis configurations and without any complex processing of the hologram, for particular applications such as 3D pattern recognition [32, 33].

We propose a new method for recording single-shot digital holograms based on the fractional Talbot effect [34]. The Talbot effect or self-imaging phenomenon is observed when a grating is illuminated by a coherent light beam [35]. Under parallel illumination, replicas of the grating appear periodically at distances multiple of the so-called Talbot distance. Fresnel images, the superposition of shifted replicas of the grating weighted by different phase factors, appear at fractions of the Talbot distance [36–40]. In fact, it is possible to design periodic amplitude gratings that, at certain fractional Talbot distances, generate a periodic phase distribution with uniform irradiance. This property has been extensively used to develop efficient Talbot array illuminators or kinoform sampling filters [37, 39, 41, 42]. In these applications, the periodic phase distribution is fabricated and used as an efficient diffractive optical element to generate the array of light spots. The Talbot effect has also been applied to develop interferometers and wavefront sensors [43, 44]. Digital holography has been used to analyze the Talbot effect and its application to spectrometry [45]. Recently, an approach to perform parallel phase-shifting digital holography by using the self-imaging effect has been proposed [46], in which a periodic phase mask has to be located at the reference beam.

Our dynamic Talbot-based digital holography system operates in reverse order with respect to Talbot array illuminators. A binary two-dimensional periodic amplitude grating codifies the reference beam and generates a periodic three-step phase distribution with uniform irradiance over the sensor plane by fractional Talbot effect. The sensor records only one shot of the interference between this reference beam and the light field scattered by an object. In this way, different pixels of the sensor record the interferogram with different phase shifts. By sampling and processing the pixelated phase-shifted interferogram, we are able to measure the amplitude and phase of the object light field. Therefore, we have an efficient way to perform dynamic phase-shifting digital holography. Moreover, neither complex pixelated polarization devices nor special phase diffractive elements are required. Once the digital hologram has been generated, images of the object are digitally reconstructed. Experimental results are presented here to prove our idea.

In the following sections, we describe the basic principle of the method to perform single-shot Talbot digital holography and report experimental results.

11.2 Talbot Digital Holography Interferometry

The Talbot digital holography layout that we propose is based on a Mach–Zehnder interferometer with a 2D binary amplitude grating codifying the reference beam, as shown in Fig. 11.1. In this optical setup, a laser beam is spatially filtered, collimated,

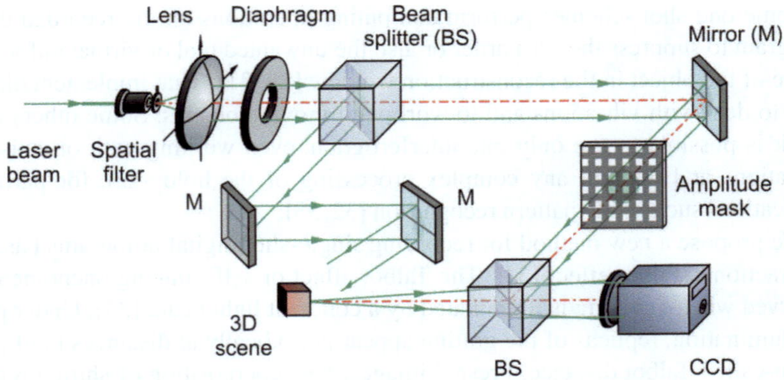

Fig. 11.1 Setup for recording digital holograms by fractional Talbot effect

and split into an object path and a reference path in order to obtain two uniform plane waves traveling in different directions. The light scattered by the 3D object travels toward the CCD detector through a second beam splitter. The complex amplitude distribution, $O(x, y, z)$, of the light field diffracted by the object at a location (x, y, z) is the superposition of the spherical waves emitted by different points of the 3D object.

Let us write the complex amplitude in the plane of the CCD, located at $z = 0$, in the following form:

$$O(x, y, 0) = A_0(x, y)\mathrm{e}^{i\varphi_0(x,y)}, \tag{11.1}$$

where $A_0(x, y)$ and $\varphi_0(x, y)$ denote the amplitude and phase of the object light field. We intend to measure both parameters to be able to reconstruct the complex amplitude distribution $O(x, y, z)$ at different distances z.

The parallel reference beam is diffracted by a 2D binary amplitude grating, and travels toward the sensor. The system is aligned in such a way that the reference beam generates a plane wave traveling perpendicular to the sensor whenever the 2D grating is absent. The amplitude transmittance $t(x, y)$ of the grating can be written as follows:

$$t(x, y) = t_c(x, y) \otimes \left[\sum_{j=-P/2}^{P/2-1} \sum_{k=-P/2}^{P/2-1} \delta(x - jd)\delta(y - kd) \right], \tag{11.2}$$

where $t_c(x, y)$ is the amplitude transmittance of the unit cell, $\otimes$ denotes convolution, d is the period of the array, $\delta(x)$ represents the Dirac delta function, and j and k are the integer numbers.

Since $t(x, y)$ is a periodic function, it exhibits self-imaging properties and, in particular, the fractional Talbot effect. Therefore, the amplitude distribution associated with the grating produces both self-images and Fresnel images by free-space propagation. Self-images are diffraction patterns that are a copy of the input distribution. Fresnel images are a superposition of a certain number, r, of copies of the input grating in each transversal dimension. These copies are displaced by a distance given by integer multiples of d/r and have different weight phase factors. Under monochromatic illumination with wavelength λ, the Fresnel images are obtained at distances [36]

$$z' = \frac{2d^2}{\lambda}\left(q + \frac{n}{m}\right) = z_t\left(q + \frac{n}{m}\right), \tag{11.3}$$

where z_t is the Talbot distance, q is an integer, and n and m are natural numbers with no common factor (with $n < m$). The ratio n/m determines the amplitude distribution at the unit cell of a particular Fresnel image. Different values of q lead to different positions of the same Fresnel-image pattern. The number of replicas, r, associated with the Fresnel image of index n/m is given by $m/2$ when m is even and by m when m is odd. By choosing the opening ratio of the grating in accordance with the index n/m of the Fresnel image, it is possible to get a uniform irradiance distribution with a periodic multilevel phase at the output plane. Several formulations have been proposed to obtain the weighting phase factors of Fresnel images for particular or general cases [36–40]. As we are interested in just three different phases at the output plane, we restrict our analysis to the particular case $n/m = 1/4$ or 3/4. However, other orders of the Fresnel image can be interesting to obtain more complex periodic phase distributions. For our case, the amplitude distribution generated by the reference beam at the output plane, at a distance given by (11.3), is given as follows:

$$R(x, y, 0) = \frac{A}{2}\left[t(x, y) + it\left(x + \frac{d}{2}, y\right) + it\left(x, y + \frac{d}{2}\right) + i^2 t\left(x + \frac{d}{2}, y + \frac{d}{2}\right)\right], \tag{11.4}$$

where A is the constant amplitude of the reference beam just before the grating, assumed to be equal to unity in the analysis that follows. To get a Fresnel image with uniform irradiance, the amplitude transmittance, $t_c(x, y)$, of the unit cell of the grating in (11.2) is given as follows:

$$t_c(x, y) = \text{rect}\left(\frac{x}{d/2}\right)\text{rect}\left(\frac{y}{d/2}\right). \tag{11.5}$$

Therefore, we are able to obtain three different phase factors, 0, $\pi/2$, and π at the unit cell distributed as shown in Fig. 11.2.

For practical purposes, it is relevant to consider the effect of using a finite grating in (11.2) in order to achieve the amplitude distribution in (11.4). It has been shown that the number of periods P of the grating and the number of periods P' of the Fresnel image must fulfill two restrictions. The first one guarantees the proper profile of the Fresnel image and the second is imposed by the paraxial approximation [47]. In our case, these restrictions require that

$$P >> 8\frac{n}{m} \quad \text{and} \quad P + P' << 4\frac{d}{\lambda}\left(q + \frac{n}{m}\right). \tag{11.6}$$

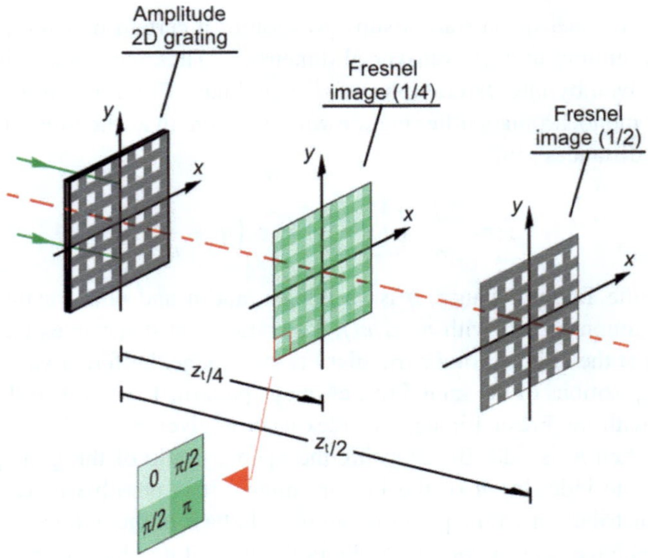

Fig. 11.2 Schematic diagram of the phase distribution at the 1/4 fractional Talbot plane provided by an amplitude 2D grating, and the corresponding irradiance distribution at the 1/2 Talbot plane. The magnified unit cell shows the different phases obtained at the 1/4 Fresnel image

When the amplitude distribution $O(x, y, 0)$ in (11.1), generated by the object beam, interferes with that of the Talbot-codified reference beam $R(x, y, 0)$ in (11.4), we obtain a pixelated interferogram with different periodic phase shifts. Two approaches are now possible to reconstruct the light field diffracted by the object. In the first approach, three interferograms with the same size of the original one, $I(x, y, 0)$, $I(x, y, \pi/2)$, and $I(x, y, \pi)$, are generated by extracting the values of the original interferogram periodically at locations with the same phase shift and using linear interpolation to allocate the empty pixels. Note that the interpolation is more accurate for the interferogram with a $\pi/2$ shift, as we have information about the two pixels of the unit cell.

In this way, it can be shown that the complex field generated by the object beam at the output plane, our digital hologram, can be written as

$$O(x, y, 0) = \frac{1}{4}\{I(x, y, 0) - I(x, y, \pi) + i[2I(x, y, \pi/2) - I(x, y, 0) - I(x, y, \pi)]\}. \tag{11.7}$$

In the second approach, (11.7) is applied to each unit cell of the original interferogram to get each discrete value of the complex amplitude distribution O. In this case, the final hologram is half the size of the previous one in each transversal dimension. We choose to use the first approach since it has been shown in other interferometry applications that it gives slightly better results than the second one [23].

The resulting complex digital hologram, $O(x, y, 0)$ allows us to numerically reconstruct the complex amplitude distribution, $O(x, y, z)$, generated by the 3D object at any plane located at a distance z from the camera sensor. The reconstruction can be obtained by computing a discrete Fresnel integral or, alternatively, by using the propagation transfer function method, i.e.,

$$O(m, n, z) = F^{-1}\left\{ F[O(m, n, 0)] \exp\left[-i\pi\lambda z \left(\frac{u^2}{(\Delta x N_x)^2} + \frac{v^2}{(\Delta y N_y)^2} \right) \right] \right\}, \tag{11.8}$$

where F denotes the fast Fourier transform, (u, v) are discrete spatial frequency variables, (m, n) are discrete transversal spatial coordinates in both the CCD plane and the output plane, and N_x and N_y are the number of samples in the x and y directions. In this approach, the resolution at the output plane is the same for any propagation distance z, and is given by the resolution at the input plane, i.e., the size of the pixel $(\Delta x, \Delta y)$ in the CCD sensor.

11.3 Experimental Results

A Talbot digital holography system, as the setup shown in Fig. 11.1, has been assembled to record the digital holograms of several 2D objects located at different distances. The light source is an Ar laser operating at 514.5 nm. The sensor is a monochrome CCD camera with 2,024 × 2,024 pixels with size equal to $9 \times 9\,\mu\text{m}^2$. The amplitude mask at the reference beam is a 2D binary square grating with size $2 \times 2\,\text{cm}^2$, period $d = 144\,\mu\text{m}$, and a 2D opening ratio of 0.25, manufactured in our laboratory by laser photolithography on a chrome photomask. The Talbot distance for this grating is $z_t = 80.6\,\text{mm}$. Figure 11.3a shows a central region of the

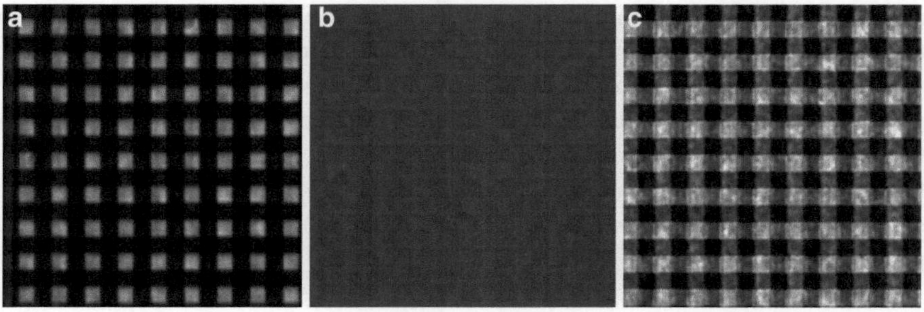

Fig. 11.3 Gray level pictures of the central region of the light distribution generated by the grating at the output plane of the system in Fig. 11.1: (**a**) irradiance distribution of the first self-image, (**b**) uniform irradiance distribution of the Fresnel image with $z = (1 + 3/4)z_t$, (**c**) interference pattern between the Fresnel image at $z = (1 + 3/4)z_t$ and a parallel object beam, showing the periodic three-step phase distribution

Fig. 11.4 (**a**, **b**) Pictures of the irradiance distribution associated to the input 2D objects located at the object beam of the optical system depicted in Fig. 11.1. Panels (**c**) and (**d**) show a partial region of the interference patterns generated for (**a**) and (**b**) objects

irradiance distribution of the first self-image of the grating recorded by the CCD. The distance from the grating to the CCD was then adjusted to obtain a Fresnel image with $q = 1$ and $n/m = 3/4$. Figure 11.3b shows the irradiance distribution (approximately uniform) generated by the Fresnel image on the CCD. Figure 11.3c is a gray level picture of the interference pattern generated when a parallel light beam is used as object beam in the interferometer. Note the periodic three-step phase distribution (with values 0, $\pi/2$, and π) associated to the Fresnel image. Each square of constant phase has a size of $72 \times 72\,\mu\text{m}^2$, corresponding to 8×8 pixels of the camera.

The requirements established in (11.6) to get high-quality Fresnel images are fulfilled by using a parallel light beam with a 15 mm diameter, which illuminates about 100 periods of the grating. The objects positioned at the object beam were 2D transparencies with an approximate size of $1.3 \times 1.3\,\text{cm}^2$ located at different distances. The first transparency codifies the binary object shown in Fig. 11.4a and was located at a distance of 32 cm from the CCD, while the second corresponds to the object in Fig. 11.4b at a distance of 37 cm.

The interferogram between the object diffraction patterns and the Talbot-codified reference beam was recorded with a single shot of the CCD. Figure 11.4c, d shows a central region of the result for the first two objects. It is possible to note the pixelated structure of the recorded interferogram because of the periodic structure of the Fresnel image. Starting from this original interferogram, a new interferogram was then obtained for each phase value of the reference beam by using the interpolation method. The digital hologram is then obtained by applying (11.7) to the three interpolated interferograms. Because of the size of each phase structure, the final digital hologram has a resolution equivalent to 8×8 pixels of the camera and, so, in our preliminary experiment, we get a hologram with only 256×256 real pixels.

Once the digital holograms were recorded, images of the object were reconstructed evaluating the Fresnel diffraction integral in the computer by using a fast Fourier transform algorithm. The results are shown in panels (a) and (b) of Fig. 11.5, corresponding to the two objects in Fig. 11.4. These results confirm the feasibility and flexibility of our method.

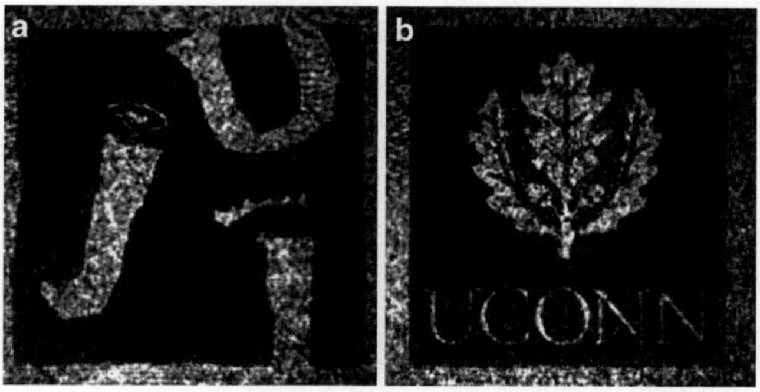

Fig. 11.5 Gray level pictures showing the result of the reconstruction of the digital holograms recorded experimentally. Panels (**a**) and (**b**) show the reconstruction of the objects in Fig. 11.4a and b, respectively

11.4 Conclusions

We have proposed a parallel phase-shifting digital holography system based on the fractional Talbot effect. The basic idea consists in using a 2D binary grating codifying the reference beam of a conventional Mach–Zehnder interferometer. Thus, a periodic three-step phase distribution with uniform irradiance is generated by the reference beam over the sensor plane by fractional Talbot effect. The interference of this diffraction pattern with the light diffracted by an object allows us to measure the Fresnel digital hologram of the object with one single shot of the light sensor. The light field diffracted by the object at different distances can then be reconstructed from the digital hologram.

As previous proposals of parallel holographic methods are based on spatial division-multiplexing, our scheme presents some limitations. For instance, the increase in time resolution is at the expense of a decrease in spatial resolution. However, this is unavoidable to develop dynamic applications of digital holography. Also, as we use binary amplitude gratings, a fraction of light is lost by absorption in the mask. Nevertheless, as the grating is at the reference beam, this missing energy can be easily compensated. In contrast, our proposal has several advantages over previous systems. We do not need special phase diffractive elements to make several copies of the object and reference beams. Neither we need complex pixelated polarization devices, which are difficult to build and hard to integrate with the CCD sensor. Our codifying mask is very simple and commercially available, as two conventional square 1D gratings can be combined in orthogonal directions to codify the reference beam. Alternatively, it can be easily fabricated by standard lithographic techniques. Moreover, Fresnel images are generated by free-space propagation avoiding the use of lenses that can introduce aberrations. Experiments were performed with 2D objects but their extension to 3D scenes will be the subject of future work. Nevertheless, preliminary experimental results confirm the feasibility and flexibility of our method.

Acknowledgments This work has been supported in part by the agreement between the Universitat Jaume I and the Fundació Caixa Castelló-Bancaixa, grant P1-1B2009-36. M.A.-E. gratefully acknowledges a grant, 08I011.36/1 from the same agreement and partial support by PROMEP PIFI 2007–33–07, Mexico Bahram Javidi is grateful for support from Guggenheim Foundation. The authors are very grateful to the Serveis Centrals d'Instrumentació Científica (SCIC) of the Universitat Jaume I for the use of the lithography system based on direct laser writing.

References

1. Schnars, U., Jüptner, W.P.O.: Digital recording and numerical reconstruction of holograms. Meas. Sci. Technol. **13**, 85–101 (2002)
2. Goodman, J.W., Lawrence, R.W.: Digital image formation from electronically detected holograms. Appl. Phys. Lett. **11**, 77–79 (1967)
3. Kronrod, M.A., Yaroslavsky, L.P.: Reconstruction of holograms with a computer. Sov. Phys. Tech. Phys. **17**, 333–334 (1972)
4. Onural, L., Scott, P.D.: Digital decoding of in-line holograms. Opt. Eng. **26**, 1124–1132 (1987)
5. Schnars, U., Jüptner, W.P.O.: Direct recording of holograms by a CCD target and numerical reconstruction. Appl. Opt. **33**, 179–181 (1994)
6. Yamaguchi, I., Zhang, T.: Phase-shifting digital holography. Opt. Lett. **22**, 1268–1270 (1997)
7. Zhang, T., Yamaguchi, I.: Three-dimensional microscopy with phase-shifting digital holography. Opt. Lett. **23**, 1221–1223 (1998)
8. Cuche, E., Bevilacqua, F., Depeursinge, C.: Digital holography for quantitative phase-contrast imaging. Opt. Lett. **24**, 291–293 (1999)
9. Dubois, F., Joannes, L., Legros, J.-C.: Improved three-dimensional imaging with a digital holography microscope with a source of partial spatial coherence. Appl. Opt. **38**, 7085–7094 (1999)
10. Indebetouw, G., Klysubun, P.: Space-time digital holography: a three-dimensional microscopic imaging scheme with an arbitrary degree of spatial coherence. Appl. Phys. Lett. **75**, 2017–2019 (1999)
11. Pedrini, G., Tiziani, H.J.: Short-coherence digital microscopy by use of a lensless holographic imaging system. Appl. Opt. **41**, 4489–4496 (2002)
12. Martínez-León, L., Pedrini, G., Osten, W.: Applications of short-coherence digital holography in microscopy. Appl. Opt. **44**, 3977–3984 (2005)
13. Frauel, Y., Naughton, T., Matoba, O., Tajahuerce, E., Javidi, B.: Three dimensional imaging and display using computational holographic imaging. Proc. IEEE, **94**, 636–654 (2006)
14. Javidi, B., Nomura, T.: Securing information by use of digital holography. Opt. Lett. **25**, 28–30 (2000)
15. Tajahuerce, E., Javidi, B.: Encrypting three-dimensional information with digital holography. Appl. Opt. **39**, 6595–6601 (2000)
16. Matoba, O., Javidi, B.: Optical retrieval of encrypted digital holograms for secure real-time display. Opt. Lett. **27**, 321–323 (2002)
17. Javidi, B., Tajahuerce, E.: Three dimensional object recognition using digital holography. Opt. Lett. **25**, 610–612 (2000)
18. Tajahuerce, E., Matoba, O., Javidi, B.: Shift-invariant three-dimensional object recognition by means of digital holography. Appl. Opt. **40**, 3877–3886 (2001)
19. Frauel, Y., Tajahuerce, E., Castro, M.-A., Javidi, B.: Distortion-tolerant 3D object recognition using digital holography. Appl. Opt. **40**, 3887–3893 (2001)
20. Meng, X.F., Cai, L.Z., Xu, X.F., Yang, X.L., Shen, X.X., Dong, G.Y., Wang, Y.R.: Two-step phase-shifting interferometry and its application in image encryption. Opt. Lett. **31**, 1414–1416 (2006)

21. Hettwer, A., Kranz, J., Schwider, J.: Three channel phase-shifting interferometer using polarization-optics and a diffraction grating. Opt. Eng. **39**, 960–966 (2000)
22. Millerd, J., Brock, N., Hayes, J., North-Morris, M., Novak, M., Wyant, J. C.: Pixelated phase-mask dynamic interferometer. Proc. SPIE **5531**, 304–314 (2004)
23. Novak, M., Millerd, J., Brock, N., North-Morris, M., Hayes, J., Wyant, J.: Analysis of a micropolarizer array-based simultaneous phase-shifting interferometer. Appl. Opt. **44**, 6861–6868 (2005)
24. Awatsuji, Y., Sasada, M., Kubota, T.: Parallel quasi-phase shifting digital holography. Appl. Phys. Lett. **85**, 1069–1071 (2004)
25. Awatsuji, Y., Fujii, A., Kubota, T., Matoba, O.: Parallel three-step phase-shifting digital holography. Appl. Opt. **45**, 2995–3002 (2006)
26. Awatsuji, Y., Tahara, T., Kaneko, A., Koyama, T., Nishio, K., Ura, S., Kubota, T., Matoba, O.: Parallel two-step phase-shifting digital holography. Appl. Opt. **47**, 183–189 (2008)
27. Nomura, T., Murata, S., Nitanai, E., Numata, T.: Phase-shifting digital holography with a phase difference between orthogonal polarizations. Appl. Opt. **45**, 4873–4877 (2006)
28. Kreis, T.M., Jüptner, W.P.O.: Suppression of the dc term in digital holography. Opt. Eng. **37**, 2357–2360 (1997)
29. Takaki, Y., Kawai, H., Ohzu H.: Hybrid-holographic microscopy free of conjugate and zero-order images. Appl. Opt. **38**, 4990–4996 (1999)
30. Cuche, E., Marquet, P., Depeursinge, C.: Spatial filtering for zero-order and twin-image elimination in digital off-axis holography. Appl. Opt. **39**, 4070–4075 (2000)
31. Zhang, Y., Lu, Q., Ge, B.: Elimination of zero-order diffraction in digital off-axis holography. Opt. Commun. **240**, 261–267 (2004)
32. Javidi, B., Kim, D.: Three dimensional object recognition by use of single-exposure on-axis digital holography. Opt. Lett. **30**, 236–238 (2005)
33. Javidi, B., Yeom, S., Moon, I.: Real-time 3D sensing, visualization and recognition of biological microorganisms. Proc. IEEE **94**, 550–567 (2006)
34. Martínez-León, L., Araiza-E, M., Javidi, B., Andrés, P., Climent, V., Lancis, J., Tajahuerce, E.: Single-shot digital holography by use of the fractional Talbot effect. Opt. Express **17**, 12900–12909 (2009)
35. Patorski, K.: The self-imaging phenomenon and its applications. In: Wolf, E. (ed.) Progress in Optics XXVII, pp. 1–108. Elsevier, Amsterdam (1989)
36. Winthrop, J.T., Worthington, C.R.: Theory of Fresnel images: I. J. Opt. Soc. Am. **55**, 373–381 (1965)
37. Arrizón, V., Ojeda-Castañeda, J.: Multilevel phase gratings for array illuminators. Appl. Opt. **33**, 5925–5931 (1994)
38. Werterholm, J., Turunen, J., Huttunen, J.: Fresnel diffraction in fractional Talbot planes: a new formulation. J. Opt. Soc. Am. A **11**, 1283–1290 (1994)
39. Zhou, C., Liu, L.: Simple equations for the calculation of a multilevel phase grating for Talbot array illumination. Opt. Commun. **115**, 40–44 (1995)
40. Arrizón, V., Rojo-Velázquez, G., Ibarra, J.G.: Fractional Talbot effect: compact description. Opt. Rev. **7**, 129–131 (2000)
41. Lohmann, A.W., Thomas, J.A.: Making an array illuminator based on the Talbot effect. Appl. Opt. **29**, 4337–4340 (1990)
42. Kolodziejczyk, A., Jaroszewicz, Z., Kowalik, A., Quintero, O.: Kinoform sampling filter. Opt. Commun. **200**, 35–42 (2001)
43. Lohmann, A., Silva, D.E.: An interferometer based on the Talbot effect. Opt. Commun. **2**, 413–415 (1971)
44. Salama, N.H., Patrignani, D., De Pasquale, L., Sicre, E.E.: Wavefront sensor using the Talbot effect. Opt. Laser Technol. **31**, 269–272 (1999)
45. De Nicola, S., Ferraro, P., Coppola, G., Finizio, A., Pierattini, G., Grilli, S.: Talbot self-image effect in digital holography and its application to spectrometry. Opt. Lett. **29**, 104–106 (2004)
46. Fajst, A., Sypek, M., Makowski, M., Suszek, J., Kolodziejczyk, A.: Self-imaging phase mask used in digital holography with phase-shifting, Proc. SPIE **7141**, 1–7 (2008)
47. Arrizón, V., Rojo-Velázquez, G.: Fractional Talbot field of finite gratings: compact analytical formulation. J. Opt. Soc. Am. A **18**, 1252–1255 (2001)

Chapter 12
Improvement of Viewing-Zone Angle and Image Quality of Digital Holograms

Takanori Nomura, Yusuke Teranishi, Eiji Nitanai, and Takuhisa Numata

Abstract A method to improve the viewing-zone angle and the image quality of a digital hologram is presented. A number of digital holograms of a central object are recorded from the position on the circumference. The holograms are used for a hologram synthesis to improve the image quality from whole viewing-zone angle. The synthesis is achieved by a correlation between a hologram and numerically propagated holograms. The large-sized synthesized digital hologram has a wide viewing-zone angle and less speckles. Some experimental results are shown to confirm our proposed method.

12.1 Introduction

Digital holography [1–8] is a useful technique for recording the fully complex field of a wavefront. The digital holography has been used for lots of applications including security [9,10], measurement [11,12], and 3D object recognition [13–17]. On the other hand, to improve the quality of the reconstructed image in digital holography is necessary, because the image is suffering from speckles. Some approaches [18–20] have been proposed to improve the quality in terms of reduced speckles. However, the viewing-zone angle defined by the size of the hologram and the distance from object to hologram are small, because the area of imaging device is much smaller than a conventional film hologram. Furthermore, it is useful for 3D observation if a digital hologram from any viewing-zone angle is obtained. To achieve the wide viewing-zone angle, a mathematical approach of superposition of holograms has been proposed [21]. In the method, some nonoverlapped digital holograms are used for superposition. However, for perfect superposition,

T. Nomura (✉)
Faculty of Systems Engineering, Wakayama Univesity, 930 Sakaedani, Wakayama, Wakayama 640-8510, Japan
e-mail: nom@sys.wakayama-u.ac.jp

T. Fournel and B. Javidi (eds.), *Information Optics and Photonics: Algorithms, Systems, and Applications*, DOI 10.1007/978-1-4419-7380-1_12,

the difficulty in knowing exactly the amount of rotation remains. In this study, we propose a straightforward method to superpose digital holograms recorded from various viewing-zone angles on concyclic positions.

12.2 Recording and Synthesis of Digital Holograms

Digital holograms of the object, which is placed on the rotational stage, from different angles are recorded by an image sensor on the concyclic positions as shown in Fig. 12.1. If we record N holograms at angular intervals of $2\pi/N$ radian, a reconstructed image is obtained at angular intervals of $2\pi/N$ radian. Therefore, wide viewing-zone angle is obtained. However, the quality of the reconstructed image is very poor, because it is suffering from the speckles due to the small extent of the hologram. To reduce the speckles, it is effective to use a hologram with a large extent [18]. For example, if we record M holograms to be synthesized into a large-extent hologram for each N digital hologram, a good quality image will be reconstructed. In this case, we have to record MN holograms. This is not practical. In the recording process, we have the concerned hologram H_n and neighboring holograms H_{n-1} and H_{n+1} as shown in Fig. 12.2. Figure 12.2 denotes two neighbors. If we use neighboring holograms to be synthesized, we have a large-extent digital holograms. However, simple synthesis between neighbor holograms is not available, because each hologram is in the different plane. To synthesize the holograms, we have to bring holograms on the same plane like H'_{n-1} and H'_{n+1} as shown in Fig. 12.2. To obtain the hologram on the same plane where the concerned hologram lays, the following procedure is introduced. It is assumed that the neighbor hologram $H_{n-1}(x', y')$ can be written as

$$\begin{aligned} H_{n-1}(x', y') &= |U_r(x', y') + U_o(x', y')|^2 \\ &= |U_r(x', y')|^2 + |U_o(x', y')|^2 \\ &\quad + U_o(x', y')U_r^*(x', y') + U_o^*(x', y')U_r(x', y'), \end{aligned} \tag{12.1}$$

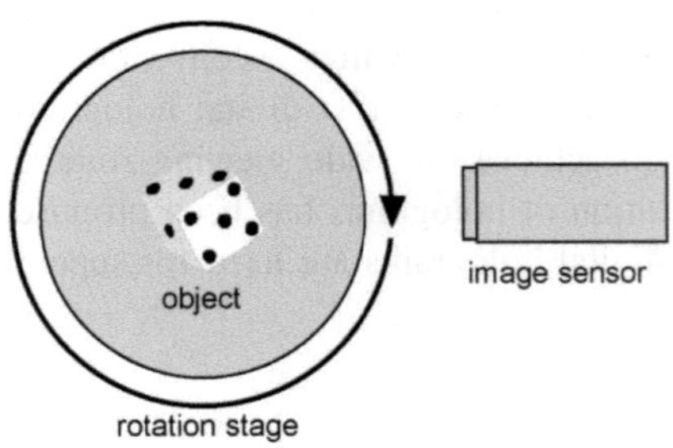

Fig. 12.1 Recording digital holograms by an image sensor on the concyclic positions

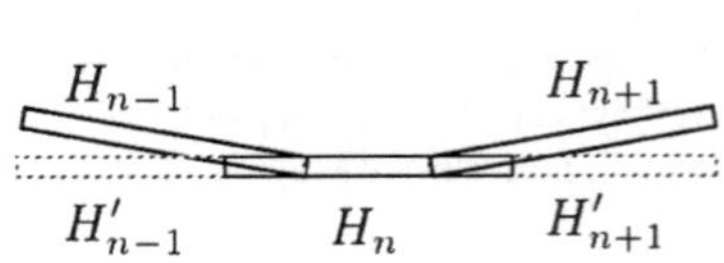

Fig. 12.2 A digital hologram and its neighbor holograms

where $U_o(x', y')$ and $U_r(x', y')$ denote the complex amplitude distributions of the object wave and the reference wave on the hologram plane (x', y'). If we use the plane wave that falls on the hologram at an angle, the third term on the right side $U_o(x', y')U_r^*(x', y')$ can be extracted using a spatial frequency filtering. After extracting the object wave information $U_o(x', y')U_r^*(x', y')$, we obtain the object wave information on the plane (x, y) numerically. For numerical calculation, we assume that wave propagates straightforward and only the phase is changed according to the propagation distance (Fig. 12.3). We can make a hologram using interference between the numerically obtained object wave on the plane (x, y) and the reference wave. After realignment of neighbor digital holograms, we can synthesize them by a correlational technique.

12.3 Simulation

To confirm the feasibility of the proposed method, we perform computer simulation for a point source. The following condition is introduced. The wavelength of the light source is 632.8 nm. The number of pixels of an imaging device is 1,024 × 1,024. The size of the pixel is 4.65 μm × 4.65 μm. The distance between the point source and the imaging device is 400 mm. A plane reference wave falls at an angle of 0.5°.

Seven digital holograms recorded at an interval of 0.2° are used for synthesis. The synthesized hologram is shown in Fig. 12.4. The size of the hologram is 2,820 × 1,025 pixels. The discontinuity of the hologram is due to the above-mentioned

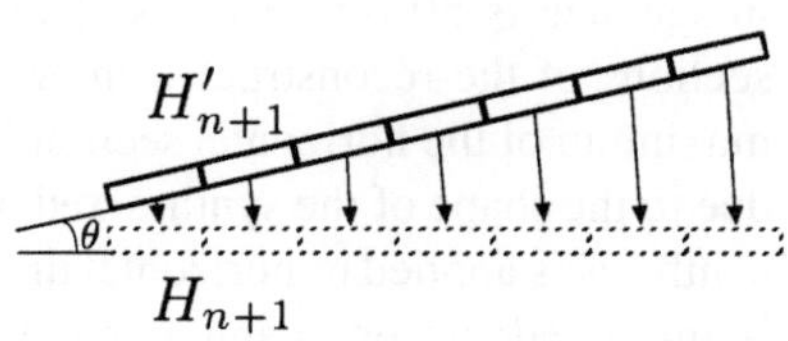

Fig. 12.3 Numerical propagation of the wavefront of the object wave

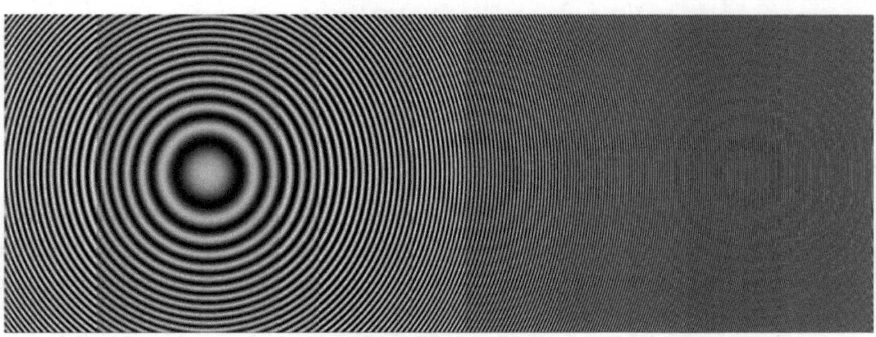

Fig. 12.4 A synthesized digital hologram using seven holograms for simulation

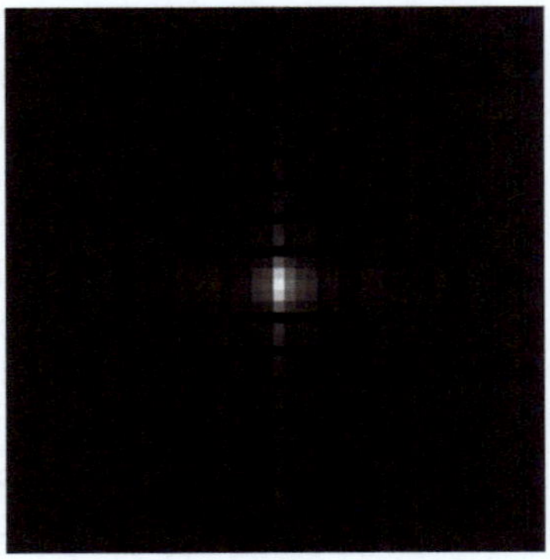

Fig. 12.5 A reconstructed image of the synthesized digital hologram

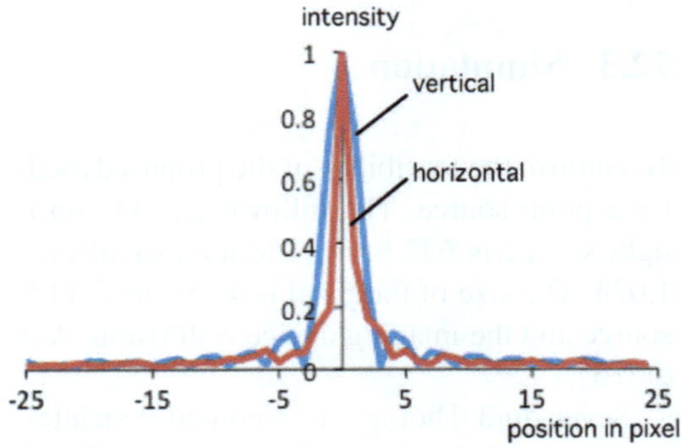

Fig. 12.6 Cross sections of the reconstructed image shown in Fig. 12.5

assumption. The reconstructed image of the hologram is shown in Fig. 12.5. The image size is 50 by 50 pixels. This is the extracted part of the center. The cross sections of the reconstructed image are shown in Fig. 12.6 The full width at half maximum of the horizontal section is narrower than that of the vertical section. It is due to the shape of the synthesized hologram shown in Fig. 12.4. In this simulation, synthesis is applied to horizontal direction only. Therefore, the horizontal resolution is improved. As one cannot see significant false peaks by the side effect due to the assumption of the propagation, it is confirmed the feasibility of the proposed propagation and synthesis method.

12.4 Experimental Results

We show the experimental results to confirm the feasibility of the proposed method. An optical experimental setup is shown in Fig. 12.7. A He–Ne laser with a wavelength of 632.8 nm is used as a coherent light source. A CCD camera is used to record the hologram. The number of pixels of the CCD camera is 1,024 × 768, and its pixel size is 4.65 μm × 4.65 μm. The off-axis configuration with tilting the mirror is used for the spatially separation of the zero-order and the conjugate image.

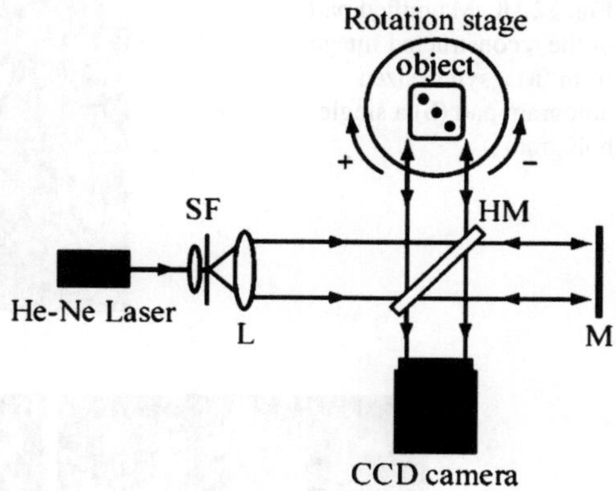

Fig. 12.7 An optical setup for digital hologram recording. *SF* spatial filter, *HM* half mirror, *M* mirror, *L* lens

Fig. 12.8 A synthesized digital hologram using 11 holograms by optical experiments

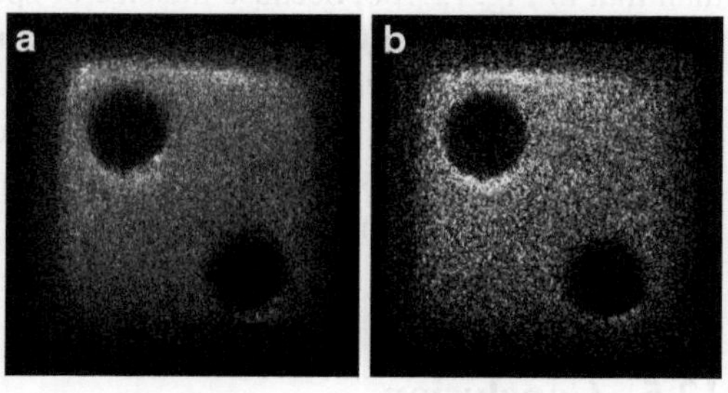

Fig. 12.9 Reconstructed images from (**a**) a synthesized hologram and (**b**) a single hologram

The distance from the object (a die; size of 8 mm × 8 mm × 8 mm) to the CCD camera is 400 mm. It is necessary to adjust the rotation angle to 0.4° or less to synthesize holograms, because two adjacent holograms are overlapped each other. In the experiments, the object was rotated from −180° to +180° at an interval of 0.2° and 1,801 holograms were recorded. Then, some of recorded holograms were processed and superposed by the proposed method.

To confirm the feasibility of superposition, we compare the reconstruction images from a single hologram and a superposed hologram. The synthesized hologram as shown in Fig. 12.8 is obtained from 11 holograms. The size of the synthesized hologram is 4,100 × 900. The roughness of the upside and downside of the hologram is due to the shakiness of the rotation stage. The reconstructed images are shown in Fig. 12.9. The magnified part of the each reconstructed images are shown in Fig. 12.10. The horizontal size of speckles in Fig. 12.10a are smaller

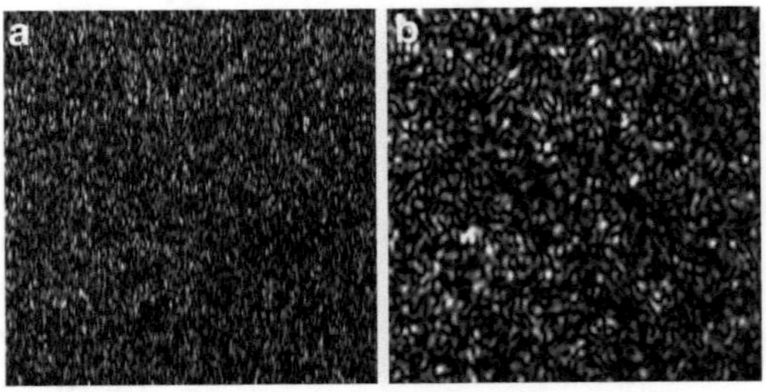

Fig. 12.10 Magnified part of the reconstructed images from (a) a synthesized hologram and (b) a single hologram

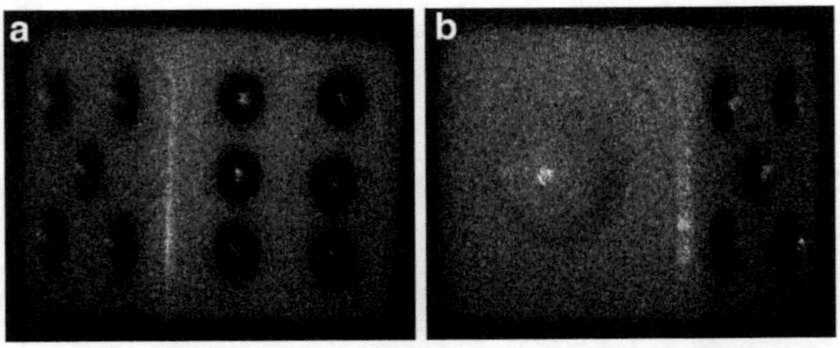

Fig. 12.11 Reconstructed images from different viewing angles. (a) From left side (60°) and (b) from right side (60°)

than that in Fig. 12.10b, because the size of speckles is in inverse proportion to the size of holograms. Furthermore, as neither blurred nor double images can be seen in Fig. 12.9a, it is shown that the synthesis is successful.

Figure 12.11 shows the reconstructed images from different viewing directions. We can see different aspects of the object in the figures. The reconstruction image pair can be used for binocular stereopsis.

12.5 Conclusion

We proposed the method to synthesize digital holograms recorded from different viewing-zone angle. Experimental results show that the synthesized digital hologram with large viewing-zone angle and arbitrary direction can be obtained by recording from whole directions. In addition, the quality of a reconstructed image is improved. For more improvement of the reconstructed image quality, our previously proposed technique [19] might be useful.

Acknowledgement This study was partly supported by a research granted form The Murata Science Foundation.

References

1. I. Yamaguchi and T Zhang, "Phase-shifting digital holography," Opt. Lett. **22**, 1268–1270 (1997).
2. W. Osten, T. Baumbach, and W. Jüptner, "Comparative digital holography," Opt. Lett. **27**, 1764–1766 (2002).
3. T. Kreis, *Handbook of Holographic Interferometry* (Wiley-VCH, Weinheim, 2005).
4. Y. Frauel, T. Naughton, O. Matoba, E. Tahajuerce, and B. Javidi, "Three-dimensional imaging and processing using computational holographic imaging," Proc. IEEE **94**, 636–653 (2006).
5. J. W. Goodman and R. Lawrence, "Digital image formation from electronically detected holograms," Appl. Phys. Lett. **11**, 77–79 (1967).
6. J. Rosen, "Three-dimensional optical Fourier transform and correlation," Opt. Lett. **22**, 964–966 (1997).
7. P. Ferraro, S. Grilli, D. Alfieri, S. D. Nicola, A. Finizio, G. Pierattini, B. Javidi, G. Coppola, and V. Striano, "Extended focused image in microscopy by digital Holography," Opt. Exp. **13**, 6738–6749 (2005).
8. Y. Awatsuji, T. Tahara, A. Kaneko, T. Koyama, K. Nishio, S. Ura, T. Kubota, and O. Matoba, "Parallel two-step phase-shifting digital holography," Appl. Opt. **47**, D183–D189 (2008).
9. E. Tajahuerce and B. Javidi, "Encrypting three-dimensional information with digital holography," Appl. Opt. **39**, 6595–6601 (2000).
10. S. Tamano, Y. Hayasaki, and N. Nishida, "Phase-shifting digital holography with a low-coherence light source for reconstruction of a digital relief object hidden behind a light-scattering medium," Appl. Opt. **45**, 953–959 (2006).
11. T. Nomura, B. Javidi, S. Murata, E. Nitanai, and T. Numata, "Polarization imaging of a 3D object by use of on-axis phase-shifting digital holography," Opt. Lett. **32**, 481–483 (2007).
12. M. Yokota, "Polarization analysis by off-axis digital holography with an improved optical system and an evaluation of its performance by simulation," Appl. Opt. **47**, 6325–6333 (2008).
13. T.-C. Poon and T. Kim, "Optical image recognition of three-dimensional objects," Appl. Opt. **38**, 370–381 (1999).
14. E. Tajahuerce, O. Matoba, and B. Javidi, "Shift-invariant three-dimensional object recognition by means of digital holography," Appl. Opt. **40**, 3877–3888 (2001).
15. B. Javidi and D. Kim, "Multiple object recognition by digital holography with a wavelength-tuned laser diode," Opt. Lett. **30**, 236–238 (2005).
16. Y. Ishii and T. Takahashi, "Three-dimensional-object recognition by use of single-exposure on-axis digital holography," Proc. SPIE **6252**, 62521M (2006).
17. T. Nomura and B. Javidi, "Object recognition by use of polarimetric phase-shifting digital holography," Opt. Lett. **32** 2146–2148 (2007).
18. R. Binet, J. Colineau, and J.-C. Lehureau, "Short-range synthetic aperture imaging at 633 nm by digital holography," Appl. Opt. **41**, 4775–4782 (2002).
19. T. Nomura, M. Okamura, E. Nitanai, and T. Numata, "Image quality improvement of digital holography by superposition of reconstructed images obtained by multiple wavelengths," Appl. Opt. **47**, D38–D43 (2008).
20. T. Nakatsuji and K. Matsushima, "Free-viewpoint images captured using phase-shifting synthetic aperture digital holography," Appl. Opt. **47**, D136–D143 (2008).
21. B. M. Hennelly, C. Mc Elhinney, Y. Frauel, T. J. Naughton, J. B. McDonald, and B. Javidi, "Superposition of digital holograms," AIP Conf. Proc. **860**, 281–289 (2006).

Chapter 13
Shot Noise in Digital Holography

Fadwa Joud, Frédéric Verpillat, Michael Atlan, Pierre-André Taillard, and Michel Gross

Abstract We discuss on noise in heterodyne holography in an off-axis configuration. We show that, for a weak signal, the noise is dominated by the shot noise on the reference beam. This noise corresponds to an equivalent noise on the signal beam of one photoelectron per pixel, for the whole sequence of images used to build the digital hologram.

13.1 Introduction

Demonstrated by Gabor [1] in the early 1950s, the purpose of holography is to record, on a 2D detector, the phase and the amplitude of the radiation field scattered by an object under coherent illumination. The photographic film used in conventional holography is replaced by a 2D electronic detection in digital holography [2] enabling quantitative numerical analysis. Digital holography has been waiting for the recent development of computer and video technology to be experimentally demonstrated [3]. The main advantage of digital holography is that, contrary to holography with photographic plates [1], the holograms are recorded by a CCD, and the image is digitally reconstructed by a computer, avoiding photographic processing [4].

Off-axis holography [5] is the oldest configuration adapted to digital holography [3, 6, 7]. In off-axis digital holography, as well as in photographic plate holography, the reference or local oscillator (LO) beam is angularly tilted with respect to the object's observation axis. It is then possible to record, with a single hologram, the two quadratures of the object's complex field. However, the object's field of view is reduced, since one must avoid the overlapping of the image with the conjugate image alias [8]. In Phase shifting digital holography, which has been introduced later

F. Joud (✉)
Laboratoire Kastler Brossel École Normale Supérieure, UMR 8552, UPMC, CNRS,
24 rue Lhomond, 75231 Paris Cedex 05, France
e-mail: joud@lkb.ens.fr

T. Fournel and B. Javidi (eds.), *Information Optics and Photonics: Algorithms, Systems, and Applications*, DOI 10.1007/978-1-4419-7380-1_13,

[9], several images are recorded with different LO beam phases. It is then possible to obtain the two quadratures of the object's field in an in-line configuration even though the conjugate image alias and the true image overlap, because aliases can be removed by taking image differences.

We have developed an alternative phase shifting digital holography technique, called heterodyne holography, that uses a frequency shift of the reference beam to continuously shift the phase of the recorded interference pattern [10]. One of the advantages of this technique is its ability to provide accurate phase shifts that allow to suppress twin images aliases [11]. This greatly simplifies holographic data handling, and improves sensitivity. Moreover, it is possible to perform holographic detection at a frequency different from illumination. One can, for example, detect "tagged photons" [12, 13] in ultrasound-modulated optical imaging [14]. One can also image vibrating objects at the frequencies corresponding to vibration sidebands [15, 16]. To the end, it is possible to perform Laser Doppler imaging [17] within microvessels [18–20].

More generally, our setup can be viewed as a multipixel heterodyne detector that is capable of recording the complex amplitude of the signal electromagnetic field $\mathscr{E}$ onto all pixels of the CCD camera in parallel. We get the map of the field over the array detector (i.e., $\mathscr{E}(x, y)$ where x and y are the pixels coordinates). Since the field is measured on all pixels at the same time, the relative phase that is measured for different locations (x, y) is meaningful. This means that the field map $\mathscr{E}(x, y)$ is an hologram that can be used to reconstruct the field $\mathscr{E}$ in any location, in particular in the object's plane.

In this chapter, we will discuss on noise in digital holography, and we will try to determine what is the ultimate noise limit both theoretically, and in real time holographic experiments. We will see that, in the theoretical ideal case, the limiting noise is the Shot Noise on the holographic reference beam. In reference to heterodyne detection, the reference beam is also called LO. We will see that the ultimate theoretical limiting noise can be reached in real life holographic experiment, by using heterodyne holography [10] in off-axis configuration. This combination makes possible to fully filter off the technical noise, whose main origin is the LO beam technical noise, opening the way to holography with ultimate sensitivity [21, 22].

13.2 Theoretical Noise

To discuss on noise in digital holography, we will consider both the case of off-axis holography, where the hologram is obtained from one frame of the CCD camera, and the case of phase shifting holography, where the holographic information is extracted from a sequence of M frames.

We will thus consider a sequence of M frames: I_0 to I_{M-1} (where $M = 1$ in the one shot, off-axis case). For each frame I_k, let us note $I_{k,p,q}$ the CCD camera signal on each pixel, where k is the frame index, and p, q the pixel indexes along the x

and y directions. The CCD signal $I_{k,p,q}$ is measured in Digital Counts (DC) units. In the typical case of the 12 bit digital camera used in experiments below, we have $0 \leq I_{k,p,q} < 4{,}096$. For each frame k, the optical signal is integrated over the acquisition time $T = 1/f_{\rm ccd}$ of the CCD camera. The pixel signal $I_{k,p,q}$ is thus defined by

$$I_{k,p,q} = \int_{t_k-T/2}^{t_k+T/2} \mathrm{d}t \iint_{(p,q)} \mathrm{d}x\mathrm{d}y \ |E(x,y,t) + E_{\rm LO}(x,y,t)|^2, \qquad (13.1)$$

where $\iint_{(p,q)} \mathrm{d}x\mathrm{d}y$ represents the integral over the pixel (p,q) area, and t_k is the recording moment of frame k. Introducing the complex representations $\mathcal{E}$ and $\mathcal{E}_{\rm LO}$ of the fields E and $E_{\rm LO}$, we get

$$E(x,y,t) = \mathcal{E}(x,y)\mathrm{e}^{j\omega_I t} + \text{c.c.} \qquad (13.2)$$

$$E_{\rm LO}(x,y,t) = \mathcal{E}_{\rm LO}(x,y)\mathrm{e}^{j\omega_{\rm LO} t} + \text{c.c.} \qquad (13.3)$$

$$I_{k,p,q} = a^2 T \left(|\mathcal{E}_{p,q}|^2 + |\mathcal{E}_{\rm LO}|^2 + \mathcal{E}_{p,q}\mathcal{E}^*_{\rm LO} \cdot \mathrm{e}^{(j(\omega_I - \omega_{\rm LO})t_k)} + \text{c.c.} \right), \qquad (13.4)$$

where a is the pixel size. To simplify the notations in (13.4), we have considered that the LO field $\mathcal{E}_{\rm LO}$ is the same in all locations (x,y), and that signal field $\mathcal{E}_{p,q}$ does not vary within the pixel (p,q). If $\mathcal{E}_{\rm LO}$ varies with location, one has to replace $\mathcal{E}_{\rm LO}$ by $\mathcal{E}_{{\rm LO},p,q}$ in (13.4).

In the single-shot, off-axis holography case, the hologram H is simply $H \equiv I_k$. To simplify the discussion in the phase shifting digital holography case [9], we will consider four phases holographic detection ($M = 4n$). In that case, the phase shift of the LO beam is equal to $\pi/2$ from one recorded frame to the next. Because of this shift, the complex hologram H is obtained by summing the sequence of M frames I_1 to I_M with the appropriate phase coefficient:

$$H \equiv \sum_{k=1}^{M} (j)^{k-1} I_k, \qquad (13.5)$$

where H is a matrix of pixel $H_{p,q}$, $M = 4n$ in the four-phases phase shifting case, and $M = 1$ in the single-shot, off-axis case. We get from (13.4):

$$H_{p,q} = \sum_{k=1}^{M} (j)^k I_{k,p,q} = 4na^2 T \mathcal{E}_{p,q}\mathcal{E}^*_{\rm LO}. \qquad (13.6)$$

The complex hologram $H_{p,q}$ is thus proportional to the object's field $\mathcal{E}_{p,q}$ with a proportionality factor that involves $\mathcal{E}^*_{\rm LO}$.

13.2.1 The Shot Noise on the CCD Pixel Signal

Because of spontaneous emission, laser emission and photodetection are random processes, the signal that is obtained on a CCD pixel exhibits a Poisson noise called "shot noise". The effect of this Poisson noise on the signal, and on the holographic images, is the Ultimate Theoretical Limiting noise, which we will study here.

We can split the signal $I_{k,p,q}$ we get for frame k and pixel (p, q) in : a noiseless average component $\langle I_{k,p,q} \rangle$ (here $\langle\ \rangle$ is the statistical average operator) and a noise component $i_{k,p,q}$:

$$I_{k,p,q} \equiv \langle I_{k,p,q} \rangle + i_{k,p,q}. \tag{13.7}$$

To go further in the discussion, we will use photoelectrons (e) Units to measure the signal $I_{k,p,q}$.

We must notice that the LO signal $\mathcal{E}_{\mathrm{LO}}$ is large, and corresponds to a large number of photoelectrons. In real life, this assumption is true. For example, if we adjust the power of the LO beam to be at the half maximum for the camera signal in DC unit (2,048 DC for a 12 bits camera), the pixel signal will be about 10^4 e for the camera used below in experiments, since the "Camera Gain" is 4.8 e per DC. This yields two consequences, which simplify the analysis. First, the signal $I_{k,p,q}$ exhibits a gaussian distribution around its statistical average. Second, both the quantization noise of the photoelectron signal ($I_{k,p,q}$ is an integer in photoelectron units), and the quantization noise of the Digital Count signal ($I_{k,p,q}$ is an integer in DC units) can be neglected. These approximations are valid, since the width of the $I_{k,p,q}$ gaussian distribution is much larger than one in both photoelectron and DC units. In the example given above, $\langle I_{k,p,q} \rangle \simeq 10^4$, and this width is $\simeq 10^2$ in photoelectron units, and $\simeq$20 in DC units. One can thus consider that $I_{k,p,q}$, $\langle I_{k,p,q} \rangle$ and $i_{k,p,q}$ are floating numbers (and not integers). Moreover, $i_{k,p,q}$ is a random Gaussian distribution, with

$$\langle i_{k,p,q} \rangle = 0, \tag{13.8}$$

$$\langle i^2_{k,p,q} \rangle = \langle I_{k,p,q} \rangle. \tag{13.9}$$

To analyse the shot noise's contribution to the holographic signal $H_{p,q}$, one of the most simple methods is to perform Monte Carlo simulation from (13.7), (13.8), and (13.9). Since $I_{k,p,q}$ is ever large in real life (about 10^4 in our experiment), $\langle I_{k,p,q} \rangle$ can be replaced by $I_{k,p,q}$ (which is measured in experiment) in the right member of (13.9). One has thus:

$$\langle i^2_{k,p,q} \rangle = \langle I_{k,p,q} \rangle \simeq I_{k,p,q}. \tag{13.10}$$

Monte Carlo simulation of the noise can be done from (13.7), (13.8), and (13.10).

13.2.2 The Object Field Equivalent Noise for One Frame

To discuss the effect of the shot noise on the heterodyne signal $\mathcal{E}_{p,q}\mathcal{E}^*_{\mathrm{LO}}$ of (13.4), let us consider the simple situation sketched on Fig. 13.1. A weak object field E, with

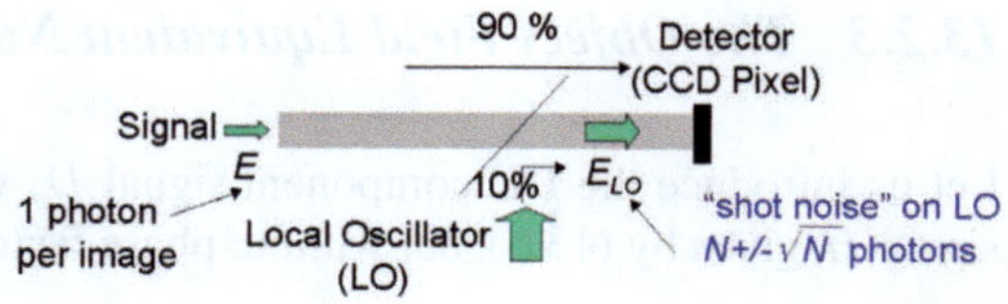

Fig. 13.1 One-photon equivalent signal (accounting Heterodyne gain) and shot noise on the holographic local oscillator beam

one photon or one photoelectron per pixel and per frame, interferes with a LO field E_{LO} with N photons, where N is large ($N = 10^4$, in the case of our experiment). Since the LO beam signal $a^2T|\mathcal{E}_{\mathrm{LO}}|^2$ is equal to N photons, and the object field signal $a^2T|\mathcal{E}_{p,q}|^2$ is one photon, we have

$$I_{k,p,q} = N + 1 + i_{k,p,q} + a^2T\mathcal{E}_{p,q}\mathcal{E}^*_{\mathrm{LO}}\mathrm{e}^{\cdots} + \mathrm{c.c.} \qquad (13.11)$$

The heterodyne signal $\mathcal{E}_{p,q}\mathcal{E}^*_{\mathrm{LO}}$ is much larger than $|\mathcal{E}_{p,q}|^2$. This is the gain effect, associated to the coherent detection of the field $\mathcal{E}_{p,q}$. This gain is commonly called "heterodyne gain", and is proportional to the amplitude of the LO field $\mathcal{E}^*_{\mathrm{LO}}$.

The purpose of the present discussion is to determine the effect of the noise term $i_{k,p,q}$ in (13.11) on the holographic signal $H_{p,q}$. Since $H_{p,q}$ involves only the heterodyne term $\mathcal{E}_{p,q}\mathcal{E}^*_{\mathrm{LO}}$ [see (13.6)], we have to compare, in (13.11), the shot noise term $i_{k,p,q}$, and the heterodyne term $\mathcal{E}_{p,q}\mathcal{E}^*_{\mathrm{LO}}$.

Consider first the shot noise term. We have

$$\langle i^2_{k,p,q}\rangle = \langle I_{k,p,q}\rangle = N + 1 \simeq N. \qquad (13.12)$$

The variance of the shot noise term is thus $N^{1/2}$. Since this noise is mainly related to the shot noise on the LO (since $N \gg 1$), one can group together, in (13.11), the LO beam term (i.e., N) with the noise term $i_{k,p,q}$, and consider that the LO beam signal fluctuates, the number of LO beam photons being thus "$N \pm N^{1/2}$", as mentioned on Fig. 13.1.

Consider now the heterodyne beat signal. Since we have N photons on the LO beam, and one photon on the object beam, we get

$$a^2T|\mathcal{E}_{p,q}\mathcal{E}^*_{\mathrm{LO}}| \equiv \left[\left(a^2T|\mathcal{E}_{p,q}|^2\right)\left(a^2T|\mathcal{E}_{\mathrm{LO}}|^2\right)\right]^{1/2} = N^{1/2}. \qquad (13.13)$$

The heterodyne beat signal $\mathcal{E}_{p,q}\mathcal{E}^*_{\mathrm{LO}}$ is thus $N^{1/2} = 100$.

The shot noise term $i_{k,p,q}$ is thus equal to the heterodyne signal $\mathcal{E}_{p,q}\mathcal{E}^*_{\mathrm{LO}}$ corresponding to one photon on the object field. This means that shot noise $i_{k,p,q}$ yields an equivalent noise of one photon per pixel, on the object beam. This result is obtained here for one frame. We will show that it remains true for a sequence of M frames, whatever $M = 4n$ is.

13.2.3 The Object Field Equivalent Noise for $M = 4n$ Frames

Let us introduce the DC component signal D, which is similar to the heterodyne signal H given by (13.5), but without phase factors:

$$D \equiv \sum_{k=1}^{M} I_k. \tag{13.14}$$

The component D can be defined for each pixel (p, q) by

$$D_{p,q} \equiv \sum_{k=1}^{M} I_{k,p,q}. \tag{13.15}$$

Since $I_{k,p,q}$ is always large in real life (about 10^4 in our experiment), the shot noise term can be neglected in the calculation of $D_{p,q}$ by (13.15). We have thus:

$$D_{p,q} \equiv \sum_{k=1}^{M} I_{k,p,q} = Ma^2T\left(|\mathcal{E}_{p,q}|^2 + |\mathcal{E}_{\mathrm{LO}}|^2\right). \tag{13.16}$$

We are implicitly interested by the low signal situation (i.e., $\mathcal{E}_{p,q} \ll \mathcal{E}_{\mathrm{LO}}$) because we focus on noise analysis. In that case, the $|\mathcal{E}_{p,q}|^2$ term can be neglected in (13.16). This means that $D_{p,q}$ gives a good approximation for the LO signal.

$$D_{p,q} \equiv \sum_{k=1}^{M} I_{k,p,q} \simeq Ma^2T|\mathcal{E}_{\mathrm{LO}}|^2. \tag{13.17}$$

We can get then the signal field $|\mathcal{E}_{p,q}|^2$ from (13.6) and (13.17):

$$\frac{|H_{p,q}|^2}{D_{p,q}} \simeq Ma^2T|\mathcal{E}_{p,q}|^2. \tag{13.18}$$

In this equation, the ratio $|H_{p,q}|^2/D_{p,q}$ is proportional to the number of frames of the sequence ($M = 4n$), This means that $|H_{p,q}|^2/D_{p,q}$ represents the signal field $|\mathcal{E}_{p,q}|^2$ summed over the all frames.

Let us calculate the effect of the shot noise on $|H_{p,q}|^2/D_{p,q}$. To calculate this effect, one can make a Monte Carlo simulation as mentioned above, but a simpler calculation can be done here. Let us develop $|H_{p,q}|$ in statistical average and noise components [as done for $I_{k,p,q}$ in (13.7)]:

$$H_{p,q} = \langle H_{p,q}\rangle + h_{p,q} \tag{13.19}$$

with

$$h_{p,q} = \sum_{k=1}^{4n} j^{\,k} i_{k,p,q}. \tag{13.20}$$

Let us calculate $\langle |H_{p,q}|^2 / D_{p,q} \rangle$ from (13.18). Since $D_{p,q} \simeq \langle D_{p,q} \rangle$, we get

$$\left\langle \frac{|H_{p,q}|^2}{D_{p,q}} \right\rangle \simeq \frac{|\langle H_{p,q}\rangle|^2 + \langle |h_{p,q}|^2\rangle + \langle \langle H_{p,q}\rangle h^*_{p,q}\rangle + \langle \langle H^*_{p,q}\rangle h_{p,q}\rangle}{\langle D_{p,q}\rangle}. \tag{13.21}$$

In (13.21), the $\langle \langle H_{p,q}\rangle h^*_{p,q}\rangle$ term is zero since $h^*_{p,q}$ is random while $\langle H_{p,q}\rangle$ is not. The two terms $\langle \langle H_{p,q}\rangle h^*_{p,q}\rangle$ and $\langle \langle H^*_{p,q}\rangle h_{p,q}\rangle$ can be thus removed. On the other hand, we get for $|h_{p,q}|^2$

$$|h_{p,q}|^2 = \sum_{k=1}^{4n} |i_{k,p,q}|^2 + \sum_{k=1}^{4n} \sum_{k'=1,k'\neq k}^{4n} j^{\,k-k'} i_{k,p,q} i_{k',p,q}. \tag{13.22}$$

Since $i_{k,p,q}$ and $i_{k',p,q}$ are uncorrelated, the $i_{k,p,q} i_{k',p,q}$ terms cancel in the calculation of the statistical average of $|h_{p,q}|^2$. We get then from (13.9)

$$\langle |h_{p,q}|^2 \rangle = \sum_{k=1}^{4n} \langle |i_{k,p,q}|^2 \rangle = \sum_{k=1}^{4n} \langle I_{k,p,q} \rangle = \langle D_{p,q} \rangle. \tag{13.23}$$

Equation 13.21 becomes thus:

$$\left\langle \frac{|H_{p,q}|^2}{D_{p,q}} \right\rangle = \frac{|\langle H_{p,q}\rangle|^2}{\langle D_{p,q}\rangle} + 1. \tag{13.24}$$

Equation 13.24 means that the average detected intensity signal $\langle |H_{p,q}|^2 / D_{p,q} \rangle$ is the sum of the square of the average object field $\langle |H_{p,q}| \rangle / (\langle D_{p,q} \rangle^{1/2})$ plus one photoelectron. Without illumination of the object, the average object field is zero, and the detected signal is one photoelectron. The equation establishes thus that the LO shot noise yields a signal intensity corresponding exactly to one photoelectron per pixel.

The 1 e noise floor, we get here, can be also interpreted as resulting from the heterodyne detection of the vacuum field fluctuations [23].

13.3 Reaching the Shot Noise in Real Life Holographic Experiment

In Sect. 13.2, we have shown that the theoretical noise on the holographic reconstructed intensity images is one photoelectron per pixel whatever the number of recorded frames is. We will now discuss the ability to reach this limit in real life holographic experiment. Since we consider implicitly a very weak object beam signal ($\mathscr{E} \ll \mathscr{E}_{\rm LO}$), the noises that must be considered are the readout noise of the CCD camera, the technical noise from laser amplitude fluctuations on the LO beam, and the LO beam shot noise, which yields the theoretical noise limit.

Consider a typical holographic experiment made with a PCO Pixelfly 12 bit digital camera. The LO beam power is adjusted in order to be at half saturation of the digital camera output. Since the camera is 12 bits, and since the camera "gain" is 4.8 e per DC, half saturation corresponds to 2,000 DC on the A/D Converter, that is, about 10^4 e on the each CCD pixel. The LO shot noise, which is about 100 e, is thus much larger than the Pixelfly Read Noise (20 e), Dark Noise (3 e/s) and A/D quantization noise (4.8 e, since 1 DC corresponds to 4.8 e). The noise of the camera can be neglected, and is not a limiting factor for reaching the noise theoretical limit.

The LO beam that reaches the camera is essentially flat field (i.e., the field intensity $|\mathscr{E}_{\rm LO}|^2$ is roughly the same for all the pixels). The LO beam technical noise is thus highly correlated in all pixels. This is in particular the case for the noise induced by the fluctuations of the main laser intensity, or by the vibrations of the mirrors within the LO beam arm. To illustrate this point, we have recorded a sequence of $M = 4n = 4$ frames I_k (with $k = 0 \ldots 3$) with a LO beam, but without signal from the object (i.e., without illumination of the object). We have thus recorded the hologram of the "vacuum field". We have calculated then the complex hologram $H(x, y)$ by (13.5), and the reciprocal space hologram $\tilde{H}(k_x, k_y)$ by Fourier transform:

$$\tilde{H}(k_x, k_y) = \text{FFT} \;\; H(x, y) \tag{13.25}$$

The reciprocal space holographic intensity $|\tilde{H}|^2$ is displayed on Fig. 13.2 in arbitrary logarithm grey scale. On most of the reciprocal space (within for example circle 1), $|\tilde{H}|^2$ corresponds to a random speckle whose average intensity

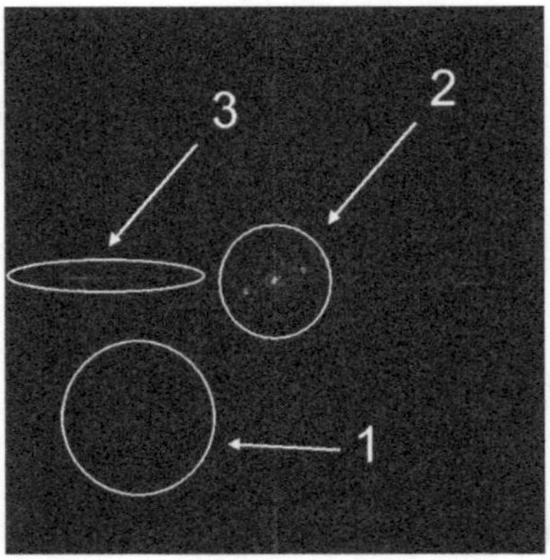

Fig. 13.2 Intensity image of $\tilde{H}(k_x, k_y, 0)$ for $M = 4n = 4$ frames without illumination of the object (no signal field $\mathscr{E}$). Three kind of noises can be identified. *Down left* (1): shot noise; *center* (2): technical noise of the CCD; *left* (3): FFT aliasing. By truncating the image and keeping only the *left down part*, the shot noise limit is reached. The image is displayed in arbitrary logarithm grey scale

is uniformly distributed along k_x and k_y. One observes nevertheless bright points within circle 2, which corresponds to $(k_x, k_y) \simeq (0,0)$. These points correspond to the technical noise, which is a flat field within the CCD plane (x, y), and which corresponds thus to low spatial frequency components gathered around the center of the (k_x, k_y) reciprocal space. One see also, on the Fig. 13.2 image, an horizontal and a vertical bright line, which corresponds to $k_y \simeq 0$ and $k_x \simeq 0$ (zone 3 on Fig. 13.2). These parasitic bright lines are related to Fast Fourier Transform aliases, that are related to the discontinuity of the signal I_k and H at edge of the calculation grid, in the (x, y) space.

We have measured $\langle|\tilde{H}|^2\rangle$ by replacing the statistical average $\langle\ \rangle$ by a spatial average over a region of the conjugate space without technical noise (i.e., over region 1). This gives a measurement of $\langle|\tilde{H}|^2\rangle$, that is, a measurement of $\langle|H|^2\rangle$, since the space average of $|\tilde{H}|^2$ and $|H|^2$ are equal, because of the FFT Parceval theorem. We have also measured D from the sequence of frames I_k with $k = 0\ldots3$ [see (13.14)]. Knowing the camera Analog Digital (A/D) conversion factor (4.8 e per DC), we have calculated the noise intensity $\langle|\tilde{H}|^2\rangle/\langle D\rangle$ in photoelectron units, and we get, within 10%, one photoelectron per pixel for the average noise within region 2, as expected theoretically for the shot noise [see (13.21)].

To verify that we have truly reached the shot noise limit, we have performed a control experiment with a camera illuminated by a tungsten lamp powered by a battery. The lamp provides here a clean white light source. The lamp voltage is adjusted to get half saturation of the camera (about 2,000 DC). Similarly to the laser experiment described above, we have recorded a sequence of $M = 4n = 4$ frames I_k with $k = 0\ldots3$, and we have calculated $H(x, y)$, and $\tilde{H}(k_x, k_y)$. The image of $|\tilde{H}(k_x, k_y)|^2$ we get is very similar to Fig. 13.2. Moreover, the average noise intensity in region 2 is exactly the same as with a laser (one photoelectron per pixel). One has thus:

$$\langle|\tilde{H}|^2\rangle/\langle D\rangle = 1. \tag{13.26}$$

This result is expected since the camera "gain" is measured by assuming that the noise obtained in clean lamp control experiment is shot noise limited [24]. Since $\langle|\tilde{H}|^2\rangle/\langle D\rangle$ is dependent of the camera "gain" (13.26) indirectly allows the determination of the "gain" in e per DC units. The control experiment made here, redoes the "gain" calibration made by the camera manufacturer (i.e., PCO). We simply get here, within 10%, the same camera gain (4.8 e per DC).

13.3.1 Experimental Validation with an USAF Target

We have verified that it is possible to perform shot noise limited holography in real life, by recording the hologram of an USAF target in transmission. The holographic setup is sketched on Fig. 13.3. We have recorded sequences of $M = 4n = 12$ frames, and we have reconstructed the image of the USAF target.

Figure 13.4 shows the holographic reconstructed images of the USAF target. The intensity of the signal illumination is adjusted with neutral density filters.

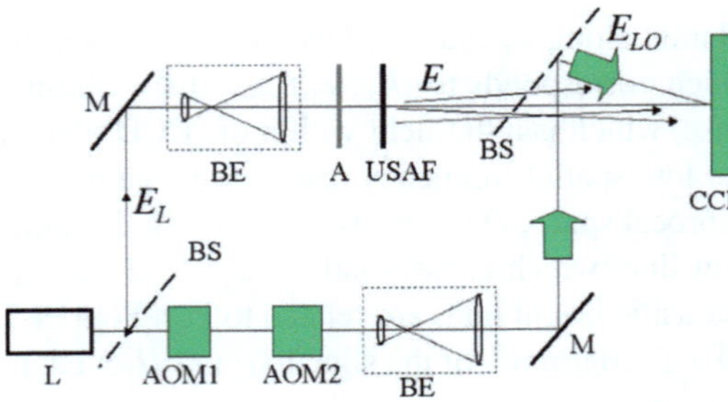

Fig. 13.3 Setup of the test experiment with USAF target. *L* main laser, *BS* beam splitter, *AOM1* and *AOM2* acousto-optic modulators, *BE* beam expander, *M* mirror, *A1 and A2* light attenuators, *USAF* transmission USAF target that is imaged, *CCD* CCD camera

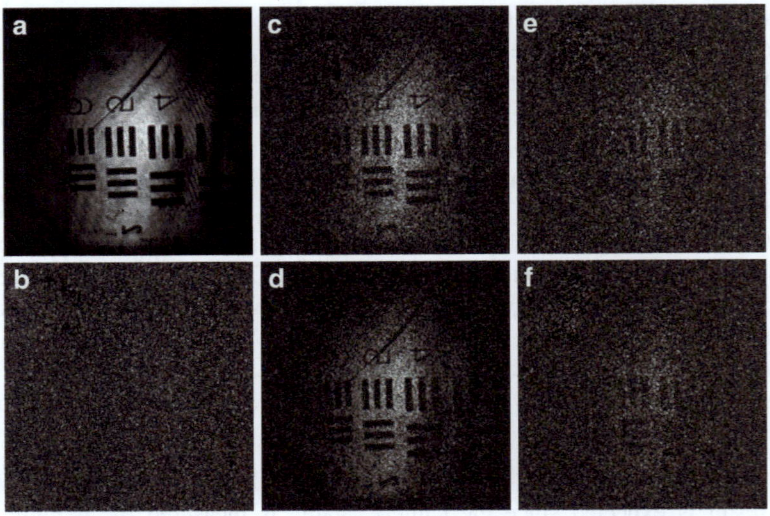

Fig. 13.4 (**a, c, d**): Reconstructions of an USAF target with different level of illumination 700 (**a**), 1 (**c**), and 0.15 e per pixel (**d**). (**b**) Simulated Shot Noise noise image. (**e, f**): Simulated reconstructed image obtained by mixing image (**a**) with weight X, and image (**b**) with weight $1 - X$. The weight X is 1/700 (**e**), and 0.15/700 (**f**). Images are displayed in arbitrary logarithmic grey scale

To filter-off the technical noise, the reconstruction is done by selecting the order 1 image of the object, within the reciprocal space [8]. Since the 400×400 pixels region that is selected is off-axis, the low spatial frequency noisy region, which correspond to the zero order contributions (region 1 on Fig. 13.2), is filtered-off.

Figure 13.4a,c,d shows the reconstructed images obtained for different illumination levels of the USAF target. For each image, we have measured the average number of photoelectrons per pixel corresponding to the object beam, within the reciprocal space region that has been selected for the reconstruction (i.e., 400 × 400 pixels). The images of Fig. 13.4 correspond to 700 (a), 1 (c), and 0.15 (d) electron per pixel for the sequence of $M = 4n = 12$ frames respectively (i.e., $700/12$, $1/12$, and $0.15/12$ e per pixel and per frame).

Here, the object beam intensity hhas been measured as follows: we have first calibrated the response of our camera with an attenuated laser whose power is known. We have then measured with the camera, at high level of signal, the intensity of the signal beam alone (without LO beam). We have decreased, to the end, the signal beam intensity by using calibrated attenuators in order to reach the low signal level of the images of Fig. 13.2a,c,d. In the case of image (a) with 700 e per pixel, we also have measured the averaged signal intensity from the data themselves by calculating $|H|^2/D$ [see (13.18)]. The two measurements gave the same result: 700 e per pixel.

On Fig. 13.4a, with 700 e per pixel, the USAF signal is much larger than the shot noise, and the Sinal to Noise Ratio (SNR) is large. On Fig. 13.4c, with 1 e per pixel, the USAF signal is roughly equal to the shot noise, and the SNR is about 1. With 0.15 e per pixel, the SNR is low on Fig. 13.4d (about 0.15), and the USAF is hardly seen.

It is nevertheless quite difficult to evaluate the SNR of an image. To perform a more quantitative analysis of the noise within the images, we have synthesized the noisy images of Fig. 13.4e,f by adding noise to the Fig. 13.4a noiseless image. We have first synthesized a pure Noise image, which is displayed on Fig. 13.4b. The Noise image, which corresponds to the image that is expected without signal, is obtained by the following way. From one of the measured frames (e.g., I_0), we have calculated the noise components $i_{k,p,q}$ by Monte Carlo drawing with the condition:

$$\langle i_{k,p,q}^2 \rangle = I_{0,p,q}. \tag{13.27}$$

This condition corresponds to (13.9) since $\langle I_{k,p,q} \rangle \simeq I_{0,p,q}$. We have synthesized the image sequence I_k in the following manner:

$$I_{k,p,q} = I_{0,p,q} + i_{k,p,q}. \tag{13.28}$$

The Shot Noise image of Fig. 13.4b is reconstructed then from the $I_{k,p,q}$ sequence with $k = 0 \dots 12$ since $M = 4n = 12$.

We have synthesized noisy images by summing the noiseless image of Fig. 13.4a with weight X, with the Noise image of Fig. 13.4b with weight $(1-X)$. The image of Fig. 13.4e is obtained with $X = 1/700$. Figure 13.4e corresponds thus to the same signal, and the same noise than Fig. 13.4c (i.e., 1 e of signal, and 1 e of noise respectively). As expected, Figs. 13.4c,e are visually very similar. The image of Fig. 13.4 is similarly obtained with $X = 0.15/700$. It corresponds to the same Signal and Noise than Fig. 13.4d (i.e., 0.15 e of signal, and 1 e of noise). As expected, Figs. 13.4d,f are visually very similar too.

Here we demonstrated our ability to synthesize a noisy image with a noise that is calculated by Monte Carlo from (13.27) and (13.28). Moreover, we have verified that the noisy image is visually equivalent to the image we have obtained in experiments. These results prove that we are able to assess quantitatively the noise, and that the noise that is obtained in experiments reaches the theoretical limit of 1 e of noise per pixel for the whole sequence of $M = 4n = 12$ frames.

13.4 Conclusion

In this chapter, we have studied the noise limits in off-axis, heterodyne digital holography. We have shown that because of the heterodyne gain of the holographic detection, the noise of the CCD camera can be neglected. Moreover, by a proper arrangement of the holographic setup, that combines off-axis geometry with phase shifting acquisition of holograms by heterodyne holography, it is possible to reach the theoretical shot noise limit. We have studied theoretically this limit, and we have shown that it corresponds to one photoelectron per pixel for the whole sequence of frame that is used to reconstruct the holographic image. This paradoxical result is related to the heterodyne detection, where the detection bandwidth is inversely proportional to the measurement time. We have verified all our results experimentally, and we have shown that is possible to image objects at very low signal levels. We have also shown that is possible to mimic the very weak illumination levels holograms obtained in experiments by Monte Carlo noise modeling.

References

1. D. Gabor. Microscopy by reconstructed wavefronts. *Proc. R. Soc. A*, 197:454, 1949.
2. A. Macovsky. Consideration of television holography. *Opt. Acta*, 22(16):1268, August 1971.
3. U. Schnars. Direct phase determination in hologram interferometry with use of digitally recorded holograms. *JOSA A.*, 11:977, July 1994.
4. J. W. Goodmann and R. W. Lawrence. Digital image formation from electronically detected holograms. *Appl. Phys. Lett.*, 11:77, 1967.
5. E. N. Leith, J. Upatnieks, and K. A. Haines. Microscopy by wavefront reconstruction. *J. Opt. Soc. Am.*, 55(8):981–986, 1965.
6. U. Schnars and W. Jüptner. Direct recording of holograms by a CCD target and numerical reconstruction. *Appl. Opt.*, 33(2):179–181, 1994.
7. T. M. Kreis, W. P. O. Juptner, and J. Geldmacher. Principles of digital holographic interferometry. *SPIE*, 3478:45, July 1988.
8. E. Cuche, P. Marquet, and C. Depeursinge. Spatial filtering for zero-order and twin-image elimination in digital off-axis holography. *Appl. Opt.*, 39(23):4070–4075, 2000.
9. I. Yamaguchi and T. Zhang. Phase-shifting digital holography. *Opt. Lett.*, 18(1):31, 1997.
10. F. LeClerc, L. Collot, and M. Gross. Numerical heterodyne holography using 2D photodetector arrays. *Opt. Lett.*, 25:716, 2000.
11. M. Atlan, M. Gross, and E. Absil. Accurate phase-shifting digital interferometry. *Opt. Lett.*, 32(11):1456–1458, 2007.
12. M. Gross, P. Goy, and M. Al-Koussa. Shot-noise detection of ultrasound-tagged photons in ultrasound-modulated optical imaging. *Opt. Lett.*, 28(24):2482–2484, 2003.
13. M. Atlan, B. C. Forget, F. Ramaz, A. C. Boccara, and M. Gross. Pulsed acousto-optic imaging in dynamic scattering media with heterodyne parallel speckle detection. *Opt. Lett.*, 30(11):1360–1362, 2005.
14. L. Wang and X. Zhao. Ultrasound-modulated optical tomography of absorbing objects buried in dense tissue-simulating turbid media. *Appl. Opt.*, 36(28):7277–7282, 1997.
15. F. Joud, F. Laloe, M. Atlan, J. Hare, and M. Gross. Imaging a vibrating object by sideband digital holography. *Opt. Exp.*, 17:2774, 2009.
16. F. Joud, F. Verpillat, F. Laloë, M. Atlan, J. Hare, and M. Gross. Fringe-free holographic measurements of large-amplitude vibrations. *Opt. Lett.*, 34(23):3698–3700, 2009.

17. M. Atlan, M. Gross, and J. Leng. Laser Doppler imaging of microflow. *J. Eur. Opt. Soc.*, 1:06025-1, 2006.
18. M. Atlan, M. Gross, B. C. Forget, T. Vitalis, A. Rancillac, and A. K. Dunn. Frequency-domain wide-field laser Doppler in vivo imaging. *Opt. Lett.*, 31(18):2762–2764, 2006.
19. M. Atlan, B. C. Forget, A. C. Boccara, T. Vitalis, A. Rancillac, A. K. Dunn, and M. Gross. Cortical blood flow assessment with frequency-domain laser Doppler microscopy. *J. Biomed. Opt.*, 12:024019, 2007.
20. M. Atlan, M. Gross, T. Vitalis, A. Rancillac, J. Rossier, and A. C. Boccara. High-speed wave-mixing laser Doppler imaging in vivo. *Opt. Lett.*, 33(8):842–844, 2008.
21. M. Gross and M. Atlan. Digital holography with ultimate sensitivity. *Opt. Lett.*, 32(8):909–911, 2007.
22. M. Gross, M. Atlan, and E. Absil. Noise and aliases in off-axis and phase-shifting holography. *Appl. Opt.*, 47(11):1757–1766, 2008.
23. H. A. Bachor, T. C. Ralph, S. Lucia, and T. C. Ralph. *A guide to experiments in quantum optics*. Wiley-VCH, New York, 1998.
24. M. Newberry. Measuring the gain of a CCD camera. *Axiom Technical Note: 1*, 1:1–9, 1998–2000.

Chapter 14
Deformation of Digital Holograms for Full Control of Focus and for Extending the Depth of Field

M. Paturzo and P. Ferraro

Abstract We show here that through an adaptive deformation of a digital hologram, it is possible to manage the depth of focus in 3D imaging reconstruction. For Fresnel holograms, a linear deformation is applied to the original hologram with the aim to change the reconstruction distance. On the other hand, by a quadratic deformation of spatial coordinates of a digital hologram acquired in Fourier configuration, it is possible to recover the extended focus image (EFI) of a tilted object.

14.1 Introduction

One of main problems in optical imaging is the limited depth of focus. All optical systems are affected by this limitation that hinder to get in focus in a single image plane objects that are located at different distances, but falling inside the same field of view. This is especially true in optical systems like microscopes where the depth of field is squeezed due to the request of a high magnification. This problem is an important issue in biological applications when a fast analysis for dynamic processes is necessary [1]. In fact, in classical optical microscopes, the Extended Focus Image (EFI) is built up by the optical longitudinal mechanical scanning and by the acquisition of a sequence of images during the scanning, called "image stack." To create an EFI, in each image of the "stack," the in-focus parts are selected and then stitched together in a single image (i.e., EFI). However, the long acquisition time required for digitizing many images prevents it to be used in case of dynamic events. To overcome this problem, some approaches have been developed for incoherent imaging system. One approach is based on the so-called "wavefront coding" using an opportune phase plate that introduces a wavefront distortion, giving the possibility to have an EFI in a single image acquisition [2]. Recently, the problem to extend the depth of field has been successfully addressed for incoherent imaging system by

P. Ferraro (✉)
CNR – Istituto Nazionale di Ottica, via Campi Flegrei 34, 80078-Pozzuoli (NA), Italy
e-mail: pietro.ferraro@inoa.it

T. Fournel and B. Javidi (eds.), *Information Optics and Photonics: Algorithms, Systems, and Applications*, DOI 10.1007/978-1-4419-7380-1_14,

using a birefringent plate [3], by means of lenses with radial and angular modulation [4], by logarithmic phase mask [5], or by amplitude and phase modulation of the pupil function [6].

The need of an EFI image is also important in 3D optical imaging systems, both for incoherent and coherent light sources [7–16]. Of course, holographic microscopes allow 3D imaging capability but suffer at the same time of a limited depth of focus. Recently, various solutions have been proposed to obtain EFI in DH systems [13,16–21]. Some of these approaches use the angular spectrum transformation; others exploit the DH capability to reconstruct image planes at different distances, or image processing during the reconstruction process. The diffraction of a tilted planar object is also studied in [22], where the authors generate a computer-generated hologram of a 3D object, composed of planar segment tilted in respect to the optical axis.

Here, we show that by means of an opportune coordinates transformation of holograms recorded both in Fresnel and Fourier configuration, it is possible to manage the depth of focus and also to obtain an EFI image of a tilted object. In Fresnel holography, a linear transformation allows to extend the depth of focus. However, such approach does not work properly for Fourier-type holograms. In the latter case, only a quadratic geometrical deformation allows to obtain a tilted object with all points in focus by a single reconstruction step. We report the theoretical analysis and the results of some specific experiments with the aim to demonstrate the validity of the method [23, 24].

14.2 Fresnel Holograms: Linear Deformation

In this section, we consider a digital hologram of an object at distance d from the CCD, acquired in Fresnel configuration, i.e., by using a plane reference wave.

The numerical reconstruction of the object field in focus is obtained through the numerical modeling of the Fresnel approximation of the Rayleigh–Sommerfeld diffraction integral, given by the formula:

$$b(x,y,d) = \frac{1}{i\lambda d}\iint h(\xi,\eta)r(\xi,\eta)\mathrm{e}^{ikd\left[1+(x-\xi)^2/2d^2+(y-\eta)^2/2d^2\right]}\mathrm{d}\xi\mathrm{d}\eta, \quad (14.1)$$

where $h(\xi,\eta)$ and $r(\xi,\eta)$ are the recorded hologram and the reference wave, respectively. If an affine geometric transformation is applied to the original recorded hologram, consisting of a simple stretching and described by $[\xi'\eta'] = [\xi \quad \eta \quad 1]\,T$ through the operator:

$$T = \begin{bmatrix} a & 0 \\ 0 & a \\ 0 & 0 \end{bmatrix},$$

we obtain the transformed hologram $h(\xi', \eta') = h(\alpha\xi, \alpha\eta)$ where α represents the scaling factor. Consequently, the propagation integral changes to

$$\begin{aligned} b(x, y, d) &= \frac{1}{i\lambda d} e^{ikd} \iint h(\alpha\xi, \alpha\eta) e^{[ik\alpha^2(x-\xi)^2]/2d\alpha^2} e^{[ik\alpha^2(y-\eta)^2]/2d\alpha^2} d\xi d\eta \\ &= \frac{1}{i\alpha^2\lambda d} e^{ikd} \iint h(\xi', \eta')\, e^{[ik(x'-\xi')^2]/2D} e^{[ik(y'-\eta')^2]/2D} d\xi' d\eta' \\ &= \frac{1}{\alpha^2} b(x', y', D) \end{aligned} \tag{14.2}$$

with $D = d/\alpha^2$ and $x' = x/\alpha$, $y' = y/\alpha$. We set $r(\xi, \eta) = 1$ since we consider that the reference wave is a collimated and normally impinging wavefront.

Such simple stretching applied to the hologram has a very interesting impact on the numerical or even optical reconstructions. In fact, from (2) it is clear that by reconstructing the stretched hologram, the distance of the in-focus plane becomes $D = d/\alpha^2$. Therefore, through subsequent reconstruction of the same hologram, stretched with a variable scaling factor α, it is possible to put in focus different object planes.

The impact that the stretching has on numerical reconstructions is shown in Fig. 14.1. The object was made of three different wires positioned at different distances from the CCD array. The horizontal wire, the twisted wire with the eyelet, and the vertical wire were set at distances of 100, 125, and 150 mm, respectively. The numerical reconstructions at the above distances give an image in which each wire

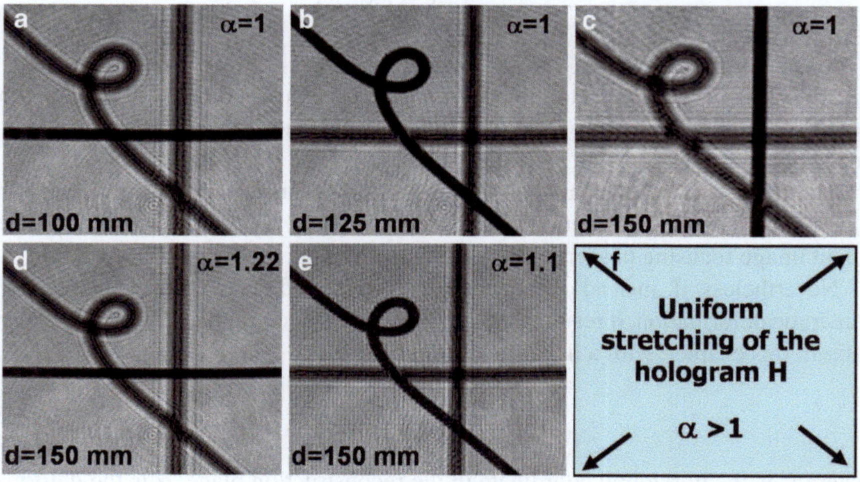

Fig. 14.1 The reconstruction of a digital hologram as recorded (no-stretching) at three different distances with each wire in focus at distance of (**a**) 100 mm, (**b**) 125 mm, and (**c**) 150 mm, respectively. When the hologram is stretched with $\alpha = 1.1$, the horizontal wire is in focus at distance of $d = 150$ mm (**d**), while for $\alpha = 1.22$, the curved wire is in focus even at $d = 150$ mm (**e**). Conceptual draw of the stretching of H (**f**) (Image from [23])

at a time is in good focus, as shown in Fig. 14.1a–c, at the corresponding recording distances. When the hologram $h(x, y)$ is uniformly stretched with an elongation factor along both dimensions (x, y) of $\alpha = 1.1$, the horizontal wire results to be in focus at a distance $d = 150$ mm (Fig. 14.1d), instead of 100 mm. If $\alpha = 1.22$, the twisted wire with the eyelet is in focus at $d = 150$ mm (Fig. 14.1e) instead that at $d = 125$ mm. Anyway, the results shown in Fig. 14.1 demonstrate that the depth of focus can be controlled by means of the uniform stretching of the holograms according to (2).

14.3 Fourier Holograms: Quadratic Deformation

For holograms recorded in Fourier configuration, the simple linear stretching gives different results in respect to the Fresnel holograms. In fact, by application of a linear transformation to a Fourier hologram, no changes can be observed about image focusing properties during reconstruction.

In fact, in the Fourier configuration, the numerical reconstruction of the hologram $f(\xi, \eta)$ is obtained simply by calculating its spatial FT $\hat{f}(x, y)$.

When the hologram is linearly stretched, we obtained a new 2D matrix, that is:

$$h(\xi, \eta) = f(a\xi, a\eta). \tag{14.3}$$

Because of the Fourier transform properties we have that:

$$\hat{h}(x, y) = \frac{1}{a}\hat{f}\left(\frac{x}{a}, \frac{y}{a}\right), \tag{14.4}$$

where $\hat{h}(x, y)$ corresponds to the new numerical reconstruction and it results to be equal to the reconstruction of the initial hologram with scaled dimensions and a scaled intensity.

Therefore, there is no change in reconstruction distance. Consequently, by application of a linear transformation to a Fourier hologram, no changes can be observed about image focusing properties during reconstruction.

Nevertheless, if, instead of a linear transformation described by (1), we apply a quadratic deformation, it results that the in-focus distance changes, as will be shown later in this chapter, with a linear law given by:

$$D = L\left(1 + 2\alpha l'\right), \tag{14.5}$$

where l' is the horizontal coordinate in the reconstruction plane, α is the deformation parameter, and L is the reconstruction distance of the recorded (not deformed) hologram. Therefore, by this kind of hologram deformations, we can put correctly in focus the numerical reconstruction of a tilted object, recovering its EFI. It important to note that we are considering the possibility to deform the hologram only along the direction of the tilt, i.e., in our case the x-axis.To demonstrate how the change of the in-focus distance depends on the quadratic deformation, we need to remind some Fourier Transform properties for a composite function $h(x) = g[f(x)]$.

If we consider the following composite function

$$h(x) = g[f(x)] = \int G(l)\mathrm{e}^{i2\pi l\cdot f(x)}\mathrm{d}l, \tag{14.6}$$

where $G(l)$ is the Fourier Transform of $g(y)$, we have that the FT of results to be:

$$\hat{h}(k) = \int \mathrm{e}^{-i2\pi k\cdot x}\int G(l)\mathrm{e}^{i2\pi l\cdot f(x)}\mathrm{d}l\,\mathrm{d}x \tag{14.7}$$

$$= \int G(l)P(k,l)\mathrm{d}l, \tag{14.8}$$

where $P(k,l) = \int \mathrm{e}^{-i2\pi k\cdot x}\mathrm{e}^{i2\pi l\cdot f(x)}\mathrm{d}x$. For a quadratic coordinate transformation of the hologram $g(x)$, we have that $f(x) = x + \alpha x^2$.

Then, it results that $P(k,l) = \sqrt{\frac{\pi}{2\pi\alpha l}}\mathrm{e}^{i\pi^2[(k-1)^2/2\pi\alpha l]}\mathrm{e}^{i(\pi/4)}$

Therefore,

$$\hat{h}(k) = \int G(l)P(k,l)\mathrm{d}l = \mathrm{e}^{i(\pi/4)}\int G(l)\sqrt{\frac{1}{2\alpha l}}\mathrm{e}^{i2\pi[(k-1)^2]/4\alpha l}\mathrm{d}l, \tag{14.9}$$

where $\hat{h}(k)$ is the reconstruction of the deformed hologram, while $G(l)$ is the reconstruction of the initial hologram. If we set $k = k'/\lambda L$ and $l = l'/\lambda L$ to change the coordinates from the spatial frequency domain to the space domain, then in (6) the exponential term $\mathrm{e}^{i2\pi[(k-l)^2/4\alpha l]}$ becomes $\mathrm{e}^{(i2\pi/\lambda)[(k'-l')^2/2d']}$ where $d' = 2L\alpha l'$.

Therefore, the final reconstruction, i.e., the reconstruction of the deformed hologram, can be interpreted as the reconstruction of the initial hologram at distance L and further propagated to a distance d' that depends linearly on the coordinate l' (x-axis in the reconstruction plane) and it is a function of the deformation parameter α and of L, that is the reconstruction distance of the not-deformed hologram (Fig. 14.2).

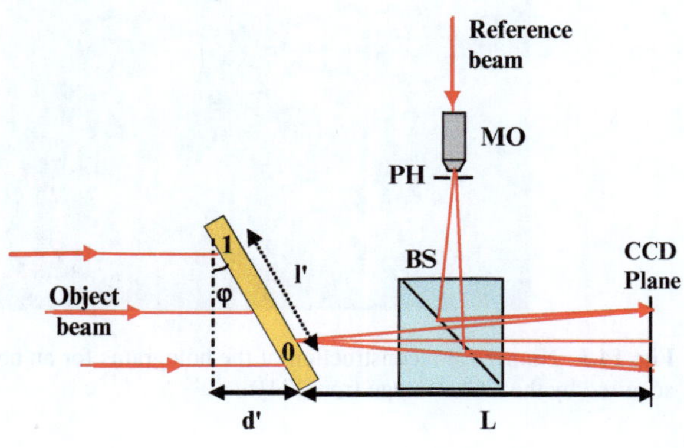

Fig. 14.2 Experimental setup: *BS* beam splitter, *MO* microscope objectives, *PH* pinhole (Image from [24])

Summarizing, we can claim that if we apply a quadratic deformation to the hologram of a tilted object along the tilting direction, we obtain an all in-focus reconstructed image of the object. In fact, the reconstruction distance changes pixel by pixel along the tilting direction l' according to (3) as experimentally demonstrated in the next section.

The quadratic deformation is applied to the original recorded hologram through the coordinates transformation $[x'y'] = \begin{bmatrix}1 & x & y & xy & x^2 & y^2\end{bmatrix} T$ where the operator is:

$$T = \begin{bmatrix} 0 & 0 \\ 1 & 0 \\ 0 & 1 \\ 0 & 0 \\ \alpha & 0 \\ 0 & 0 \end{bmatrix}. \tag{14.10}$$

For the experimental validation, we made an experiment with a Mach–Zehnder interferometer in Fourier configuration as depicted in Fig. 14.2.

The object was an USAF target positioned in a tilted way in respect to the laser light illumination direction. The Fourier holographic configuration is such that the curvature of the reference beam matches that of the light scattered by the left side of the object (where there is the number "0"), i.e., the part of the target closer to the CCD (at a distance L). Consequently, this region results to be in focus in the numerical reconstruction, i.e., the hologram Fourier Transform. On contrary, the right side of the target (where there is the number "1") results to be completely out of focus in the numerical reconstruction. In Fig. 14.3, we show the numerical reconstruction of the acquired hologram for an object tilted with an angle $\varphi = 55°$. It is clear that the left part of the object, where there is the number "0," results to be in focus, while the right part, with the number "1," is out of focus and that the focusing gradually worsens as going from left to right side.

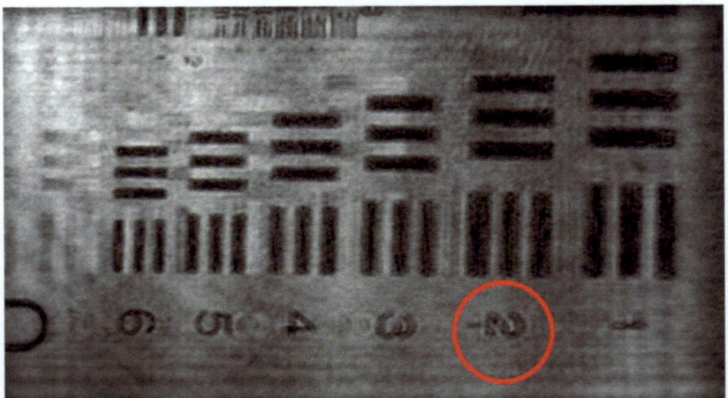

Fig. 14.3 Numerical reconstruction of the holograms for an object tilted with an angle of 55°, as acquired by the CCD (Image from [24])

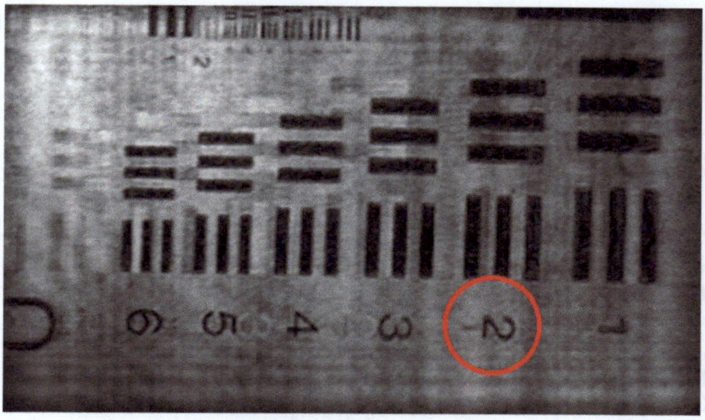

Fig. 14.4 Numerical reconstruction of the holograms for an object tilted with an angle of 55°, after the quadratic deformation. (Image from [24])

If we apply the quadratic deformation before performing the reconstruction (with an opportune choice of the deformation parameter α), we obtain the image shown in Fig. 14.4. In this case, the α value is 2.1×10^{-5}. In the reconstructions of the deformed objects, all details of the target result to be in focus. We encircled the number "2" in all the reconstructions to better guide the reader to visualize the comparison between the corrected and uncorrected images. In all the images, some replicas of the target elements appear. They are due to the multiple reflections of the laser light on the target surfaces. It is clear that the replicas are not due to the applied deformation as they are visible in the reconstructions of the not-deformed holograms.

14.4 Conclusion

We have demonstrated that by an opportune deformation (numerical transformation) of a digital hologram, it is possible to control the reconstruction distance in case of the Fresnel-type holograms and to obtain an all in-focus image (i.e., EFI) for tilted objects in case of the Fourier-type holograms. The experimental results are in agreement with the theoretical model. It has been demonstrated that the EFI can be recovered for very high tilted angles. The approach is completely new in respect to the previous methods developed to obtain EFI both for incoherent and coherent imaging systems. The method is simple because it requires just a single recorded hologram dispensing from the need to record multiple images and avoiding mechanical scanning. The method is easily applied to digital holograms where the coordinate deformation is performed numerically. Moreover, for holograms recorded on a specific material (physically recorded on a photosensitive media), it is possible to foresee to apply an adaptive mechanical deformation to extend the depth of focus even in that case.

Acknowledgments This research is funded by the European Community's Seventh Framework Program FP7/2007–2013 under grant agreement 216105.

References

1. Perry, D. J., Venkatachalam, V., Liang, L., Hall, B. E., Frost, K., Basiji, D.A.: Extended depth of field imaging for high speed cell analysis. Cytometry A **71A**, 215–231 (2007)
2. Dowski, E. R. Jr., Cathey, W.T.: Extended depth of field through wavefront coding. Appl. Opt. **34**, 1859–1866 (1995)
3. Zalevsky, Z., Ben-Yaish, S.: Extended depth of focus imaging with birefringent plate. Opt. Express **15**, 7202–7210 (2007)
4. Mikula, G., Jaroszewicz, Z., Kolodziejczyk, A., Petelczyc, K., Sypek, M.: Imaging with extended focal depth by means of lenses with radial and angular modulation. Opt. Express **15**, 9184–9193 (2007)
5. Zhao, H., Li, Q., Feng, H.: Improved logarithmic phase mask to extend the depth of field of an incoherent imaging system. Opt. Lett. **33**, 1171–1173 (2008)
6. Bagheri, S., Javidi, B.: Extension of depth of field using amplitude and phase modulation of the pupil function. Opt. Lett. **33**, 757–759 (2008)
7. Kanka, M., Riesenberg, R., Kreuzer, H. J.: Reconstruction of high-resolution holographic microscopic images. Opt. Lett. **34**, 1162–1164 (2009)
8. Wang, D., Zhao, J., Zhang, F., Pedrini, G., Osten, W.: High-fidelity numerical realization of multiple-step Fresnel propagation for the reconstruction of digital holograms. Appl. Opt. **47**, D12–D20 (2008)
9. Seok Hwang, Y., Hong, S.-H., Javidi, B.: Free view 3-D visualization of occluded objects by using computational synthetic aperture integral imaging. J. Display Technol. **3**, 64–70 (2007)
10. Takaki, Y., Ohzu, H.: Hybrid holographic microscopy: visualization of three-dimensional object information by use of viewing angles. Appl. Opt. **39**, 5302–5308 (2000)
11. Malkiel, E., Abras, J. N., Katz, J.: Automated scanning and measurements of particle distributions within a holographic reconstructed volume. Meas. Sci. Technol.**15**, 601–612 (2004)
12. Ferraro, P., Coppola, G., De Nicola, S., Finizio, A., Pierattini, G.: Digital holographic microscope with automatic focus tracking by detecting sample displacement in real time. Opt. Lett. **28**, 1257–1259 (2003)
13. Antkowiak, M., Callens, N., Yourassowsky, C., Dubois, F.: Extended focused imaging of a microparticle field with digital holographic microscopy. Opt. Lett. **33**, 1626–1628 (2008)
14. Dubois, F., Schockaert, C., Callens, N., Yourassowsky, C.: Focus plane detection criteria in digital holography microscopy by amplitude analysis. Opt. Express **14**, 5895–5908 (2006)
15. Liebling, M., Unser, M.: Autofocus for digital Fresnel holograms by use of a Fresnelet sparsity criterion. J. Opt. Soc. Am. A **21**, 2424–2430 (2004)
16. Pieper, R. J., Korpel, A.: Image processing for extended depth of field. Appl. Opt. **22**, 1449–1453 (1983)
17. Ferraro, P., Grilli, S., Alfieri, D., De Nicola, S., Finizio, A., Pierattini, G., Javidi, B., Coppola, G., Striano, V.: Extended focused image in microscopy by digital holography. Opt. Express **13**, 6738–6749 (2005)
18. De Nicola, S., Finizio, A., Pierattini, G., Ferraro, P., Alfieri, D.: Angular spectrum method with correction of anamorphism for numerical reconstruction of digital holograms on tilted planes. Opt. Express **13**, 9935–9940 (2005)
19. Jeong, S. J., Hong, C. K.: Pixel-size-maintained image reconstruction of digital holograms on arbitrarily tilted planes by the angular spectrum method. Appl. Opt. **47**, 3064–3071 (2008)
20. Matsushima, K.: Formulation of the rotational transformation of wave fields and their application to digital holography. Appl. Opt. **47**, D110–D116 (2008)
21. McElhinney, C. P., Hennelly, B. M., Naughton, T. J.: Extended focused imaging for digital holograms of macroscopic three-dimensional objects. Appl. Opt. **47**, D71–D79 (2008)

22. Leseberg, D., Frère, C.: Computer-generated holograms of 3-D objects composed of tilted planar segments. Appl. Opt. **27**, 3020–3024 (1988)
23. Ferraro, P., Paturzo, M., Memmolo, P., Finizio, A.: Controlling depth of focus in 3D image reconstructions by flexible and adaptive deformation of digital holograms. Opt. Lett. **34**, 2787–2789 (2009)
24. Paturzo, M., Ferraro, P.: Creating an extended focus image of a tilted object in Fourier digital holography. Opt. Express **17**, 20546–20552 (2009)

Chapter 15
Zoom Algorithms for Digital Holography

Bryan M. Hennelly, Damien P. Kelly, David S. Monaghan, and Nitesh Pandey

Abstract Digital Holography is an imaging modality made up of two parts: (1) using a digital camera to record the interference pattern between a field scattered from an object and a known reference field so that the complex object wavefield can be obtained and (2) replaying or reconstructing the hologram on a computer by simulating the propagation of the object wavefield back to the object plane. Thus an image is obtained. The most commonly used algorithms for the reconstruction algorithm are the so-called direct method and the spectral method of calculating the Fresnel Transform, which describes free space propagation in the paraxial approximation. These algorithms differ in the output range that they display. For the direct method, the output image size is proportional to the distance making it more appropriate for large objects at large distances. The spectral method has an output image size equal to the size of the CCD. We show how to adapt this latter algorithm in a simple way to allow it to generate any output range and in any location making it far more versatile for zooming in on specific regions of our reconstructed image.

15.1 Introduction to Digital Holography

Holography is a technique for recording and reconstructing a complex electromagnetic wavefield that was invented by Gabor [1]. This initial invention concerned itself with electron microscopy and predated the invention of the laser. With the onset of the laser, E. Leith and J. Upatnieks [2, 3] appended the holographic principle with the introduction of the offset reference wave. This enabled the spatial separation of the object wavefield from the other components that are generated in the optical recording process, namely the intensities of the object and reference wavefields and also the so-called "ghost" or conjugate image. Using photosensitive recording materials to record holograms is costly and inflexible. Digital

B.M. Hennelly (✉)
National University of Ireland, Maynooth, Ireland
e-mail: bryanh@cs.nuim.ie

T. Fournel and B. Javidi (eds.), *Information Optics and Photonics: Algorithms, Systems, and Applications*, DOI 10.1007/978-1-4419-7380-1_15,

holography [4–7] (DH) is the technique of using discrete electronic devices, such as CCDs to record the hologram. In this case, reconstruction is performed numerically by simulating the propagation of the wavefield back to the plane of the object [8–13].

In recent years, DH has been demonstrated to be a useful method in many areas of optics, in particular in microscopy [14, 15], and this branch has become known as digital holographic microscopy (DHM). In this chapter, we investigate the application of the zoom algorithm to digitally zooming in on small regions of interest in a large object. Though we limit ourselves in this chapter to digital holograms of macroscopic images, we believe that the analysis presented herein may be applied to DHM for investigating digital zoom.

Numerical reconstruction requires some kind of fast algorithms to numerically calculate the Fresnel Transform in $O(N \log N)$ calculations. In general, these algorithms are often based around the use of the Fast Fourier Transform (FFT) [16] algorithm developed for $O(N \log N)$ calculation of the Fourier Transform. Various different algorithms exist such as the well-known direct and spectral methods [12] to simulate the Fresnel Transform and thereby reconstruct digital holograms to obtain the object image. The differences between these algorithms lies in the range of distances of the Fresnel Transform that can be simulated and the relationship between the input and output field of view. In order to allow for a more flexible range of values for the output field of view, a number of Fresnel algorithms have been developed that allow for the user to define the output pitch size arbitrarily [17, 18]. The first of these algorithms is based on application of the chirp z transform to the discretized Fresnel kernel and the second is based on two applications of the direct method. In this chapter, we derive a zoom algorithm based on the spectral method and we compare it the previous two methods. We show that all three algorithms amount to the same fundamental algorithm comprising three chirps and two FFT algorithms.

In Sect. 15.2, we provide a brief overview of digital hologram recording and the two most common algorithms for DH reconstruction. In Sect. 15.3, we show how to develop a zoom algorithm based on the spectral method. In Sect. 15.4, we show how to adapt the zoom algorithm so that we can zoom into smaller regions on larger objects. In Sect. 15.5, we discuss the relationship of our zoom algorithm with existing zoom algorithms used in DH. In Sect. 15.6, we offer a brief conclusion.

15.2 Digital Hologram Recording and Reconstructing

15.2.1 Digital Hologram Recording

In Fig. 15.1, we illustrate a typical optical setup that is used to record digital holograms, similar to that used to record the holograms in this chapter. In our experiments, we use a 785 nm CrystalLaser (coherence length 1 m) and a 1,392 by 1,040 pixel AVT Dolphin F-145B camera (pixel pitch, $T = 6.45\,\mu\text{m}$). The laser

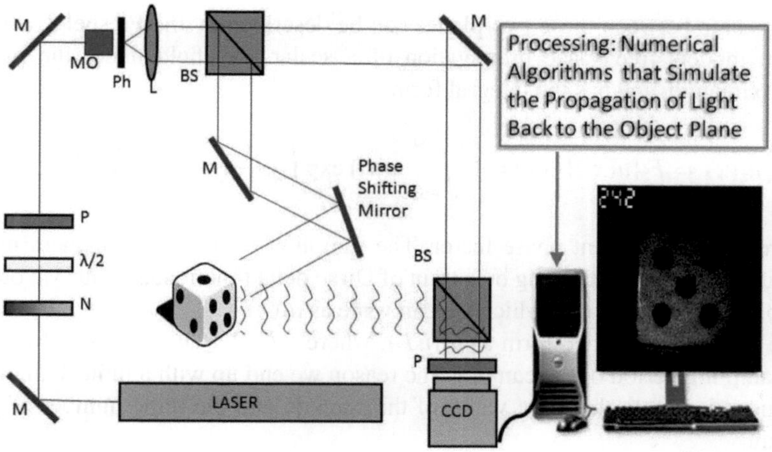

Fig. 15.1 Digital Holography In-line Architecture. The presence of the phase retarders allow us to make multiple captures with different constant phase shifts of the reference beam. We can therefor implement the Phase Shifting Interferometry algorithm to extract the complex wavefront: *MO* Microscope objective, *Ph* Pinhole, *N* Neutral density filter, *P* Polarizer, *M* Mirror, *L* Lens, *BS* Beam splitter, $\lambda/2$ Half wave plate

beam is first passed through a neutral density filter that allows control of the input power to the system. This is immediately followed by a half wave plate and a linear polarizer. Having been spatially filtered using a pinhole of 10 μm, the beam is expanded and it now passes through a polarizing beam splitter. By using the half wave plate, the linear polarizer, and the polarizing beam splitter, we can effectively change the ratio of powers in the two arms that emerge from this beam splitter. One arm is aimed directly at an object, or in the case of transmissive objects it is aimed through the object. The light coming from the object is incident on a CCD via a second polarizing beam splitter and a linear polarizer. The second arm is the reference beam that takes an approximately equidistant path to the CCD via the same latter two elements. The final linear polarizer in the setup serves to force both the reference and object wavefields into the same state of polarization such that the diffraction efficiency of the systems is maximized.

We write the interference pattern recorded by the camera as

$$I(x) = |u_z(x)|^2 + |r(x)|^2 + u_z(x)r^*(x) + u_z^*(x)r(x), \tag{15.1}$$

where $u_z(x)$ represents the signal that has propagated a distance z from the object to the object to the CCD and $r(x)$ denote the reference wavefield at the plane of the CCD. The superscript $*$ denote complex conjugate. In our setup, the phase shifting mirror shown in Fig. 15.1 is moved and four captures allow us to separate out $u_z(x)$ [19].

We denote the original signal by $u(x')$, where x' refers to the coordinate space in the object plane. In the plane of the CCD, this signal is denoted by $u_z(x)$. The

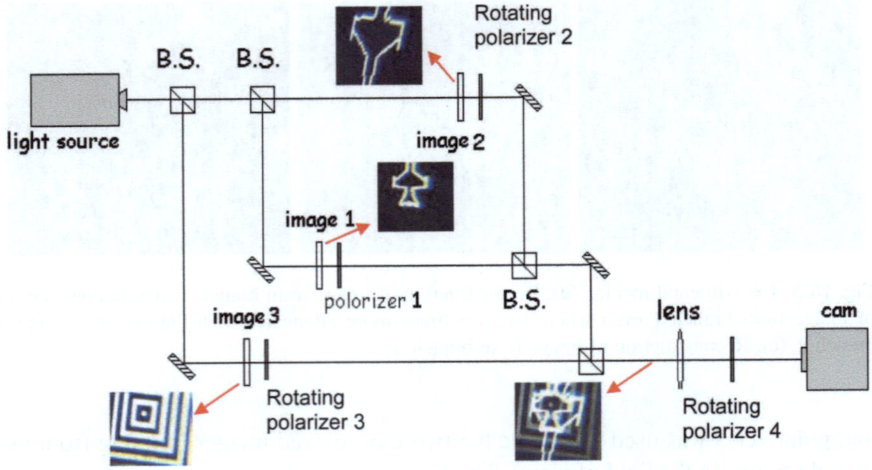

Fig. 18.4 The experimental setup with coding of three regions

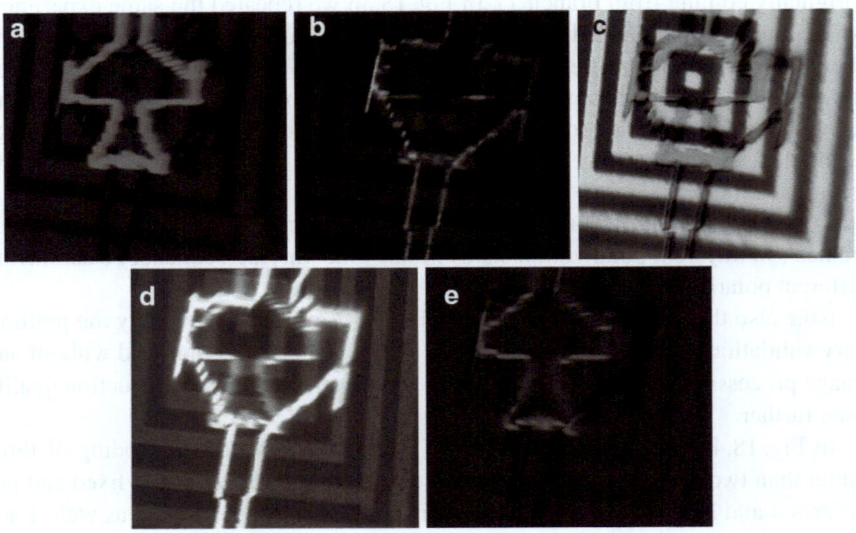

Fig. 18.5 Experimental results. (**a**). Reconstruction of image from branch 1. (**b**). Reconstruction of image from branch 2. (**c**). Reconstruction of image from branch 3. (**d**). The case when the input image from branch 2 is four times stronger than the rest of the images. (**e**). The reconstruction of image from branch 1 for in the input case of (**d**)

seen in Fig. 18.5e). One may see that although image from region 1 was attenuated more than the image from region 2, we yet could reconstruct it. This experiment shows that the orthogonal depolarization sequence allows reconstruction even when the required spatial information is lower in its signal-to-noise ratio in comparison to the undesired information (regarded as noise).

relationship between these two planes can be described by the Fresnel Transform, which models free space propagation of a scalar wavefield under the paraxial approximation, and has the integral form

$$u_z(x) = F_z\{u(x')\}(x) = A_z \int_{-\infty}^{\infty} u(x') \exp\left(j\frac{\pi}{\lambda z}\left(x - x'\right)^2\right)\mathrm{d}x', \qquad (15.2)$$

where A_z is a constant phase factor. The sampling of the digital camera may be accounted for by multiplying by a train of Dirac delta functionals. Thus we obtain a sampled version of $u_z(x)$, which we can write as $u_z(x) \sum_{n=-\infty}^{\infty} \delta(x - nT)$. We can write this in a shortened form as $u_z(nT)$, where $-N/2 \leq n \leq N/2 - 1$ and T is the sampling period of the camera. The reason we end up with a finite sum of delta functions is due to the finite width of the camera, and the finite number of pixels therein.

15.2.2 Reconstruction of Digital Holograms

15.2.2.1 The Direct Method

In Sect. 15.2.1, we showed how the recording process allows one to obtain $u_z(nT)$. Discretizing the input and output variables of the Fresnel Transform, that is, setting $x- > nT$ and $x'- > mT''$, we get

$$\begin{aligned}\mathrm{recon}(mT'') = A_z \exp\left(\frac{j2\pi\,(mT'')^2}{\lambda z}\right) \sum_{n=-N/2}^{N/2-1} \left[u_z(nT) \exp\left(\frac{j2\pi\,(nT)^2}{\lambda z}\right)\right] \\ \times \exp\left(\frac{-j2\pi nmTT''}{\lambda z}\right).\end{aligned} \qquad (15.3)$$

Focusing on the last term in (15.3), if we fix the output sampling (in the image plane) period as follows

$$T'' = \frac{\lambda z}{NT} = \frac{\lambda z}{W_{\mathrm{CCD}}} \qquad (15.4)$$

and if we interest ourselves with only a range of x'', x'' denotes the image plane, where corresponding to the range of samples defined by $-N/2 \leq m \leq N/2-1$, then (15.3) may be reduced into the form of the Discrete Fourier Transform (DFT),

$$\begin{aligned}\mathrm{recon}(mT'') &= \mathrm{DIRECT}\{u_z(nT), T, z, \lambda\} \\ &= A_z \exp\left(\frac{j\pi\,(mT'')^2}{\lambda z}\right) \\ &\quad \times \mathrm{DFT}\left\{u_z(nT) \exp\left(\frac{j\pi\,(nT)^2}{\lambda z}\right)\right\}(mT'').\end{aligned} \qquad (15.5)$$

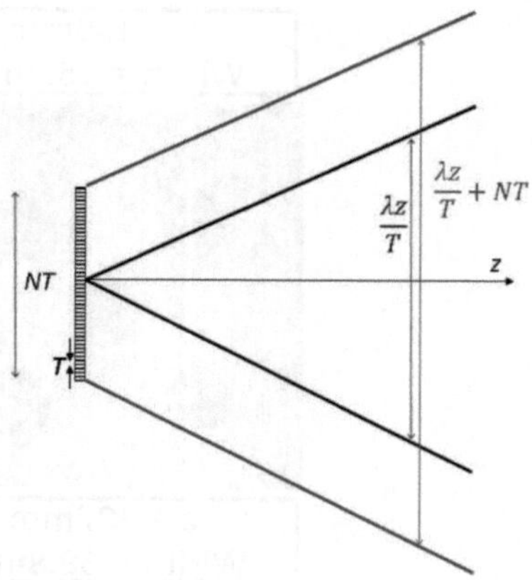

Fig. 15.2 Input and output widths of images obtained by reconstructing holograms using the direct method for simulating the Fresnel transform

It is very useful to have this equation in the form of the DFT because the DFT may be implemented in a computationally efficient manner ($N \log N$ computations compared with N^2) by employing a FFT algorithm [10]. The resulting numerical algorithm, derived from (15.5) is known as the direct method [9] and is made up of three parts: (1) multiply the hologram by a discrete chirp function, (2) implement the DFT using a FFT algorithm, and (3) multiply the result by a second discrete chirp function.

The output distribution spans a width of NT'', which we can calculate to be $\lambda z/T$. In Fig. 15.2, we illustrate the dependence of the output reconstruction width on the distance parameter. The inner lines show the window obtained by the direct method for a range of different distances. The outer lines show the maximum possible width of a discrete signal with these characteristics using the Fresnel approximation. In Fig. 15.3, we further illustrate this point by showing the reconstructed images of digital holograms of the same object recorded at four different distances.

15.2.2.2 The Spectral Method

Another method for simulating the Fresnel transform is known as the spectral method. This method is based on the fact that the Fresnel transform is a convolution and may be written as a product with a chirp function in the frequency domain. The Fresnel transform may be written as

$$u(x) = A_z \Im_{-1} \left\{ \Im \left\{ u_z(x) \exp\left(j2\pi\lambda z k^2 \right) \right\} \right\}, \tag{15.6}$$

where $\Im$ is the Fourier transform and transforms the signal into the frequency, k, domain. The following algorithm for simulating the Fresnel transform may be derived from this expression.

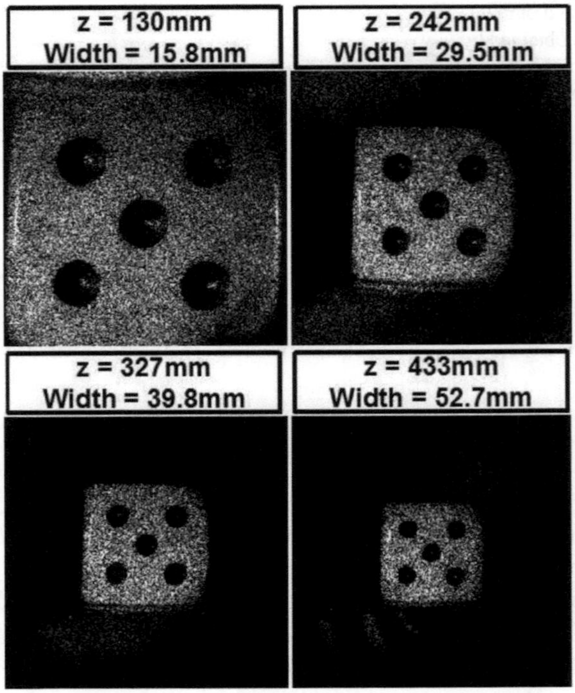

Fig. 15.3 Reconstructions from direct method when the object is placed at four different distances and reconstructed using the appropriate distance parameter. Clearly as we increase the value for z we also increase the output reconstruction width

$$\begin{aligned}\text{recon}(mT'') &= \text{SPECTRAL}\,\{u_z(nT), T, z, \lambda\}\\ &= A_z\text{IDFT}\left\{\text{DFT}\,\{u_z(nT)\}\exp\left(j2\pi\lambda z\left(\frac{l}{NT}\right)^2\right)\right\},\end{aligned} \quad (15.7)$$

where the integers n, m, and l all have ranges $-N/2 \leq n, m, l \leq N/2 - 1$. The result is that we obtain samples of our reconstruction with a pitch $T'' = T$. Unlike in the previous case of the direct method the output spatial width is not proportional to the reconstruction distance z and remains at a constant value of $NT'' = W_{\text{CCD}}$. In Fig. 15.4, we illustrate this point showing the constancy of the output window width from the spectral method, regardless of distance. An important point arises; if the actual object size is greater than the output window we must zeropad the input hologram up to a size equal to that of the object. For the dice object used in the current experiments, we have $N = 1{,}392$, $T = 6.45\,\mu\text{m}$ and $W_{\text{OBJ}} = 1.5\,\text{cm}$. Thus, our hologram size (and therefore our output window size) is $W_{\text{CCD}} = 9\,\text{mm}$ and is considerably smaller than the object size. We therefore zeropad our hologram up to a value of $N = 2{,}700$ and we reconstruct this new enlarged hologram. For the sake of comparison, we do this to the four holograms previously reconstructed by the direct

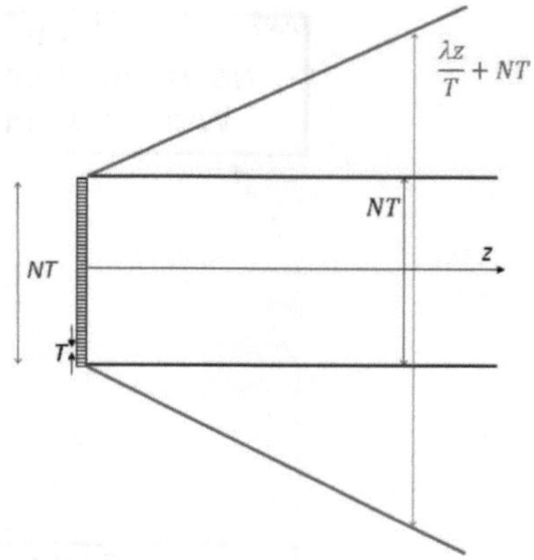

Fig. 15.4 Input and output widths of images obtained by reconstructing holograms using the spectral method for simulating the Fresnel transform

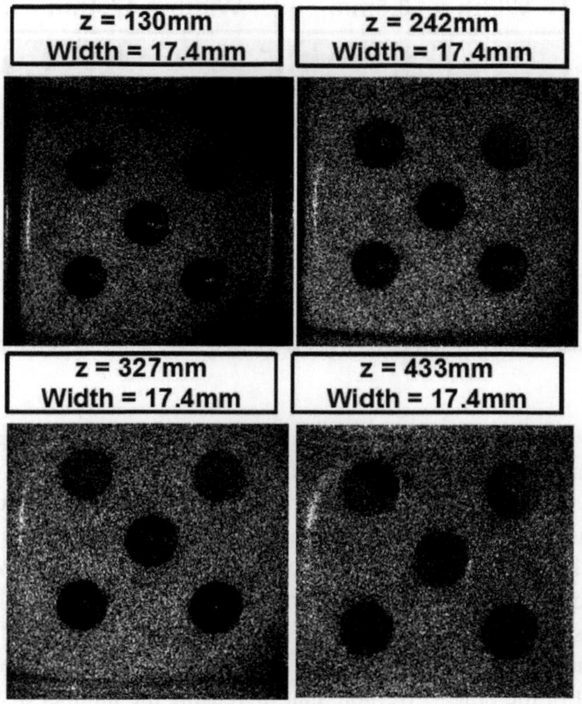

Fig. 15.5 Reconstructions from spectral method when the object is placed at four different distances and reconstructed using the appropriate distance parameter

method in Fig. 15.3 corresponding to the same object recorded at four different distances. The result is shown in Fig. 15.5. We expect to see the same output window size in all four cases. However, it is noticeable that the object appears to be different sizes in all four holograms and appears to be larger for increasing recording

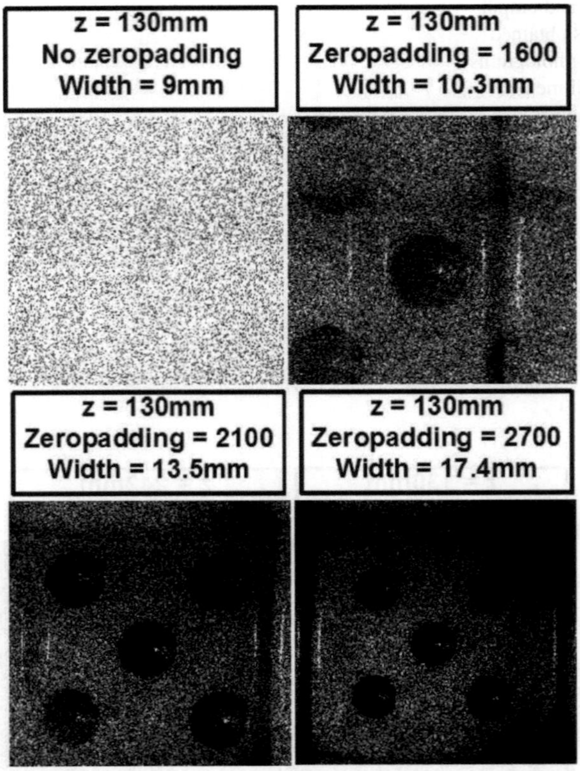

Fig. 15.6 Reconstructions from spectral method using varying amounts of zeropadding

distance. This is attributed to the presence of a slight spherical curvature on the reference beam which brings about an effective magnification in reconstruction. We now examine the impact of failing to zeropad the hologram sufficiently and we reconstruct the dice hologram (at distance $z = 130\,\text{mm}$) for varying amounts of zeropadding. Four different reconstructions are shown in Fig. 15.6. We can see that for no zeropadding the object is not reconstructed at all. The aliased copies begin to separate out as we increase the amount of zeropadding up to the point where we have full separation when the zeropadding is sufficient to bring the effective hologram size up to that of the object. We note that increasing the number of samples in this way has a detrimental effect on the time taken to reconstruct. For this reason, the direct method is often the method of choice if there is a better match between its output window width at the appropriate reconstruction distance and the physical size of the object. In the next section, we show how to adapt the spectral method so that a significantly reduced amount of zeropadding is required. In addition, we show how this adaptation allows us to reconstruct the object at any magnification.

15.3 Zooming Algorithm

15.3.1 Zooming Out

In this section, we develop an algorithm that allows us to effectively zoom such that the output window size is approximately equal the size of the object. This algorithm will allow us to arbitrarily change the output pixel size and therefore the output window size as we see fit. The basic principle is based on the following property of the Fresnel transform.

$$u_z(x/M) = \exp\left(j\frac{2\pi}{\lambda f_2}x'^2\right)F_{z_n}\left\{u(x')\exp\left(j\frac{2\pi}{\lambda f_1}x'^2\right)\right\}, \tag{15.8}$$

where M is a magnification parameter that we are free to choose and this imposes the conditions:

$$z_n = zM, \tag{15.9}$$

$$f_1 = z_n/\left(M-1\right), \tag{15.10}$$

$$f_2 = z_n - f_1. \tag{15.11}$$

Equation (15.8) implies that to obtain a magnified image of our reconstructed image, our digital hologram can be multiplied by a discrete chirp signal with parameter f_1 [defined using the desired magnification according to (15.10)], and then reconstructed using the spectral method where we propagate a distance z_n defined by (15.9) and finally we multiply by another discrete chirp function. For a discussion on the subject of discrete Fresnel propagation of chirped signals such as the equation above, the reader may consult the following reference [20]. The overall algorithm may be written as follows:

$$\begin{aligned}\text{recon}(mT'') = {} & \exp\left(j\frac{2\pi}{\lambda f_2}(mT'')^2\right)\\ & \times \text{SPECTRAL}\left\{u_z(nT)\exp\left(j\frac{2\pi}{\lambda f_1}(nT)^2\right), T, z_n, \lambda\right\}.\end{aligned} \tag{15.12}$$

In this case, m takes the range $-N/2 \le m \le N/2 - 1$ and the actual pixel pitch at the output is now

$$T'' = T/M. \tag{15.13}$$

The user can define T'' arbitrarily, which will in turn define values for M, z_n, f_1, and f_2. It is interesting to note that in the case of the direct method we cannot apply this idea successfully to zoom. The reason for this becomes clear when we consider that the actually propagated distance is given by $z_n = Mz$. We recall from earlier discussion that the direct method will in this case have an output window size given

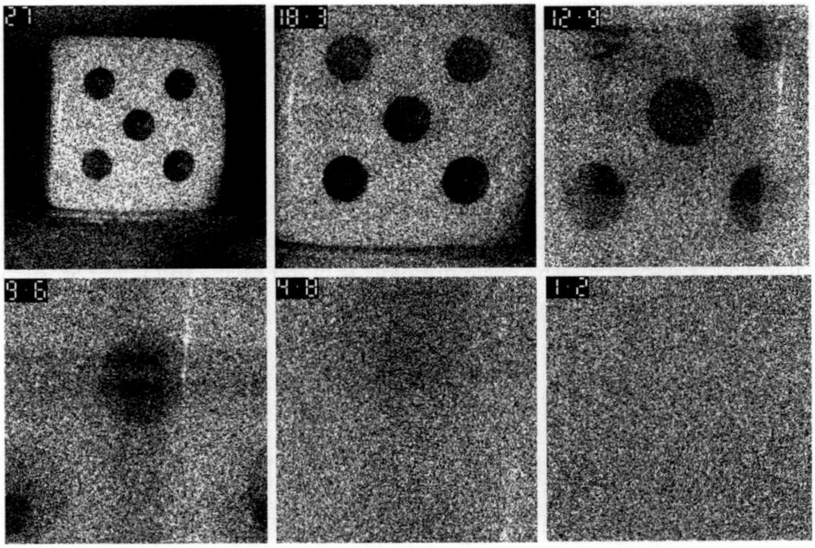

Fig. 15.7 Reconstructions from the zoom algorithm. The numbers on the figure indicate the width of the image in millimeter. Clearly the zoom algorithm fails if we choose our output window size to be less than the size of the object

by $\lambda z_n / T = \lambda z M / T$. Since we know the output window is scaled according to (15.13), we can scale this value by M to get the true output window size. Thus there is a cancelation of the magnification term and we find that the direct method produces identical results as if there were no zoom.

Our algorithm may be broken down into five numerical steps; three chirp multiplications separated by two FFT operations. We use the following notation to summarise the order of these five operations; $CFCFC$ where C and F represent a chirp multiplication and an FFT operation, respectively. The rightmost operator is the first one that we apply. Using the same notation we could describe the direct method and the spectral method as CFC and FCF, respectively.

In Fig. 15.7, we show some results of our zooming algorithm. All six parts of the figure show the reconstruction of the same hologram as before, but for different zooms. We have not employed any zeropadding. In the first two parts of the figure, we have chosen the output window size (visible in millimeters in the top left corner of each image) such that the full object is in view without any aliasing. We recall that when we investigated the spectral method we saw that this object could not be reconstructed without aliasing unless we applied considerable zeropadding and thereby significantly increased the computation time. Our zoom algorithm shows that zeropadding is no longer necessary. The remaining figures in Fig. 15.7 show what happens when we choose our output window to be smaller than the object size. From the second image, we can see that the object is approximately 18 mm in size in both the x and y dimensions. In the third part of the figure, we have shown the result of choosing an output window of 12.9 mm. We can clearly see the wrap around of the image. Choosing an output window of 9.6 mm results in a greater wrap around of the image. We can see in the fifth part of the figure, that the choice of output window of 4.8 mm results in more than one wrap around of the image and the quality of the image is further reduced. In the final part of the figure, the output window is 1.2 mm. In this case the image has wrapped around multiple times and now appears a uniform noise. All image detail has been lost. The aliasing becomes worse with smaller output window sizes. We can conclude that the algorithm works well when we zoom out to a sufficient window size but it fails when we zoom in to a window smaller that the object size. In Sect. 15.4, we design an adaptation to our algorithm that permits zooming in.

As of yet we have given no consideration to the sampling conditions that we must impose at each stage in our zoom algorithm. The first one is obvious from the discussion above; the magnification M must be chosen such that the output window is at least as large as the object itself. The second is that the signal $u_z(nT)\exp\left(j\frac{2\pi}{\lambda f_1}(nT)^2\right)$ is well sampled in the Nyquist sense. The reason for this is that we do not want aliasing to occur. If aliasing occurs, then multiple copies of the Fourier Transform will appear in the output of the DFT. Since we are multiplying the output of this DFT by a discrete chirp function we do not want this chirp function to be multiplying by more than one copy of the signal of interest. We want it to multiply by only one replica. We can state that these two conditions are a sufficient set of conditions for our algorithm to work successfully. Whether they constitute a necessary set of conditions or not remains a subject for debate. We give this subject further consideration in Sect. 15.5.4.

15.4 Adaptation of Algorithm to Permit Zooming In

It was previously discussed in Sect. 15.3.1 that choosing out magnification to produce an output window size less than the size of our object causes aliasing to occur. In this case, copies of our image wrap around and invade the output window. The smaller the output window size the more dramatic the effect and the image may wrap around many times to completely obscure the detail. In Fig. 15.7, we showed four examples of this. Clearly our zoom algorithm fails when we attempt to zoom in to see fine details in our image. We may conceptualize this effect as follows: Since every step in our zoom algorithm is unitary, that is, that energy is conserved through FFT and chirp multiplication, all of the energy in the hologram plane must end up in our reconstructed image regardless of our choice of magnification. Since the hologram contains information about every point in the image, then this information must appear in some way in our reconstructed image. We now propose a method that does allow for zooming in a series of steps. Step 1: The first step in the method is to apply our zoom algorithm choosing the output window size to be approximately equal to the object size. In other words, we zoom out before we zoom in. Step 2: The second step is to select a region of pixels at the center of our image, corresponding to the region we wish to magnify. Step 3: The third step is to

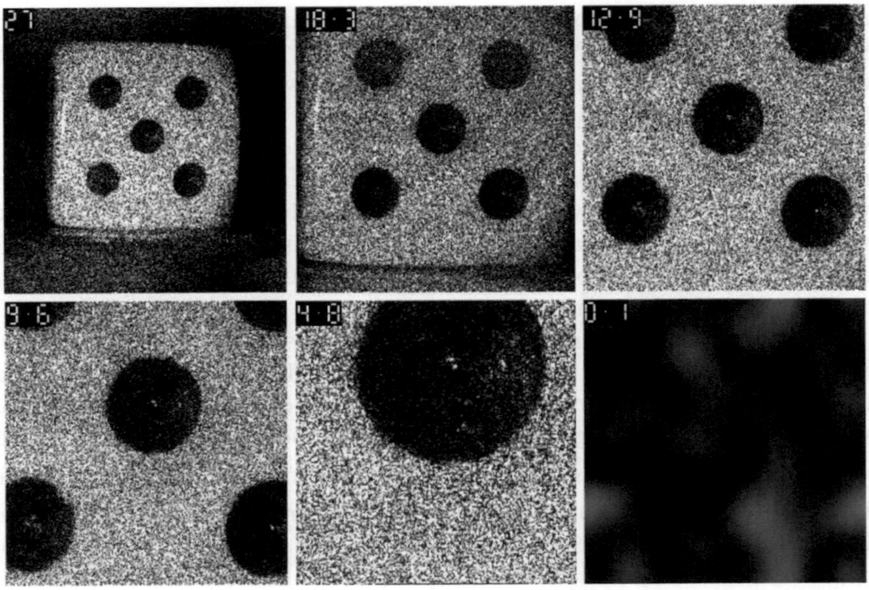

Fig. 15.8 Zooming in with the adapted zoom in technique. The numbers on the figure indicate the width of the image in millimeter

carry out an FFT algorithm on this small block of pixels. Step 4: The fourth step is to zeropad this DFT up to a number of samples that we would like. Step 5: The final step is an inverse FFT algorithm. In Fig. 15.8, we show the equivalent results for those shown in Fig. 15.7. In this case, we have selected the appropriate section of pixels and zeropadded up to the original object size in the Fourier domain. We can see a remarkable improvement. There is clearly no wrap around and no reduction in image quality. In this case, each part of the figure shows the center of the image having been magnified by a different amount. We are not limited to magnifying the center of our image. We can magnify any section by simply choosing that area of pixels that interests us and carrying out the same procedure as above. We show some results of scanning and zooming in Fig. 15.9.

15.5 Relationship of This Algorithm to Previous Work

15.5.1 The Rhodes Light Tube

We note that the relationship given in (15.8) was first used to create a zooming algorithm for the Fresnel transform, known as the "Rhodes Light Tube" in [21, 22], where it was shown that by multiplying by an appropriate chirp the light emanating from the discrete signal as shown by the green lines in Fig. 15.4 can be mapped

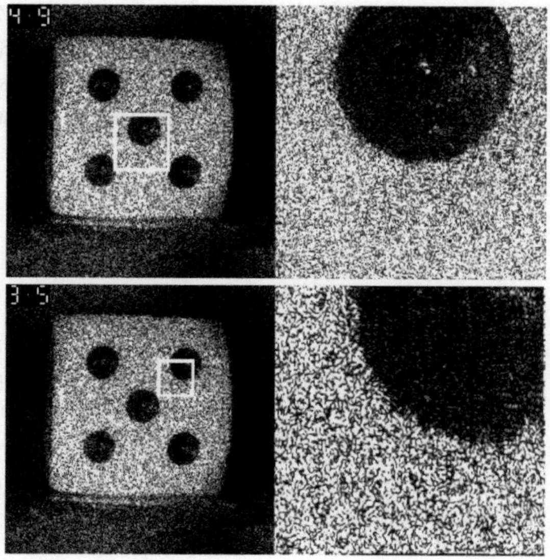

Fig. 15.9 Zooming in with the adapted low pass filtering technique implemented before applying the zoom algorithm. The numbers on the figure indicate the width of the image in millimeter

down to the blue lines, that is, to the actual output width of the spectral method for all distances. Using the same notation as earlier the Rhodes Light Tube method may be described as $CFCFC$, which is identical to the form of our zoom algorithm. The difference is that in our algorithm we account for an arbitrary choice of magnification for any distance, and so the chirp functions are different depending on our choice of magnification.

15.5.2 The Double Direct Method

It is also interesting to note the relevance of the other previous zoom algorithms in the literature [17, 18] to this chapter. In [17], a method was outlined to zoom based on two applications of the direct method applied in sequence. By choosing z_1 and z_2 appropriately, a desired magnification could be implemented. This becomes clear when we reconsider (15.4). By making use of the additive property of the Fresnel Transform, we can break it up into two Fresnel Transforms applied in sequence, where $z = z_1 + z_2$. In this case, there are two sampling periods to consider. The first is in the plane at distance z_1 and the second is at a further distance z_2. Respectively, they are

$$T'' = \frac{\lambda z_1}{NT}, \quad T''' = \frac{\lambda z_2}{NT''} = T\frac{z_2}{z_1}. \tag{15.14}$$

Therefore, a magnification is achieved of z_2/z_1. In [17], the authors state that the size of the output plane NT''' must be greater than the object size to avoid aliasing. However, they make no mention of aliasing happening in the intermediate plane and they impose no other conditions on the use of their algorithm. We note that using the same notation as earlier the double direct method may be described as $CFCCFC$. Since two chirp multiplications in sequence may be described as a single chirp multiplication, we can rewrite this as $CFCFC$. This is identical to the form of our zoom algorithm.

15.5.3 The Chirp Z Transform Method

In [18], a method was outlined to zoom based on the chirp z transform being applied to the Fourier part of the Fresnel kernel. The approach is based on taking (15.3) above (recall that this was the first equation in our derivation of the direct method but we had not yet set the output sampling interval T'' in order to create a DFT) and in the rightmost exponential function we set $nm = -\frac{1}{2}\left[(n-m)^2 - n^2 - m^2\right]$. This gives

$$\begin{aligned}\text{recon}(mT'') &= A_z \exp\left(\frac{j\pi m^2\left(T''^2 - TT''\right)}{\lambda z}\right) \\ &\quad \times \sum_{n=-N/2}^{N/2-1}\left[u_z(nT)\exp\left(\frac{j\pi n^2\left(T^2 - TT''\right)}{\lambda z}\right)\right] \\ &\quad \times \exp\left(\frac{j\pi\left(n-m\right)^2 TT''}{\lambda z}\right). \end{aligned} \tag{15.15}$$

This equation is recognized as a discrete convolution between two signals; $v(nT) = (u_z(nT)\exp(j\pi n^2(T^2 - TT'')/\lambda z))$ and $\exp(j\pi nT^2T''/\lambda zT)$. This can be solved by calculating the DFTs of both discrete functions, multiplying them and applying an inverse DFT. However, a more efficient method is found by observing that the Fourier Transform of the function $\exp(j\pi x^2T''/\lambda zT)$ can be solved analytically to be $\exp(j\pi\lambda zT{f_x}^2/T'')$, where we have omitted a constant factor. In this case, we carry out the DFT of $v(nT)$ and then multiply it by the discrete function $\exp(j\pi\lambda z(m/N)^2/TT'')$. Clearly, the value of T'' can now be chosen arbitrarily in this algorithm. Furthermore, by zeropadding N up a new value M, the size of the output window may be controlled in another manner. The authors in [18] state that the signal $\exp(j\pi\lambda z(m/L)^2/TT'')$ must be well sampled in the Nyquist sense and so L can be chosen such that $\lambda z/TT''L^2 \ll 1$. The authors further state that given that a convolution is now performed between two discrete functions, one of size N and the other of size L, the value of the zeropadded function, $M = N + L$. Indeed, the authors make one further assumption in their paper; they state that the chirped function $v(n)$ must be chosen to have N samples such that $v(n)$ is well sampled in the Nyquist sense. In their paper, the authors were free to choose the input sampling period freely as they were dealing with simulated signals. Using the same notation as earlier, this method may be described as $CFCFC$, which is identical to the form of our zoom algorithm.

15.5.4 The Correct Conditions of Use

During the course of our study, we observed that from an algorithmic point of view, the two algorithms presented in [17, 18] and the one presented in this paper based on the spectral method are identical. Not only do all three algorithms share the same form, $CFCFC$, in terms of the operations involved, but the actual numerical values corresponding to the C operations are identical for the same choice of magnification M. All three algorithms appear at first glance to be completely different because they are all derived using fundamentally different approaches and each of these approaches has its own considerations on the conditions of use of the algorithms. Nevertheless, all three algorithms are in fact identical and there can only therefore exist one legitimate set of conditions of use. The correct set of necessary conditions remain as of yet unclear and require further study, though we have defined a set of sufficient conditions above. The difficulty in uncovering the necessary conditions lies in the fact that a triple sampling has occurred: first there is sampling in the space domain, then in the Fourier and then again in the space domain. In addition, there are chirp multiplications in each domain. We believe that the true conditions may be revealed when we apply the rules of generalized sampling such as that given in [23, Chap. 10]. We believe that such a study will reveal for the first time the impulse response from double FFT methods such as the spectral method and our zoom algorithm.

15.5.5 Other Zoom Algorithms

We note that there have been other papers recently written on the subject of zooming in digital holography [24–26]. We now briefly discuss this work in the context of our algorithm. In [24], the authors present a straightforward method to obtain more samples within in a reconstruction obtained using the direct method. They simply zeropad the input and then implement the direct method. Recalling the relationship between the input and output sampling intervals for the direct method $T'' = \lambda z/NT$, we can see that by introducing zeros we will increase N and effectively reduce T'' to get a finer sampling in the image plane. However, the output window size NT'' remains the same for that value of z. Thus one cannot zoom out and zooming in requires magnifying the entire image up to a higher resolution. This is extremely time intensive compared with our algorithm. In [25], the authors propose a method of first applying the angular spectrum method (this is the nonparaxial version of the spectral method and also requires two FFTs) and then applying a form of the direct method onto a tilted plane. The disadvantage of this method when compared to our own is that it requires more FFTs, does not include a zoom in mechanism and it is often no desirable to recover a signal on a tilted plane. In the final paper [26], the authors perform an algorithm very similar to our own except the initial chirp is in the form of a nonparaxial Rayleigh–Sommerfeld (RS) spherical wave and the spectral method is replaced by the nonparaxial angular spectral method. They make similar decisions on the conditions of use that we do in order to avoid aliasing in the DFT domain before multiplying by the chirp like term associated with the RS transfer function. It would appear that this algorithm is a nonparaxial version of that derived in this chapter. They did not consider zooming in.

15.6 Conclusion

In this chapter, we have reviewed the most common algorithms used to reconstruct digital holograms, namely the direct and spectral methods of simulating the Fresnel transform. We have discussed the limitations of these two methods in terms of their

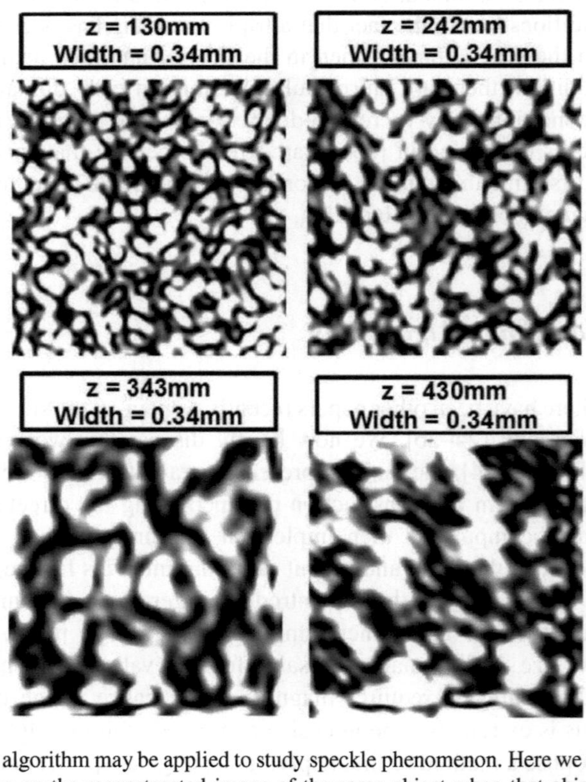

Fig. 15.10 The algorithm may be applied to study speckle phenomenon. Here we show the speckle pattern appearing on the reconstructed image of the same object when that object was placed at four different distances from the camera

output window widths. For the direct method, this output window expands linearly with z and for the spectral method it remains at a constant size equal to the size of the camera that recorded the hologram. We have derived a zoom algorithm based on an adaptation of the spectral method, where the output window size may be defined arbitrarily by the user and we have shown how this algorithm performs for an experimentally recorded digital hologram. The algorithm works well when we choose the output window size to be greater than the size of the object. It fails, however, when we try to zoom into windows smaller than the object size. We proceeded to develop an adaptation of our zoom algorithm that overcomes this limitation. The first step in this adapted algorithm is to zoom out sufficiently to see the entire object and then to employ an FFT based interpolation technique on the region of the image that we wish to magnify. We defined two sufficient conditions of use for our algorithm. The first was based on choosing a sufficiently large output window and the second to avoid aliasing in the DFT domain, where we must apply a chirp multiplication. We discovered in the process of our investigation that our algorithm is exactly the same as two previously derived algorithms in the literature. The first uses a double direct method approach and the second using a chirp z transform. These previous papers had used fundamentally different approaches but had ended up with identical algorithms as our own. Each had derived different conditions of use. We discussed these conditions of use and pointed out that more work must be done in understanding double and triple sampling in order to fully understand the effect of these double FFT type algorithms. We believe that our algorithm has widespread application in microscopy as well as the study of speckle phenomenon. To demonstrate this point, in Fig. 15.10 we show the speckle pattern obtained in the image of our dice, when it was recorded at four different distances. The zoomed in images show different size speckle in each case.

Acknowledgement This research is funded from the European Community's Seventh Framework Programme FP7/2007–2013 under grant agreement 216105 "Real 3D."

References

1. D. Gabor (1948), A new microscope principle, Nature, 161:777–778
2. E. N. Leith, J. Upatnieks (1962), Reconstructed wavefronts and communication theory, J. Opt. Soc. Am., 52:1123
3. E. N. Leith, J. Upatnieks (1963), Wavefront reconstruction with continuous-tone objects, J. Opt. Soc. Am., 53:1377
4. J. W. Goodman, R. Lawrence (1967), Digital image formation from electronically detected holograms, Appl. Phys. Lett., 11:77–79
5. L. P. Yaroslavskii, N. S. Merzlyakov (1980), Methods of Digital Holography, Consultants Bureau, New York
6. T. M. Kreis, M. Adams, W. P. O. Juptner (1997), Methods of digital holography: a comparison, Proc. SPIE, 3098:224–233
7. U. Schnars, W. P. O. Juptner (2002), Digital recording and numerical reconstruction of holograms, Meas. Sci. Technol., 13:85–101

8. D. P. Kelly, B. M. Hennelly, W. T. Rhodes, J. T. Sheridan (2006), Analytical and numerical analysis of linear optical systems, Opt. Eng., 45:008201
9. B. M. Hennelly, J. T. Sheridan (2005), Generalizing, optimizing, and inventing numerical algorithms for the fractional Fourier, Fresnel, and linear canonical transforms, J. Opt. Soc. Am. A, 22:917–927
10. B. M. Hennelly, J. T. Sheridan (2005), Fast algorithm for the linear canonical transform, J. Opt. Soc. Am. A, 22:928–937
11. D. Mendlovic, Z. Zalevsky, N. Konforti, (1997) Computation considerations and fast algorithms for calculating the diffraction integral, J. Mod. Opt., 44:407–414
12. D. Mas, J. Garca, C. Ferreira, L. M. Bernardo, F. Marinho (1999), Fast algorithms for free-space diffraction patters calculation, Opt. Commun., 164:233–245
13. D. Mas, J. Prez, C. Hernndez, C. Vzquez, J. J Miret, C. Illueca (2003), Fast numerical calculation of Fresnel patterns in convergent systems, Opt. Commun., 227:245–258
14. E. Cuche, P. Marquet, C. Depeursinge (1999), Simultaneous amplitude-contrast and quantitative phase-contrast microscopy by numerical reconstruction of Fresnel off-axis holograms, Appl. Opt., 38:6994–7001
15. P. Marquet, B. Rappaz, P. J. Magistretti, E. Cuche, Y. Emery, T. Colomb, C. Depeursinge (2005), Digital holographic microscopy: a noninvasive contrast imaging technique allowing quantitative visualization of living cells with subwavelength axial accuracy, Opt. Lett., 30: 468–470
16. J. W. Cooley, J. W. Tukey (1965), An algorithm for the machine calculation of complex Fourier series, Math. Comput., 19:297–301
17. X. Deng, B. Bihari, J. Gan, F. Zhao, R. T. Chen (2000), Fast algorithm for chirp transforms with zooming-in ability and its applications, J. Opt. Soc. Am. A, 17:762–771
18. F. Zhang, I. Yamaguchi, L. P. Yaroslavsky (2004), Algorithm for reconstruction of digital holograms with adjustable magnification, Opt. Lett., 29:1668–1670
19. I. Yamaguchi, T. Zhang (1997), Phase-shifting digital holography, Opt. Lett., 22:1268–1270
20. D. P. Kelly, B. M. Hennelly, J. T. Sheridan, W. T. Rhodes (2006), Finite-aperture effects for Fourier transform systems with convergent illumination, Part II, Opt. Commun. 263:180–188
21. W. T. Rhodes (2002), Light tubes, Wigner diagrams and optical signal propagation simulation, in Optical Information Processing: A Tribute to Adolf Lohmann, H. J. Caulfield, ed., SPIE Press, Bellingham, WA, pp. 343–356
22. W. T. Rhodes (2001), Numerical simulation of Fresnel-regime wave propagation: the light tube model, in Wave-Optical Systems Engineering, F. Wyrowski, ed., Proc. SPIE 4436:21–26
23. M. Testorf, B. Hennelly, J. Ojeda-Castaneda (2009), Phase-Space Optics, McGrawHill, New York
24. P. Ferraro, S. De Nicola, G. Coppola, A. Finizio, D. Alfieri, G. Pierattini (2004), Controlling image size as a function of distance and wavelength in Fresnel-transform reconstruction of digital holograms, Opt. Lett., 29:854–856
25. L. Yu, M. K. Kim (2006), Pixel resolution control in numerical reconstruction of digital holography, Opt. Lett., 31:897–899
26. J. Li, P. Tankam, Z. Peng, P. Picart (2009), Digital holographic reconstruction of large objects using a convolution approach and adjustable magnification, Opt. Lett., 34:572–574

Part III
Polarimetric Imaging

Chapter 16
Partial Polarization of Optical Beams: Temporal and Spectral Descriptions

Tero Setälä, Ferdinando Nunziata, and Ari T. Friberg

Abstract A fluctuating, statistically stationary, polychromatic electromagnetic field can, at any instant of time, be regarded as fully polarized but on time average or in the frequency space its state of polarization may be quite different. In this chapter, we analyze the variations between the degrees of polarization of optical beams in time and frequency domains. We point out the origins of the differences and illustrate the physical consequences by several examples.

16.1 Introduction

The temporal coherence properties of fluctuating, statistically stationary, polychromatic optical fields generally differ from the coherence properties characterizing the various frequency components of the random field. For scalar waves, an explicit relationship between the complex degrees of coherence in the space–time and space–frequency domains exists, indicating that the two quantities are connected but have, in general, different values. Further, a polychromatic field considered at a single frequency is often confused with a monochromatic wave that, by definition, is fully deterministic and therefore completely coherent and polarized. Under usual conditions, a randomly fluctuating polychromatic electromagnetic field can, at any instant of time, be regarded as fully polarized, represented by an instantaneous polarization state. The temporal variation of this state then leads conveniently to the notion of "polarization time", which characterizes the period over which the polarization state on average remains essentially unchanged. The fluctuations of the polarization state can be analyzed, in a geometric picture, on the Poincaré sphere [1] or, more physically, using directly the traditional Jones vectors [2]. In a stationary field, the time average of the instantaneous polarization state then leads to the usual (time-domain) degree of polarization [3], which in

T. Setälä (✉)
Department of Applied Physics, Aalto University, P. O. Box 13500, FI-00076 Aalto, Finland
e-mail: tero.setala@tkk.fi

T. Fournel and B. Javidi (eds.), *Information Optics and Photonics: Algorithms, Systems, and Applications*, DOI 10.1007/978-1-4419-7380-1_16,

general differs from unity. Likewise, a random optical field evaluated at a single temporal frequency is normally not fully polarized. In this chapter, we analyze the differences between the degrees of partial polarization of electromagnetic beam-like wave fields on time average and in the frequency domain. We explore the origin of the variations in these quantities and illustrate the differences and physical consequences by several practical examples.

16.2 Partial Coherence in Temporal and Spectral Domains

Let us begin by considering the partial coherence properties of random optical fields within the scalar-wave theory. Traditionally, the complex degree of coherence of such an optical field in the space–time domain is defined as [3, 4]

$$\gamma_{12}(\tau) = \frac{\Gamma_{12}(\tau)}{[\Gamma_{11}(0)\Gamma_{22}(0)]^{1/2}}, \tag{16.1}$$

where

$$\Gamma_{12}(\tau) = \langle V^*(\mathbf{r}_1, t)V(\mathbf{r}_2, t+\tau)\rangle \tag{16.2}$$

is the mutual coherence function, $V(\mathbf{r}, t)$ is the complex analytic signal of the field, $\mathbf{r}_1$ and $\mathbf{r}_2$ are position vectors of two points, and t is time. The angle brackets can denote either time or ensemble average, assuming the field is statistically stationary and ergodic. The quantity $\gamma_{12}(\tau)$ is normalized in magnitude between 0 and 1; the lower limit indicates that field $V(\mathbf{r}, t)$ is completely uncorrelated and the upper limit that it is fully correlated at points $\mathbf{r}_1$ and $\mathbf{r}_2$.

In many instances, such as in interactions with material media, it is often more convenient to work in the space–frequency domain. Use of the generalized Wiener-Khintchine theorem [3, 4]

$$W_{12}(\omega) = \frac{1}{2\pi}\int_{-\infty}^{\infty} \Gamma_{12}(\tau)e^{i\omega\tau}d\tau \tag{16.3}$$

yields the cross-spectral density function $W_{12}(\omega)$ at $\mathbf{r}_1$ and $\mathbf{r}_2$, at frequency ω. Normalization of this function according to [5]

$$\mu_{12}(\omega) = \frac{W_{12}(\omega)}{[W_{11}(\omega)W_{22}(\omega)]^{1/2}} \tag{16.4}$$

leads to the complex degree of spectral coherence, which too is bounded in absolute value as $0 \leq |\mu_{12}(\omega)| \leq 1$. The limits 0 and 1 correspond to full incoherence and complete coherence, respectively, at $\mathbf{r}_1$ and $\mathbf{r}_2$ but now at frequency ω of the fluctuating polychromatic field.

While $\gamma_{12}(\tau)$ and $\mu_{12}(\omega)$ characterize different properties of the field, they are nonetheless connected. Indeed, in view of (16.3), one readily finds that [6]

$$\gamma_{12}(\tau) = \int_0^\infty [w_1(\omega)w_2(\omega)]^{1/2}\,\mu_{12}(\omega)\mathrm{e}^{-\mathrm{i}\omega t}\mathrm{d}\omega, \tag{16.5}$$

where

$$w_j(\omega) = \frac{W_{jj}(\omega)}{\int_0^\infty W_{jj}(\omega)\mathrm{d}\omega}, \quad (j = 1, 2), \tag{16.6}$$

are the normalized spectra of the field at the two points. The ω integration in (16.5) and (16.6) starts from 0 due to the complex analytic signal representation.

It readily follows from (16.5) that $\gamma_{12}(\tau)$ and $\mu_{12}(\omega)$, though both are measures of coherence of the field at $\mathbf{r}_1$ and $\mathbf{r}_2$, are in general not the same, nor are they Fourier transforms of each other. This shows that coherence in the time domain is different from coherence in the frequency domain. It is also important to note that whereas a strictly monochromatic field is completely coherent both in time and in frequency [7], a random polychromatic field considered at some given frequency need not be fully coherent [4, 5]. For instance, when narrow-band filters are placed at the two pinholes in a Young's interferometer and the pass bands are made increasingly thin, the visibility of the interference fringes (in long-time integration) does not become unity, but instead approaches the spectral degree of coherence at the filters' center frequency, as has been demonstrated theoretically [8] and experimentally [9, 10].

16.3 Partial Polarization of Electromagnetic Beams

To account for polarization, we consider a randomly fluctuating, statistically stationary electromagnetic wave propagating primarily in the positive z direction. The assumed beam-like nature implies that the z component of the electric field is negligible, thereby allowing a two-component representation for the electric-field vector. Hence the realization of the electric vector of such a stochastic field, at an instant of time t and at a point $\mathbf{r}$, is denoted by a zero-mean complex analytic signal $\mathbf{E}(\mathbf{r}, t) = [E_x(\mathbf{r}, t), E_y(\mathbf{r}, t)]^{\mathrm{T}}$, where the superscript T denotes the transpose. We suppose further that the beam is uniformly polarized, so that the polarization features are independent of the position $\mathbf{r}$ within the beam. Therefore, from now on, the argument $\mathbf{r}$ will be suppressed and we analyze the partial polarization properties of the optical beam at some arbitrary point separately in the time and the frequency domains [11–14].

16.3.1 Degree of Polarization in Time Domain

The time-domain polarization characteristics of the field are contained in the 2×2 equal-time coherence (or polarization) matrix of the form [3, 4]

$$\mathbf{J} = \begin{pmatrix} J_{xx} & J_{xy} \\ J_{yx} & J_{yy} \end{pmatrix}, \tag{16.7}$$

where

$$J_{ij} = \langle E_i^*(t)E_j(t)\rangle, \quad (i,j) = (x,y), \tag{16.8}$$

are the intensities and mutual intensities of the electric field components $E_x(t)$ and $E_y(t)$. The angle brackets may denote time average over a single field realization or statistical average over an ensemble of realizations, since the field is taken ergodic, and the asterisk stands for complex conjugation. The matrix in (16.7) is Hermitian, $\mathbf{J}^\dagger = \mathbf{J}$, and it is also non-negative definite, i.e., $\mathbf{a}^\dagger \mathbf{J}\,\mathbf{a} \geq 0$ for any two-component complex vector $\mathbf{a}$ (the dagger denotes the adjoint). From this it follows, for instance, that the determinant $\det \mathbf{J} \geq 0$. We note that the diagonal elements of $\mathbf{J}$ are non-negative, since they are intensities, while no such restriction applies to the off-diagonal elements.

The degree of polarization characterizes the state of partial polarization of the beam. Conceptually, it is obtained by first dividing the field, unambiguously, into its fully polarized and fully unpolarized parts and then finding the ratio of the intensity of the polarized component to that of the whole beam. The degree of polarization can be expressed in many different, but mathematically identical, forms. One of the most frequently encountered formulas is obtained from the equal-time coherence matrix in (16.7) as [4]

$$P_\mathrm{t} = \left[1 - \frac{4\det \mathbf{J}}{\mathrm{tr}^2\mathbf{J}}\right]^{1/2}, \tag{16.9}$$

where tr denotes the trace and the subscript t refers to the time domain. The value of the degree of polarization is bounded within the interval $0 \leq P_\mathrm{t} \leq 1$, with the lower and upper limits corresponding to completely unpolarized and fully polarized beams, respectively. From (16.9), we observe at once that if $\det \mathbf{J} = 0$, then $P_\mathrm{t} = 1$, whereas for $\det \mathbf{J} > 0$ the beam is only partially polarized and $P_\mathrm{t} < 1$. For stationary beams, the degree of polarization P_t is constant, just one number.

16.3.2 Mutual Coherence Matrix

The equal-time coherence matrix $\mathbf{J}$ in (16.7) does not contain any information about the temporal coherence properties of the electric field. Consequently, the time-domain degree of polarization P_t also is independent of the beam's spectral structure. To include the coherence and spectral features and to eventually assess the spectral polarization properties of the beam, we employ the 2×2 electric mutual coherence matrix defined as [4, 15]

$$\mathbf{G}(\tau) = \begin{pmatrix} G_{xx}(\tau) & G_{xy}(\tau) \\ G_{yx}(\tau) & G_{yy}(\tau) \end{pmatrix}, \tag{16.10}$$

where

$$G_{ij}(\tau) = \langle E_i^*(t)E_j(t+\tau)\rangle, \quad (i,j) = (x,y). \tag{16.11}$$

Clearly $\mathbf{G}(0) = \mathbf{J}$ and hence $\mathbf{G}(\tau)$ is more general than $\mathbf{J}$. The matrix $\mathbf{G}(\tau)$ is Hermitian in the sense that $\mathbf{G}^{\dagger}(-\tau) = \mathbf{G}(\tau)$, and additionally it satisfies various non-negative definiteness conditions [4]. From the Schwartz inequality, it follows at once that $|G_{xy}(\tau)|^2 \leq G_{xx}(0)G_{yy}(0)$. We note also that the diagonal elements of $\mathbf{G}(\tau)$ are auto-correlation functions, which are subject to stringent mathematical properties that we use later, while the off-diagonal elements are cross-correlation functions with much weaker requirements.

16.3.3 Degree of Polarization in Frequency Domain

We are now in a position to introduce the spectral polarization characteristics of the random beam. To this end, we make use of a matrix version (at a single point) of the generalized Wiener–Khintchine theorem in (16.3), i.e.,

$$\mathbf{M}(\omega) = \frac{1}{2\pi} \int_{-\infty}^{\infty} \mathbf{G}(\tau) \mathrm{e}^{\mathrm{i}\omega\tau} \mathrm{d}\tau. \tag{16.12}$$

As the equal-time coherence matrix $\mathbf{J}$, the spectral coherence matrix $\mathbf{M}(\omega)$ is Hermitian and non-negative definite. In addition, its diagonal elements are non-negative since they represent spectral densities. In terms of this frequency-domain coherence matrix, we obtain the spectral degree of polarization as [3, 13, 16]

$$P_{\mathrm{f}}(\omega) = \left[1 - \frac{4 \det \mathbf{M}(\omega)}{\mathrm{tr}^2\, \mathbf{M}(\omega)}\right]^{1/2}, \tag{16.13}$$

where the subscript f emphasizes that $P_{\mathrm{f}}(\omega)$ is defined in the frequency domain. Like the time-domain quantity P_{t}, the value of the spectral degree of polarization is bounded as $0 \leq P_{\mathrm{f}}(\omega) \leq 1$, with the lower and upper limit indicating complete unpolarization and full polarization at frequency ω. Since $\mathbf{M}(\omega)$ is a function of ω, it is evident that $P_{\mathrm{f}}(\omega)$ can take on different values at different frequencies.

On inverting (16.12) and setting $\tau = 0$ we readily find that

$$\mathbf{G}(0) = \mathbf{J} = \int_{0}^{\infty} \mathbf{M}(\omega) \mathrm{d}\omega. \tag{16.14}$$

This expression, together with (16.9) and (16.13), indicates at once that the time-domain degree of polarization, P_{t}, and the spectral degree of polarization, $P_{\mathrm{f}}(\omega)$, may in general have quite different values. This is one of the main results of this chapter. But unlike for the temporal and spectral degrees of coherence of scalar waves [see (16.5)], there does not appear to exist an explicit relationship between the degrees of polarization in the space–time and space–frequency domains for statistically stationary electromagnetic beams. Physically we observe, however, in view of (16.12), that the spectral degree of polarization accounts for the temporal coherence properties of the field, while as evidenced by (16.7), the time-domain degree of polarization is independent of the beam's temporal coherence.

16.4 Examples

To learn more about the relationships between the time-domain and frequency-domain degrees of polarization, as well as of their practical implications, we next consider several illustrative examples.

16.4.1 Quasi-monochromatic Beams

For a quasi-monochromatic electromagnetic field with a central frequency at ω_0, the electric mutual coherence matrix assumes, in an approximation [3], the form $\mathbf{G}(\tau) = \mathbf{J}e^{-i\omega_0\tau}$, from which it follows that $\mathbf{M}(\omega) = \mathbf{J}\delta(\omega - \omega_0)$.[1] This result shows, in effect, that $P_f(\omega_0) = P_t$; in other words, the spectral degree of polarization at the center frequency of a quasi-monochromatic electromagnetic beam is equal to the time-domain degree of polarization. This conclusion is somewhat analogous to the one that follows from (16.5) between the degrees of coherence in the time and frequency domains for a quasi-monochromatic scalar field. Since the equal-time coherence matrix $\mathbf{J}$ can be arbitrary, the degrees of polarization P_t and $P_f(\omega_0)$ may assume any (equal) values between 0 and 1.

16.4.2 Beams with Arbitrary Uniform Coherence

A generalization of the quasi-monochromatic beam is an electromagnetic wave, for which the temporal behavior of all elements in the electric mutual coherence matrix is arbitrary but the same, i.e.,

$$\mathbf{G}(\tau) = \mathbf{J}g(\tau), \tag{16.15}$$

where $g(\tau)$ is a (normalized) auto-correlation function with the properties $g(0) = 1$ and $g(-\tau) = g^*(\tau)$. For lack of a better term, we call such fields here "uniform coherence" beams. Making use of (16.12), we find that

$$\mathbf{M}(\omega) = \mathbf{J}\tilde{g}(\omega), \tag{16.16}$$

[1] In a more rigorous analysis, the strict exponential coherence should be replaced by a narrow band of exponential functions, leading to sharp but finite spectral coherence distribution in place of the Dirac delta function.

where the tilde denotes the Fourier transform. In view of Bochner's theorem, we have $\tilde{g}(\omega) \geq 0$. Equation (16.16) implies that $P_\mathrm{f}(\omega) = P_\mathrm{t}$ for all ω. Thus, for a field represented by (16.15), the spectral degree of polarization is the same at all frequencies and, moreover, equal to the time-domain degree of polarization. This result is independent of the form of the function $g(\tau)$ specifying the temporal coherence.

Let us consider a beam field that is completely polarized in the time domain. A light beam is fully polarized if, and only if, its orthogonal electric-field components are proportional to the same random process [17], i.e., $E_y(t) = cE_x(t)$, where c is a constant. Computation of the mutual coherence matrix shows that it is of the form of (16.15), with $g(\tau) = \langle E_x^*(t)E_x(t+\tau)\rangle / \langle |E_x(t)|^2\rangle$ and $\det \mathbf{J} = 0$. Hence, a field which is fully polarized in the time domain is necessarily fully polarized also at every frequency; $P_\mathrm{t} = P_\mathrm{f}(\omega) = 1$.

However, (16.15) is not restricted to fully polarized fields. For example, if we allow $E_y(t) = cE_x(t)$, where c now is a random variable (independent of field), we find that $\mathbf{G}(\tau)$ is of the form of (16.15) with $g(\tau)$ given as above but $\det \mathbf{J} = \langle |c|^2\rangle - \langle c\rangle\langle c^*\rangle \geq 0$. As a typical example, we may choose Gaussian functions, $G_{ij}(\tau) = J_{ij}\exp(-\tau^2/2\tau_\mathrm{c}^2)$, where $(i,j) = (x,y)$ and τ_c is the "coherence time". Comparisons with the random variable c indicate that in this case $J_{xx} = 1$, $J_{xy} = \langle c\rangle$, $J_{yx} = \langle c^*\rangle$, and $J_{yy} = \langle |c|^2\rangle$, in full agreement with the conditions discussed in Sect. 16.3.2. The ensuing degree of polarization $P_\mathrm{t} = P_\mathrm{f}(\omega)$ is plotted in Fig. 16.1 for three values of $a = J_{yy}/J_{xx}$ as a function of $|J_{xy}|/J_{xx}$, which may range from 0 to $\sqrt{a}$. We see that if the intensities of the x and y polarizations are equal ($a = 1$), the degree of polarization increases linearly from 0 to 1. However, if the intensities are not equal, for instance $a = 0.5$ or $a = 1.5$, the beam is partially polarized already when $|J_{xy}| = 0$ and the degree of polarization grows monotonously to 1 as $|J_{xy}|/J_{xx}$ increases towards its respective maximum value.

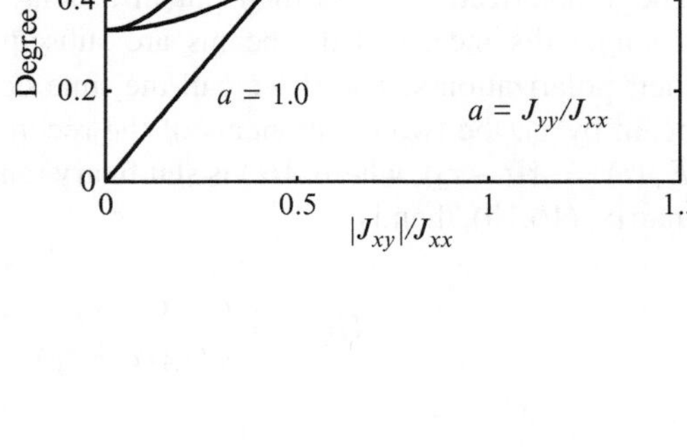

Fig. 16.1 Degree of polarization $P_\mathrm{t} = P_\mathrm{f}(\omega)$ for "uniform coherence" electromagnetic beams of the type of (16.15). The plots are for three values of the ratio of the intensities of the y and x components, J_{yy}/J_{xx}, as a function of the x-intensity normalized correlation coefficient, $|J_{xy}|/J_{xx}$

16.4.3 Superposition of Uncorrelated Orthogonal Beams

As a third example, we consider a field that consists of a superposition of two orthogonal, linearly polarized beams (electric fields along the x and y axes) which are mutually uncorrelated. The beams have intensities J_x and J_y and their temporal coherence properties are given by the normalized autocorrelation functions $g_x(\tau)$ and $g_y(\tau)$, respectively. The mutual coherence matrix for this combined wave then is

$$\mathbf{G}(\tau) = \begin{pmatrix} J_x g_x(\tau) & 0 \\ 0 & J_y g_y(\tau) \end{pmatrix}. \tag{16.17}$$

Setting $\tau = 0$ in this equation, we readily find that the temporal degree of polarization of this field is $P_\mathrm{t} = |J_x - J_y|/(J_x + J_y)$. The beam is unpolarized in the time domain if the intensities J_x and J_y are equal. On the other hand, making use of (16.12) we find for the spectral coherence matrix

$$\mathbf{M}(\omega) = \begin{pmatrix} J_x \tilde{g}_x(\omega) & 0 \\ 0 & J_y \tilde{g}_y(\omega) \end{pmatrix}, \tag{16.18}$$

from which it follows at once that the spectral degree of polarization is $P_\mathrm{f}(\omega) = |J_x \tilde{g}_x(\omega) - J_y \tilde{g}_y(\omega)|/[J_x \tilde{g}_x(\omega) + J_y \tilde{g}_y(\omega)]$. This shows that even though the field may be fully unpolarized in the time domain ($P_\mathrm{t} = 0$, if $J_x = J_y$), it as a rule is partially polarized in the frequency domain, $0 \leq P_\mathrm{f}(\omega) \leq 1$. In general, the field is unpolarized at those frequencies for which $J_x \tilde{g}_x(\omega) = J_y \tilde{g}_y(\omega)$, and fully polarized at frequencies for which either $\tilde{g}_x(\omega)$ or $\tilde{g}_y(\omega)$ is zero. This occurs, for instance, when the spectral distributions $\tilde{g}_x(\omega)$ and $\tilde{g}_y(\omega)$ have no overlap.

16.4.4 Delayed Orthogonal Beams

As a final illustration of the polarization properties, we consider a beam which is a superposition of two fields: one that is linearly polarized in the x direction and its delayed replica polarized in the y direction. Such a field can be constructed by starting from a usual 45° linearly polarized laser beam and dividing it into two equal but orthogonally polarized parts by use of a polarizing beam splitter [2]. The y-polarized beam is then put in a delay loop, i.e., it is made to propagate a longer distance, and the beams are subsequently recombined without altering their polarization states. If we call the time delay experienced by the y-polarized beam by τ_d, the two components of the recombined beam are $E_x(t) = A(t)$ and $E_y(t) = A(t - \tau_\mathrm{d})$, where $A(t)$ is stationary random process. The mutual coherence matrix, (16.10), then is

$$\mathbf{G}(\tau) = \begin{pmatrix} G_A(\tau) & G_A(\tau - \tau_\mathrm{d}) \\ G_A(\tau + \tau_\mathrm{d}) & G_A(\tau) \end{pmatrix}, \tag{16.19}$$

where $G_A(\tau) = \langle A^*(t)A(t+\tau)\rangle$. Setting $\tau = 0$, we obtain the equal-time coherence matrix, $\mathbf{J}$, from which the degree of polarization in the time domain is readily computed to be $P_t = |\gamma_A(\tau_d)|$, where $\gamma_A(\tau) = G_A(\tau)/G_A(0)$ is the correlation coefficient of the random process $A(t)$. Several interesting observations are made. If the time delay, τ_d, is much longer than the coherence time of the process $A(t)$, the combined beam is completely unpolarized, $P_t = 0$. This technique has been employed to produce unpolarized laser beams [18]. On the other hand, when $\tau_d = 0$, the field is naturally fully polarized, $P_t = 1$. This means that by adjusting the time delay τ_d, we may construct, at will, a beam whose degree of polarization in the time domain is anything between 0 and 1.

However, whatever is the value of P_t, the combined field is fully polarized at all frequencies ω. This can be seen by first using (16.12) to obtain

$$\mathbf{M}(\omega) = \begin{pmatrix} 1 & e^{i\omega\tau_d} \\ e^{-i\omega\tau_d} & 1 \end{pmatrix} \tilde{G}_A(\omega), \tag{16.20}$$

and then (16.13) to deduce that $P_f(\omega) = 1$ independently of ω. This result can be understood since the x and y components of the superposition beam are the same random process, even though delayed, and therefore, apart from a phase factor, their Fourier transformations are the same.

16.5 Conclusions

Inspired by the developments in scalar coherence theory, we have considered randomly fluctuating but statistically stationary electromagnetic beams, in particular their polarization properties. The object of interest was the degree of polarization, which is a traditional quantity in the time domain [19] but which has only quite recently begun to be studied in the frequency domain. The time-domain degree of polarization for stationary beams is just one number, whereas in the spectral domain one has a degree of polarization at every frequency. One of the principal results of this chapter is that we bring into evidence the differences between the time-domain and spectral-domain degrees of polarization. We point out that unlike the time-domain degree of polarization P_t, the spectral degree of polarization $P_f(\omega)$ depends on the temporal coherence properties of the field (both the x and y components, as well as their correlation). Consequently, the degrees of polarization in the two domains assume, in general, different values, as do the spectral degrees of polarization at different frequencies.

To demonstrate the polarization properties and their differences in the time and spectral domains, we examined several specific examples of electromagnetic beams. The results are summarized in Table 16.1. For example, we see from Table 16.1 that a field which is fully unpolarized in the time domain may be fully polarized in the spectral domain, even at every frequency. However, a field that is fully polarized in the time domain is also completely polarized in the frequency domain.

Table 16.1 Polarization properties of various electromagnetic beams

Beam type	Time domain	Spectral domain	Comments
Monochromatic	$P_t = 1$	$P_f = 1$	Idealization
Quasi-monochromatic	$0 \leq P_t \leq 1$	$P_f(\omega_0) = P_t$	At center frequency
Time-domain polarized	$P_t = 1$	$P_f(\omega) = 1$	For all frequencies
Uniform coherence (16.15)	$0 \leq P_t \leq 1$	$P_f(\omega) = P_t$	For all frequencies
Uncorrelated orthogonal waves, equal intensities	$P_t = 0$	$0 \leq P_f(\omega) \leq 1$	P_f depends on spectra
Uncorrelated orthogonal waves, unequal intensities	$0 < P_t < 1$	$0 \leq P_f(\omega) \leq 1$	P_f depends on spectra
Delayed orthogonal waves	$0 \leq P_t \leq 1$ [a]	$P_f(\omega) = 1$	For all frequencies

[a] P_t depends on time delay ($P_t = |\gamma_A(\tau_d)|$)

Acknowledgements This research was supported by the Academy of Finland. A.T. Friberg also acknowledges funding from the Swedish Foundation for Strategic Research (SSF). The authors thank A. Shevchenko for useful discussions.

References

1. T. Setälä, A. Shevchenko, M. Kaivola, and A.T. Friberg, Phys. Rev. A **78**, 033817 (2008).
2. A. Shevchenko, T. Setälä, M. Kaivola, and A.T. Friberg, New J. Phys. **11**, 073004 (2009).
3. E. Wolf, *Introduction to the Theory of Coherence and Polarization of Light* (Cambridge University Press, Cambridge, UK, 2007).
4. L. Mandel and E. Wolf, *Optical Coherence and Quantum Optics* (Cambridge University Press, Cambridge, UK, 1995).
5. L. Mandel and E. Wolf, J. Opt. Soc. Am. **66**, 529–535 (1976).
6. A.T. Friberg and E. Wolf, Opt. Lett. **20**, 623–625 (1995).
7. M. Beran and G.B. Parrent, *Theory of Partial Coherence* (Prentice-Hall, Englewood Cliffs, NJ, 1963).
8. E. Wolf, Opt. Lett. **8**, 250–252 (1983).
9. L. Basano, P. Ottonello, G. Rottigni, and M. Vicari, Appl. Opt. **42**, 6239–6244 (2003).
10. B. Kanseri and H.C. Kandpal, Opt. Lett. **35**, 70–72 (2010).
11. T. Setälä, F. Nunziata, and A.T. Friberg, in *Partial Electromagnetic Coherence and 3D Polarization* (Koli Workshop, University of Joensuu, 2009), pp. 66–67.
12. T. Setälä, F. Nunziata, and A.T. Friberg, J. Phys. Conf. Ser. **206**, 012001 (2010).
13. T. Setälä, F. Nunziata, and A.T. Friberg, Opt. Lett. **34**, 2924–2926 (2009).
14. M. Lahiri, Opt. Lett. **34**, 2936–2938 (2009).
15. C. Brosseau, *Fundamentals of Polarized Light: A Statistical Optics Approach* (Wiley, New York, 1998).
16. T. Setälä, A. Shevchenko, M. Kaivola, and A.T. Friberg, Phys. Rev. E **66**, 016615 (2002).
17. K. Lindfors, T. Setälä, M. Kaivola, and A.T. Friberg, J. Opt. Soc. Am. A **22**, 561–568 (2005).
18. K. Lindfors, A. Priimagi, T. Setälä, A. Shevchenko, A.T. Friberg, and M. Kaivola, Nature Phot. **1**, 228–231 (2005).
19. E. Wolf, Nuovo Cimento **13**, 1165–1181 (1959).

Chapter 17
A Degree of Freedom and Metrics Approach for Nonsingular Mueller Matrices Characterizing Passive Systems

V. Devlaminck and P. Terrier

Abstract In this paper, we are interested to define and analyze a general parametric form of nonsingular Mueller matrices. Starting from previous results about the nondepolarizing matrices, we generalize the method to any nonsingular Mueller matrices. We addressed this problem from the point of view of metrics and degrees of freedom in order to introduce a physical admissible solution to solve this question. Generators of this group are used to address the issue of the dimension of the full space in order to obtain an adequate number of degrees of freedom of the related Mueller subspace.

17.1 Introduction

The literature on characterization of Mueller matrices is well furnished [1–13] but various definitions are used. One definition [11], henceforth called first definition, is to define a Mueller matrix $\mathbf{M}$ as a real 4×4 matrix mapping all real vectors $\mathbf{S} = (S_0, S_1, S_2, S_3)^{\mathrm{T}}$ satisfying $S_0^2 \geq S_1^2 + S_2^2 + S_3^2$ into real vectors $\mathbf{S}' = (S'_0, S'_1, S'_2, S'_3)^{\mathrm{T}}$ satisfying $S'^2_0 \geq S'^2_1 + S'^2_2 + S'^2_3$. A second definition [13] is to define a Mueller matrix as a convex sum of so-called Mueller–Jones matrices or pure Mueller matrices or nondepolarizing Mueller matrices, which are obtained by writing the equation $\mathbf{E}' = \mathbf{J}\mathbf{E}$ mapping an input electric field vector $\mathbf{E}$ into an output electric field vector $\mathbf{E}'$ by means of the (generally complex) 2×2 Jones matrix $\mathbf{J}$, in terms of a relation between the corresponding vectors of Stokes parameters. For either definition there exist complete characterizations of the class of real 4×4 matrices $\mathbf{M}$. For instance, Cloude [12] has proved that a real 4×4 matrix $\mathbf{M}$ is a Mueller matrix with respect to the second definition if and only if the (complex hermitian) 4×4 coherency matrix constructed from $\mathbf{M}$ by linear operations has

---

V. Devlaminck (✉)
LAGIS FRE CNRS 3303, Université Lille 1, Sciences et Technologies
59655 Villeneuve d'Ascq, France
e-mail: vincent.devlaminck@univ-lille1.fr

T. Fournel and B. Javidi (eds.), *Information Optics and Photonics: Algorithms, Systems, and Applications*, DOI 10.1007/978-1-4419-7380-1_17,

only nonnegative eigenvalues. Further, it has been shown by Givens and Kostinski [7], also reported by van der Mee [2], that a real 4×4 matrix $\mathbf{M} = (m_{ij})$ with nonnegative m_{00} element is a Mueller matrix with respect to the first definition if $\mathbf{A} = \mathbf{G}\mathbf{M}^{\mathrm{T}}\mathbf{G}\mathbf{M}$, with $\mathbf{G} = \mathrm{diag}(1, -1, -1, -1)$, has only real eigenvalues, the largest eigenvalue in absolute value is nonnegative, and there is an eigenvector $\mathbf{S} = (S_0, S_1, S_2, S_3)^{\mathrm{T}}$ satisfying $S_0^2 \geq S_1^2 + S_2^2 + S_3^2$ corresponding to this eigenvalue. Actually, this is just a sufficient condition since there is a whole class of Mueller matrices according to the first definition, that does not satisfy this condition, as pointed out by Gopala Rao et al. [4] (the type II matrices following Gopala Rao terminology). But these latter matrices are very special so the Givens–Kostinski conditions are generally accepted as necessary and sufficient ones.

The first definition leads to a larger class of real 4×4 matrices than the second one. Analysing the Cloude condition on passivity, Takakura ans Stoll [14] claim that the Cloude condition is a sufficient condition that is not necessary because of some restrictions on the mutual probabilities associated with elements of the target coherency matrix expansion. They exhibit a Mueller matrix that does not satisfy this condition but still correpond to an optical system.

Then, the description of such sets of matrices is always questionable, thought a part of the solution is already well-known: nonsingular nondepolarizing Mueller matrices are the positive multiples of Lorentz group matrices[1] with positive left upper corner element and positive determinant with respect to the definition of Mueller matrices as an ensemble average of pure Mueller matrices. The nonsingular nondepolarizing matrices with respect to the first definition are the positive multiples of Lorentz group matrices with positive left upper corner element. Then, corresponding subgroup is given [15, 16] by:

$$\{SO(3,1)_e\} \cup \{\mathbf{G}{\cdot}SO(3,1)_e\} \tag{17.1}$$

with $\mathbf{G} = \mathrm{diag}(1, -1, -1, -1)$. So, the characterization of matrices of $SO(3,1)_e$ completely describes the two subgroups.

Extension of this result to the whole set of nonsingular Mueller matrices is not straightforward since this set will be expected to form only a semigroup [17]– the inverse matrix of a pure depolarizer for instance is not a Mueller matrix.

In a recent paper [16], regarding mathematical framework to define and analyze a general parametric form of an arbitrary nonsingular Mueller matrix, we proposed to address this problem in a space with more than four dimensions in order to introduce a set of transformations with a group structure. We consider a group of transformations continuously connected to the identity transformation and preserving a N-vectors norm such as the proper Lorentz transformations with the four-vectors of the Minkowski space.

---

[1] Lorentz group is the union of four disconnected pieces [15]. The restriction to those transformations which are continuously connected to the identity transformation is called the *proper orthochronous* Lorentz transformations and denoted $SO(3,1)_e$ as the identity component of $SO(3,1)$, the associated special orthogonal group.

17.2 Appropriate Norm and Dimension

The problems with such approach are then: how do we determine N, the space dimension, and how do we chose the appropriate norm?

17.2.1 *Appropriate Norm*

The proposed space should contain a Minkowskian subspace associated with the four first components for instance. Then, this subspace can be assimilated with the polarization space – it means: $(x^0, x^1, x^2, x^3) = (S_0, S_1, S_2, S_3)$ – and the associated metric should be defined by the metric tensor $g_{\mu\nu}$ with the signature $(1, -1, -1, -1, g_{44}, g_{55}, \ldots, g_{(N-1)(N-1)})$. The fundamental relation for the entries of such a transformation Γ is derived from the property of preservation of the *length* of the N-vectors:

$$g_{\mu\nu}\,\Gamma^{\mu}_{\lambda}\,\Gamma^{\nu}_{\sigma} = g_{\lambda\sigma}. \tag{17.2}$$

Considerations about properties of the polarizance vector and (17.2) for $\mu = \nu = 0$ or diattenuation and (17.2) for $\lambda = \sigma = 0$ allow us [16] to demonstrate $g_{44} = g_{55} = \ldots = g_{(N-1)(N-1)} = -1$. So, we are now considering the group $O(N-1,1)$ and more precisely $SO(N-1,1)_e$, its identity component [2]

17.2.2 *Appropriate dimension*

For a N-dimensions space preserving a N-vectors norm, the general rule gives the number of degrees of freedom (DOF):

$$D = N^2 - \frac{N\,(N+1)}{2} = \frac{N\,(N-1)}{2}. \tag{17.3}$$

Nevertheless, what is under interest is the number of DOF associated with the polarization four-dimensions subspace not with the full N-dimensions space.

As $SO(N-1,1)_e$ is a matrix Lie group with N degrees of freedom, we can focus our attention on the N generators of the group rather than the infinite number of group elements. These generators are elements of the Lie algebra [18] of the group $SO(N-1,1)_e$. Since we must have 15 DOF, the smallest value of N to consider is $N = 6$–see (17.3) – but we will prove that $N = 7$ is the actual solution.

---

[2] Matrices with unit determinant and positive left upper corner element are the $SO(N-1,1)_e$ subgroup elements.

Under this hypothesis, we now have a potential of 21 DOF. So how many DOF are actually present in the resulting Mueller matrix? We can figure out the response when analyzing the generators of $SO(6,1)_e$. The expressions of these matrices are listed in Appendix.

We first divide these generators into four sets and consider the DOF contribution to Mueller matrix subspace of each set and the DOF added by the interaction between these sets under the matrix multiplication rule.

The first set is composed by J_{23}, J_{13}, J_{12}, J_{01}, J_{02}, and J_{03}. If we consider

$$\mathbf{J}_1 = i\,(\alpha J_{01} + \beta J_{02} + \gamma J_{03} + \delta J_{12} + \varepsilon J_{13} + \eta J_{23}) \tag{17.4}$$

the 6 DOF matrix has the following expression:

$$\mathbf{J}_1 = \begin{pmatrix} \mathbf{A} & [\mathbf{0}]_{43} \\ [\mathbf{0}]_{34} & [\mathbf{0}]_{33} \end{pmatrix} \quad \text{with } \mathbf{A} = \begin{pmatrix} 0 & \otimes & \otimes & \otimes \\ * & 0 & \otimes & \otimes \\ * & * & 0 & \otimes \\ * & * & * & 0 \end{pmatrix} \quad \begin{matrix} * \equiv \text{DOF} \\ \otimes \neq \text{DOF} \end{matrix}, \tag{17.5}$$

where $[\mathbf{0}]_{mn}$ is for null matrix of $m \times n$ dimensions. Then,

$$\mathbf{J}_1{}^k = \begin{pmatrix} \mathbf{A}^k & [\mathbf{0}]_{43} \\ [\mathbf{0}]_{34} & [\mathbf{0}]_3 \end{pmatrix}. \tag{17.6}$$

As the exponential of $\mathbf{J}$ is defined as the usual power series:

$$\exp(\mathbf{J}) = \sum_{m=0}^{+\infty} \frac{J^m}{m!} \tag{17.7}$$

the six generators give associated Mueller submatrices with a total of 6 DOF. In fact these Mueller matrices are the nonsingular, nondepolarizing Mueller matrices (product of $\mathbf{M}_{\mathrm{R}}$, the retardance matrix and $\mathbf{M}_{\mathrm{D}}$ the diattenuation one [16, 19]), since these generators are translation in seven dimensions of Lorentz group generators.

The same approach with J_{46}, J_{56}, J_{45} gives a $\mathbf{J}_2$ matrix with the following expression:

$$\mathbf{J}_2 = \begin{pmatrix} [\mathbf{0}]_{44} & [\mathbf{0}]_{43} \\ [\mathbf{0}]_{34} & \mathbf{A} \end{pmatrix} \text{with } \mathbf{A} = \begin{pmatrix} 0 & \otimes & \otimes \\ * & 0 & \otimes \\ * & * & 0 \end{pmatrix}. \tag{17.8}$$

Computing of $\mathbf{J}_2{}^k$ and (17.7), show that no DOF can be directly added to the Mueller subspace or by the block matrix multiplication rule either.

We now consider the $\mathbf{J}_3$ matrix defined by nine generators:

$$\mathbf{J}_3 = i\left(\sum_{m=1}^{3} \alpha_m J_{m4} + \sum_{n=1}^{3} \beta_n J_{n5} + \sum_{p=1}^{3} \gamma_p J_{p6}\right) \tag{17.9}$$

$\mathbf{J}_3$ is a 9 DOF matrix and from the expressions of the generators, it has the following form:

$$\mathbf{J}_3 = \begin{pmatrix} [\mathbf{0}]_{44} & -\mathbf{A}^T \\ \mathbf{A} & [\mathbf{0}]_{33} \end{pmatrix} \text{ with } \mathbf{A} = \begin{pmatrix} 0 & * & * & * \\ 0 & * & * & * \\ 0 & * & * & * \end{pmatrix}. \tag{17.10}$$

It is straightforward from a recursive computation to prove that the $2k$-power and $(2k+1)$-power of $\mathbf{J}_3$ matrix have the following expressions:

$$\begin{cases} \mathbf{J}_3^{2k} = \begin{pmatrix} (-1)^k \left(\mathbf{A}^{\mathrm{T}}\mathbf{A}\right)^k & [\mathbf{0}]_{3,4}^{\mathrm{T}} \\ [\mathbf{0}]_{3,4} & (-1)^k \left(\mathbf{A}\mathbf{A}^{\mathrm{T}}\right)^k \end{pmatrix} \\ \mathbf{J}_3^{2k+1} = \begin{pmatrix} [\mathbf{0}]_{44} & (-1)^{k+1} \left(\mathbf{A}^{\mathrm{T}}\mathbf{A}\right)^k \mathbf{A}^{\mathrm{T}} \\ (-1)^k \left(\mathbf{A}\mathbf{A}^{\mathrm{T}}\right)^k \mathbf{A} & [\mathbf{0}]_{33} \end{pmatrix} \end{cases}, \tag{17.11}$$

where $\mathbf{A}^{\mathrm{T}}\mathbf{A}$ is a symmetric square matrix (with the entries of the first raw and first column equal to 0). Then, as a summation of symmetric matrices with the entries of the first raw and first column equal to 0 (except for $m = 0$ since $\mathbf{J}_3{}^0 = \mathrm{Id}$), the resulting submatrix of $\exp(\mathbf{J}_3)$ associated with the polarization space has the form:

$$\begin{pmatrix} 1 & [\mathbf{0}] \\ [\mathbf{0}]^{\mathrm{T}} & m \end{pmatrix}, \tag{17.12}$$

where m denotes a symmetric 3×3 matrix. So, this set of generators give an amount of only 6 DOF and the matrix as the following DOF distribution:

$$\exp(\mathbf{J}_3) = \left[\begin{array}{cccc|ccc} 1 & 0 & 0 & 0 & 0 & 0 & 0 \\ 0 & * & \otimes & \otimes & * & \otimes & \otimes \\ 0 & * & * & \otimes & \otimes & * & \otimes \\ 0 & * & * & * & \otimes & \otimes & * \\ \hline 0 & \otimes & \otimes & \otimes & \otimes & \otimes & \otimes \\ 0 & \otimes & \otimes & \otimes & \otimes & \otimes & \otimes \\ 0 & \otimes & \otimes & \otimes & \otimes & \otimes & \otimes \end{array}\right] = \begin{bmatrix} \mathbf{M}_3 & \mathbf{G}_3 \\ \mathbf{H}_3 & \mathbf{N}_3 \end{bmatrix} \tag{17.13}$$

We arbitrary place the DOF of $\mathbf{G}_3$ on the diagonal entries, as we see further, it is not important. $\mathbf{G}$ or $\mathbf{H}$ or $\mathbf{N}$ can only have 3 DOF since 6 DOF are kept by $\mathbf{M}$, that is what worth noticing is.

The last set of three generators gives the matrix $\mathbf{J}_4$:

$$\mathbf{J}_4 = i\left(\sum_{m=4}^{6} \delta_m J_{0m}\right) \tag{17.14}$$

with the following form:

$$\mathbf{J}_4 = \begin{pmatrix} [\mathbf{0}]_4 & \mathbf{B}^{\mathrm{T}} \\ \mathbf{B} & [\mathbf{0}]_3 \end{pmatrix} \text{ with } \mathbf{B} = \begin{pmatrix} * & 0 & 0 & 0 \\ * & 0 & 0 & 0 \\ * & 0 & 0 & 0 \end{pmatrix}. \tag{17.15}$$

It is straightforward to demonstrate the powers of $\mathbf{J}_4$ matrix have a very similar expression of (17.11) (with a substitution of $\mathbf{A}$ by $\mathbf{B}$ and without the power of -1). The result is the following form for $\exp(\mathbf{J}_4)$:

$$\exp(\mathbf{J}_4) = \begin{bmatrix} \boxed{\begin{matrix} \cdot & 0 & 0 & 0 \\ 0 & 0 & 0 & 0 \\ 0 & 0 & 0 & 0 \\ 0 & 0 & 0 & 0 \end{matrix}} & \boxed{\begin{matrix} * & * & * \\ 0 & 0 & 0 \\ 0 & 0 & 0 \\ 0 & 0 & 0 \end{matrix}} \\ \boxed{\begin{matrix} \cdot & 0 & 0 & 0 \\ \cdot & 0 & 0 & 0 \\ \cdot & 0 & 0 & 0 \end{matrix}} & \boxed{\begin{matrix} \cdots \\ \cdots \\ \cdots \end{matrix}} \end{bmatrix} = \begin{bmatrix} \mathbf{M}_2 & \mathbf{G}_2 \\ \mathbf{H}_2 & \mathbf{N}_2 \end{bmatrix} \tag{17.16}$$

and no DOF can be directly added to the Mueller subspace by these generators. But it is clear that DOF can be added to the Mueller subspace by the block matrix multiplication rule between $\exp(\mathbf{J}_3)$ and $\exp(\mathrm{J}_4)$ since we have the following expressions:

$$\mathbf{G}_4\mathbf{H}_3 = \begin{bmatrix} 0 & * & * & * \\ 0 & 0 & 0 & 0 \\ 0 & 0 & 0 & 0 \\ 0 & 0 & 0 & 0 \end{bmatrix}, \quad \mathbf{H}_4\mathbf{G}_3 = \begin{bmatrix} 0 & 0 & 0 & 0 \\ * & 0 & 0 & 0 \\ * & 0 & 0 & 0 \\ * & 0 & 0 & 0 \end{bmatrix} \tag{17.17}$$

These matrix products of (17.17) add a diattenuation vector or a polarizance vector [19] with 3 DOF to the Mueller matrix subspace.

Eventually, we obtain a total amount of 15 DOF for the resulting Mueller matrix subspace.

The same approach proves we only have an amount of 14 DOF for the Mueller matrix subspace when dealing with a six-dimensions space. Then $SO(6,1)_e$ is a physical admissible solution to solve the question of definition of arbitrary nonsingular Mueller matrices. The previously proposed solution in [16] was just a subset of this one.[3]

It is worth noticing that the relation between the both definitions of Mueller matrices and the different connected components of $O(3,1)$ introduced for the nondepolarizing matrices (see (17.1)) is not so straightforward for the case of depolarizing matrices since $SO(6,1)_e$ is a group of matrices with unit determinant but submatrices of this group can have negative determinant.

---

[3] It is possible to identify missing matrices as ones with equal depolarization factor along the three principal axis with a non-null polarizance or diattenuation vector.

17.3 Conclusion

Starting from previous results about the nondepolarizing Mueller matrices, we generalized the method to any nonsingular Mueller matrices. Considering the set of nonsingular Mueller matrices as a set of linear operators on an embedded subspace of a more general space, we take metrics and degrees of freedom in consideration in order to specify this space and the group of associated linear transformations. Generators of this group are used to address the issue of the dimension of the full space and a seven-dimensions Minkowski space gives a physical admissible solution. Arbitrary nonsingular Mueller matrices are then submatrices of $SO(6,1)_e$.

Appendix

$\mathbf{J}_{mn}$ are the matrices of the 21 generators of $SO(6,1)_e$ group:

$$\mathbf{J}_{01} = \begin{pmatrix} 0&i&0&0&0&0&0\\ i&0&0&0&0&0&0\\ 0&0&0&0&0&0&0\\ 0&0&0&0&0&0&0\\ 0&0&0&0&0&0&0\\ 0&0&0&0&0&0&0\\ 0&0&0&0&0&0&0 \end{pmatrix} \quad \mathbf{J}_{02} = \begin{pmatrix} 0&0&i&0&0&0&0\\ 0&0&0&0&0&0&0\\ i&0&0&0&0&0&0\\ 0&0&0&0&0&0&0\\ 0&0&0&0&0&0&0\\ 0&0&0&0&0&0&0\\ 0&0&0&0&0&0&0 \end{pmatrix} \quad \mathbf{J}_{03} = \begin{pmatrix} 0&0&0&i&0&0&0\\ 0&0&0&0&0&0&0\\ 0&0&0&0&0&0&0\\ i&0&0&0&0&0&0\\ 0&0&0&0&0&0&0\\ 0&0&0&0&0&0&0\\ 0&0&0&0&0&0&0 \end{pmatrix} \quad \mathbf{J}_{04} = \begin{pmatrix} 0&0&0&0&i&0&0\\ 0&0&0&0&0&0&0\\ 0&0&0&0&0&0&0\\ 0&0&0&0&0&0&0\\ i&0&0&0&0&0&0\\ 0&0&0&0&0&0&0\\ 0&0&0&0&0&0&0 \end{pmatrix}$$

$$\mathbf{J}_{05} = \begin{pmatrix} 0&0&0&0&0&i&0\\ 0&0&0&0&0&0&0\\ 0&0&0&0&0&0&0\\ 0&0&0&0&0&0&0\\ 0&0&0&0&0&0&0\\ \mathrm{i}&0&0&0&0&0&0\\ 0&0&0&0&0&0&0 \end{pmatrix} \quad \mathbf{J}_{06} = \begin{pmatrix} 0&0&0&0&0&0&i\\ 0&0&0&0&0&0&0\\ 0&0&0&0&0&0&0\\ 0&0&0&0&0&0&0\\ 0&0&0&0&0&0&0\\ 0&0&0&0&0&0&0\\ i&0&0&0&0&0&0 \end{pmatrix} \quad \mathbf{J}_{12} = \begin{pmatrix} 0&0&0&0&0&0&0\\ 0&0&-i&0&0&0&0\\ 0&i&0&0&0&0&0\\ 0&0&0&0&0&0&0\\ 0&0&0&0&0&0&0\\ 0&0&0&0&0&0&0\\ 0&0&0&0&0&0&0 \end{pmatrix} \quad \mathbf{J}_{23} = \begin{pmatrix} 0&0&0&0&0&0&0\\ 0&0&0&0&0&0&0\\ 0&0&0&-i&0&0&0\\ 0&0&i&0&0&0&0\\ 0&0&0&0&0&0&0\\ 0&0&0&0&0&0&0\\ 0&0&0&0&0&0&0 \end{pmatrix}$$

$$\mathbf{J}_{34} = \begin{pmatrix} 0&0&0&0&0&0&0\\ 0&0&0&0&0&0&0\\ 0&0&0&0&0&0&0\\ 0&0&0&0&-i&0&0\\ 0&0&0&i&0&0&0\\ 0&0&0&0&0&0&0\\ 0&0&0&0&0&0&0 \end{pmatrix} \quad \mathbf{J}_{45} = \begin{pmatrix} 0&0&0&0&0&0&0\\ 0&0&0&0&0&0&0\\ 0&0&0&0&0&0&0\\ 0&0&0&0&0&0&0\\ 0&0&0&0&0&-i&0\\ 0&0&0&0&i&0&0\\ 0&0&0&0&0&0&0 \end{pmatrix} \quad \mathbf{J}_{56} = \begin{pmatrix} 0&0&0&0&0&0&0\\ 0&0&0&0&0&0&0\\ 0&0&0&0&0&0&0\\ 0&0&0&0&0&0&0\\ 0&0&0&0&0&0&0\\ 0&0&0&0&0&0&-i\\ 0&0&0&0&0&i&0 \end{pmatrix} \quad \mathbf{J}_{13} = \begin{pmatrix} 0&0&0&0&0&0&0\\ 0&0&0&i&0&0&0\\ 0&0&0&0&0&0&0\\ 0&-i&0&0&0&0&0\\ 0&0&0&0&0&0&0\\ 0&0&0&0&0&0&0\\ 0&0&0&0&0&0&0 \end{pmatrix}$$

$$\mathbf{J}_{24} = \begin{pmatrix} 0&0&0&0&0&0&0\\ 0&0&0&0&0&0&0\\ 0&0&0&0&i&0&0\\ 0&0&0&0&0&0&0\\ 0&0&-i&0&0&0&0\\ 0&0&0&0&0&0&0\\ 0&0&0&0&0&0&0 \end{pmatrix} \quad \mathbf{J}_{35} = \begin{pmatrix} 0&0&0&0&0&0&0\\ 0&0&0&0&0&0&0\\ 0&0&0&0&0&0&0\\ 0&0&0&0&0&i&0\\ 0&0&0&0&0&0&0\\ 0&0&0&-i&0&0&0\\ 0&0&0&0&0&0&0 \end{pmatrix} \quad \mathbf{J}_{46} = \begin{pmatrix} 0&0&0&0&0&0&0\\ 0&0&0&0&0&0&0\\ 0&0&0&0&0&0&0\\ 0&0&0&0&0&0&0\\ 0&0&0&0&0&0&i\\ 0&0&0&0&0&0&0\\ 0&0&0&0&-i&0&0 \end{pmatrix} \quad \mathbf{J}_{14} = \begin{pmatrix} 0&0&0&0&0&0&0\\ 0&0&0&0&-i&0&0\\ 0&0&0&0&0&0&0\\ 0&0&0&0&0&0&0\\ 0&i&0&0&0&0&0\\ 0&0&0&0&0&0&0\\ 0&0&0&0&0&0&0 \end{pmatrix}$$

$$\mathbf{J}_{25} = \begin{pmatrix} 0&0&0&0&0&0&0\\ 0&0&0&0&0&0&0\\ 0&0&0&0&0&-i&0\\ 0&0&0&0&0&0&0\\ 0&0&0&0&0&0&0\\ 0&0&i&0&0&0&0\\ 0&0&0&0&0&0&0 \end{pmatrix} \mathbf{J}_{36} = \begin{pmatrix} 0&i&0&0&0&0&0\\ i&0&0&0&0&0&0\\ 0&0&0&0&0&0&0\\ 0&0&0&0&0&0&i\\ 0&0&0&0&0&0&0\\ 0&0&0&0&0&0&0\\ 0&0&0&i&0&0&0 \end{pmatrix} \mathbf{J}_{15} = \begin{pmatrix} 0&0&0&0&0&0&0\\ 0&0&0&0&0&i&0\\ 0&0&0&0&0&0&0\\ 0&0&0&0&0&0&0\\ 0&0&0&0&0&0&0\\ 0&-i&0&0&0&0&0\\ 0&0&0&0&0&0&0 \end{pmatrix} \mathbf{J}_{26} = \begin{pmatrix} 0&0&0&0&0&0&0\\ 0&0&0&0&0&0&0\\ 0&0&0&0&0&0&i\\ 0&0&0&0&0&0&0\\ 0&0&0&0&0&0&0\\ 0&0&0&0&0&0&0\\ 0&0&-i&0&0&0&0 \end{pmatrix}$$

$$\mathbf{J}_{16} = \begin{pmatrix} 0&0&0&0&0&0&0\\ 0&0&0&0&0&0&-i\\ 0&0&0&0&0&0&0\\ 0&0&0&0&0&0&0\\ 0&0&0&0&0&0&0\\ 0&0&0&0&0&0&0\\ 0&i&0&0&0&0&0 \end{pmatrix}$$

References

1. Z. F. Xing, "On the deterministic and non-deterministic Mueller matrix", J. Mod. Opt. **39**, 461–484 (1992)
2. V. M. van der Mee, "An eigenvalue criterion for matrices transforming Stokes parameters", J. Math. Phys. **34**, 5072–5088 (1993)
3. J. W. Hovenier, "Structure of a general pure Mueller matrix", App. Opt. **33**, 8318–8324 (1994)
4. A. V. Gopala Rao, K. S. Mallesh, and J. Sudha, "On the algebraic characterization of a Mueller matrix in polarization optics. I. Identifying a Mueller matrix from its N matrix", J. Mod. Opt. **45**, 955–987 (1998)
5. M. S. Kumar and R. Simon, "Characterization of Mueller matrices in polarization optics", Opt. Commun. **88**, 464–470 (1992)
6. A. B. Kostinski, C. R. Given, and J. M. Kwiatkowski, "Constraints on Mueller matrices of polarization optics", Appl. Opt. **32**, 1646–1651 (1993)
7. C.R. Givens and B. Kostinski, "A simple necessary and sufficient condition on physically realizable Mueller matrices", J. Mod. Opt. **40**, 471–481 (1993)
8. R. Simon, "The connection between Mueller and Jones matrices of polarization optics", Opt. Com. **42**, 293–297 (1982)
9. D. G. M. Anderson and R. Barakat, "Necessary and sufficient conditions for a Mueller matrix to be derivable from a Jones matrix", J. Opt. Soc. Am. A **11**, 2305–2319 (1994)
10. J. J. Gil, "Characteristic properties of Mueller matrices", J. Opt. Soc. Am. A **17**, 328–334 (2000)
11. E.S. Fry and G.W. Kattawar, "Relationships between elements of the Stokes matrix", Appl. Opt. **20**, 2811–2814 (1981)
12. S.R. Cloude, "Group theory and polarisation algebra", Optik **75**, 26–36 (1986)
13. K. Kim, L. Mandel, and E. Wolf, "Relationship between Jones and Mueller matrices for random media", J. Opt. Soc. Am. A **4**, 433–437 (1987)
14. Y. Takakura and M.-P. Stoll, "Passivity test of Mueller matrices in the presence of additive Gaussian noise", Appl. Opt. **48**, 1073–1083 (2009)
15. D. H. Sattiger and O. L. Weaver, *Lie Group and Algebras with Applications to Physics, Geometry and Mechanics*, (Springer Verlag, New York, 1991)
16. V. Devlaminck and P. Terrier, "Definition of a parametric form of nonsingular Mueller matrices", J. Opt. Soc. Am. A **25**, 2636–2643 (2008)
17. R. Sridhar and R. Simon, "Normal form for Mueller matrices in polarization optics", J. Mod. Opt. **41**, 1903–1915 (1994)
18. A. A. Sagle and R. E. Walde, *Introduction to Lie groups and Lie algebras*, (Academic Press, New York 1973)
19. S. Y. Lu and R. A. Chipman, "Interpretation of Mueller matrices based on polar decomposition", J. Opt. Soc. Am. A **13**, 1106–1113 (1996)

Chapter 18
Resolution-Enhanced Imaging Based upon Spatial Depolarization of Light

D. Sylman, Z. Zalevsky, V. Micó, and J. García

Abstract In this paper, we present a new approach allowing the surpassing of the diffraction-based limitation for the achievable resolution provided by imaging systems. It is based on an encoding–decoding process of various spatial pixels or regions in the field of view of the imaged object by orthogonal and differently time-varying polarization states. The reconstruction of the original spatial information is obtained by applying a decoding process in a way similar to the encoding one. Although all the spatial information is summed and mixed together by the system, the decoding provides super-resolved imaging since in every spatial position the undesired spatial information having time-varying polarization dependence, which is uncorrelated to the decoding sequence applied on that specific spatial position, is averaged to zero and, on the other hand, the information which corresponds to that specific spatial region is being reinforced. The proposed approach can be used not only for super-resolved imaging but also for imaging module that maintains the same spatial resolution while providing enlarged field of view.

18.1 Introduction

Super resolution [1, 2] is the field in which coding spatial degrees of freedom by other domains as time [3,4], wavelength [5–8], code or field of view [9,10], grey levels [11], coherence [12, 13], and polarization [14, 15] allows overcoming resolution limits of imaging systems posed by the diffraction effect. The spatial information is encoded by one of the other domains and after being transmitted through the imaging system it is decoded to reconstruct the high resolution image.

In recent work the coherence property of light was used to increase resolution or to obtain the same resolution over larger field of view [12, 13]. The idea included synthesizing the mutual coherence of the illumination source and then multiplexing

Z. Zalevsky (✉)
School of Engineering, Bar-Ilan University, Ramat-Gan 52900, Israel
e-mail: zalevsz@macs.biu.ac.il

T. Fournel and B. Javidi (eds.), *Information Optics and Photonics: Algorithms, Systems, and Applications*, DOI 10.1007/978-1-4419-7380-1_18,

the spatial information while using this apriori known encoding illumination, transferring the information through the imaging system and eventually demultiplexing it and reconstructing the high resolution image. At first the encoding was done by shaping the transversal coherence distribution and using it to improve the lateral resolution [12] and later on by the axial coherence [13] where the authors presented how to preserve the same resolution over an increased field of view.

In this paper we follow the line presented in Ref. [13] while the capability for increasing the field of view is obtained not by encoding each region in the field of view by different axial delays (while each is larger than the coherence distance of the illumination source), but rather by proper source depolarization.

In the proposed concept the illumination source is depolarized in such a way that various spatial regions of the field of view are illuminated by differently time-varying polarization states. The depolarization is such that the polarization variations in the different regions of the illumination spot are orthogonal to each other, and thus time averaging may allow separating the various parts of the illumination spot (that coded different regions of the field of view) even if they were summed or mixed together (due to the low resolution blurring). As in the case of field-of-view coding by the axial coherence of the illumination, here as well the time-averaging duration that is required in order to orthogonally separate the multiplexed lateral regions is very short.

Note that in contrast to polarization-related super-resolving approach presented in [15], here we do not seek to obtain the reconstruction by matrix inversion, but rather just by time averaging of the properly decoded depolarization of the encoded illumination source.

The theory is presented in Sect. 18.2 while in Sect. 18.3 we present experimental validation of the proposed approach. The paper is concluded in Sect. 18.4.

18.2 Theoretical Description

We assume that the illumination source is depolarized such that various spatial regions within the illumination spot have different and orthogonal time-varying state of polarization. We denote by $P(x,t)$ the spatial distribution of the temporal variation of the polarization state in the illumination spot.

We assume that the input object, which is denoted by $g_{\text{in}}(x)$, is illuminated by the time-varying polarization:

$$g_{in}(x) = \sum_n g_{in}{}^{(n)}(x - n\Delta x), \tag{18.1}$$

where Δx is the size of the lateral regions that we aim to multiplex and $g_{\text{in}}{}^{(n)}$ is the spatial distribution of the input object in those lateral regions.

After illuminating the input object with the depolarized source one has

$$g_{in}{}^{(p)}(x) = \sum_n P(x,t) g_{in}{}^{(n)}(x - n\Delta x). \tag{18.2}$$

Right after the various lateral regions are being multiplexed together and passed through laterally narrow region of regard, Δx:

$$g_{\text{multiplex}}(x) = \text{rect}\left(\frac{x}{\Delta x}\right) \times \sum_n g_{in}{}^{(p)}(x - n\Delta x) = \sum_n g_{in}{}^{(n)}(x) P(x - n\Delta x, t). \tag{18.3}$$

where $\text{rect}\,(x/\Delta x)$ is the rectangular mathematical function that equals to one within the region of $-\Delta x/2 < x < \Delta x/2$ and zero elsewhere. The demultiplexing process involving extracting the original wide field-of-view information by laterally replicating the transmitted spatially multiplexed distribution and interfering it with the same depolarized source. Alternatively, the decoding may also be obtained by passing the transmitted illumination through space and time-varying polarizer having spatial–temporal distribution identical to the illumination source $P(x,t)$. The lateral replication using (18.3) is given as follows:

$$\sum_n g_{\text{multiplex}}(x - n\Delta x) = \sum_n \sum_m g_{in}{}^{(n)}(x - m\Delta x) P(x - (n+m)\Delta x, t), \tag{18.4}$$

then the decoding which involves multiplication by $P(x,\ t)$ and time-averaging yields:

$$\begin{aligned} g_{\text{rec}}(x) &= \int \sum_n g_{\text{multiplex}}(x - n\Delta x) P(x,t)\text{d}t, \\ &= \int \sum_n \sum_m g_{in}{}^{(n)}(x - m\Delta x) P(x - (n+m)\Delta x, t) P(x,t)\text{d}t, \\ &= \sum_n \sum_m g_{in}{}^{(n)}(x - m\Delta x) \times \int P(x - (n+m)\Delta x, t) P(x,t)\text{d}t. \end{aligned} \tag{18.5}$$

Due to the orthogonality property one has

$$\int P(x - (n+m)\Delta x, t) P(x,t)\text{d}t = \delta[n+m] \tag{18.6}$$

where δ denotes Kronecker delta and thus the reconstruction equals to

$$g_{\text{rec}}(x) = \sum_n \sum_m g_{in}{}^{(n)}(x - m\Delta x)\delta[n+m] = \sum_n g_{in}{}^{(n)}(x - n\Delta x) = g_{in}(x). \tag{18.7}$$

Thus, reconstruction of the original object is obtained.

18.3 Experimental Validation

In the first step, we constructed a numerical simulation in which we aim to demonstrate the usage of the proposed approach for super-resolved imaging. The idea is to encode a 2D image into a single pixel value and to transmit it through resolution-limiting system and eventually to reconstruct the original object.

Our object in the simulation was an image of 16 × 16 pixels that is presented in Fig. 18.1a. It was encoded using the proposed approach. Each one of the 16 × 16 pixels was encoded with different time-varying orthogonal polarization sequence. Then all the pixels were added (i.e., mixed) into a single pixel value. This imitates the low resolution system having the resolution limitation we aim to overcome. Then, the one value was replicated 16 × 16 times and for each one of the replications proper time-varying polarization sequence was applied.

The reconstruction after DC subtraction is presented in Fig. 18.1b. To obtain the presented result we did 100,000 averages, i.e., we used 100,000 different polarization states in our depolarization sequence. One may see that indeed the reconstruction is very similar to the original object.

The next step is to obtain optical experimental validation. The experimental setup may be seen in Fig. 18.2. At our first experiment the intension was to multiplex two regions of the field of view, to transmit them through the imaging system and then being able to decode and to demultiplex the two regions and to construct the original image of the full field.

In Fig. 18.2, we imitate the depolarization of the illumination source by using time-varying polarizers. Two images were multiplexed. Polarizer 1 is a fixed polarizer and polarizers 2 and 3 are randomly varied in time with possibility to control for each of its own random sequence. Polarizers 1 and 2 were the encoding polarizers

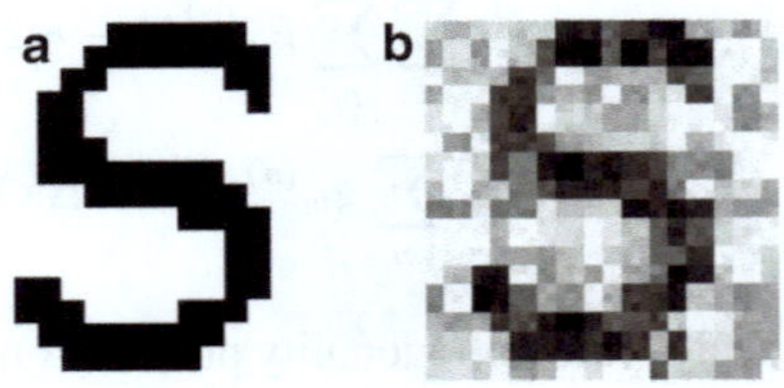

Fig. 18.1 Numerical simulation. (**a**) The original object. (**b**) The obtained reconstruction after transmitting the 2D object as a single pixel value

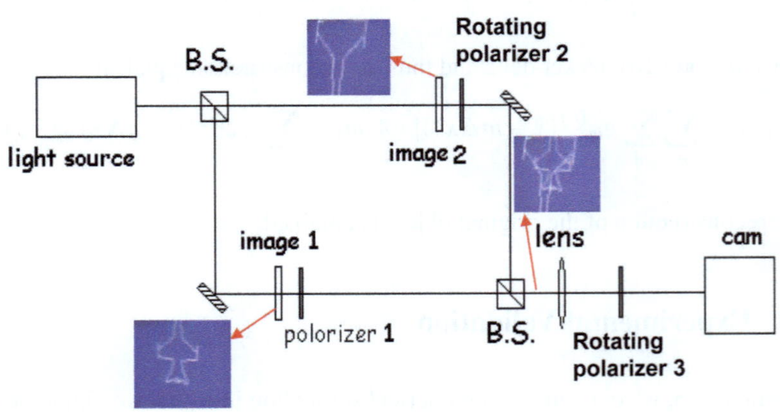

Fig. 18.2 The experimental setup with coding of two regions

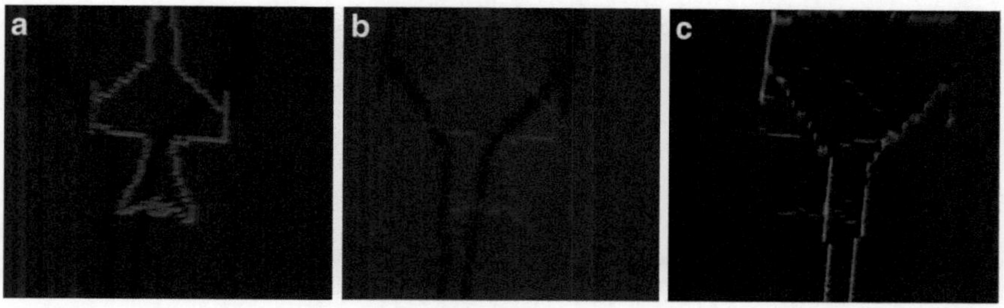

Fig. 18.3 Experimental results. (**a**). Reconstruction of image from branch 1. (**b**). Reconstruction of image from branch 1 even when it is five times more attenuated than the original image of branch 2. (**c**). Reconstruction of image from branch 2

and polarizer 3 was used to decode the two multiplexed images (coming from two lateral regions in the field of view). The mixing of the two images was done using a beam splitter (BS).

The obtained results are seen in Fig. 18.3. In Fig. 18.3a we reconstructed image 1 (originally coming from branch 1). In Fig. 18.3b we repeated the same experiment when image 1 is attenuated five times more than the undesired image 2. This is to show that the orthogonal depolarization sequence allows reconstruction even when the required spatial information is lower in its signal-to-noise ratio in comparison to the undesired information (regarded as noise). In Fig. 18.3c we reconstructed image 2. In all cases, one may see the reconstruction of the original images as appearing in the sketch of Fig. 18.2.

In order to do this reconstruction we applied a sequence of 36 time-averaged frames (36 different polarization states in our time-varying sequence) each having different polarization state.

Note also that the results presented in Fig. 18.3 aimed to show only the preliminary validation of the proposed approach and thus they were obtained without any image processing algorithm, which might have improved the reconstruction quality even further.

In Fig. 18.4 we upgraded the setup of Fig. 18.2 to encoding/decoding of three rather than two spatial regions. In this case, encoding polarizer 1 was fixed and polarizers 2 and 3 were rotating. The decoding polarizer 4 was rotating as well. Each rotating polarizer was shifted between six angular positions (each at an angular difference of 30°). Thus, for the encoding there were 36 ($= 6\times6$) different polarization states. The obtained preliminary results, after averaging all captured images, while applying in the decoding polarizer (polarizer 4) the required decoding sequence, are seen in Fig. 18.5.

One may see that all three multiplexed images (coming from three different spatial regions) were separated from each other. In Fig. 18.5a–c, we present the capability of separating the multiplexed objects of region 1, 2, and 3, respectively. In Fig. 18.5d, we repeat the same experiment when the image of region 2 is amplified four times more than the images of the rest of the regions while we try to reconstruct image from region 1 (the result of the reconstruction for the input of Fig. 18.5d is

18.4 Conclusions

In this chapter, we have demonstrated a super resolving and a field-of-view multiplexing approach in which various lateral regions of the field of view are encoded by synthesizing the depolarization of the illumination source. By interfering the output with the encoding source or by time-varying polarizers following the same depolarization sequence, allows to separate the multiplexed regions of the field of view and to reconstruct the original image.

Although experimentally demonstrated for field-of-view multiplexing (i.e., having larger field of view with the same resolution), one may use the same approach for resolution improvement configuration where each spatial pixel that we aim to transmit will be encoded by orthogonal time varying polarization sequence (as demonstrated by the numerical simulation).

The proposed approach was numerically and experimentally validated and showed good reconstruction capabilities even when the lateral region to be extracted has lower signal-to-noise ratio in comparison to the undesired lateral regions.

References

1. Z. Zalevsky and D. Mendlovic "Optical Super Resolution," Springer, New york (2002).
2. Z. Zalevsky, D. Mendlovic, and A. W. Lohmann, "Optical system with improved resolving power," *Progress in optics*, Vol. XL, Ch. 4 (1999).
3. M. Francon, "Amelioration de resolution d'optique," *Nuovo Cimento Suppl.* **9**, 283–290 (1952).
4. W. Lukosz, "Optical systems with resolving powers exceeding the classical limits," *J. Opt. Soc. Am.* **56**, 1463–1472 (1967).
5. A. I. Kartashev, "Optical systems with enhanced resolving power," *Opt. Spectrosc.* **9**, 204–206 (1960).
6. J. D. Armitage, A. W. Lohmann, and D. P. Parish, "Superresolution image forming systems for objects with restricted lambda dependence" *Jpn. J. Appl. Phys.* **4**, 273–275 (1965).
7. S. A. Alexandrov and D. D. Sampson, "Spatial information transmission beyond a system's diffraction limit using optical spectral encoding of the spatial frequency," *J. Opt. A: Pure Appl. Opt.* **10**, 025304 (5pp) (2008).
8. D. Mendlovic, J. Garcia, Z. Zalevsky, E. Marom, D. Mas, C. Ferreira, and A. W. Lohmann, "Wavelength multiplexing system for a single mode image transmission," *Appl. Opt.* **36**, 8474–8480 (1997).
9. J. Solomon, Z. Zalevsky, and D. Mendlovic, "Geometrical superresolution using code division multiplexing," *Appl. Opt.* **42**, 32–40 (2005).
10. W. Lukosz, "Optical systems with resolving powers exceeding the classical limits II,"*J. Opt. Soc. Am.* **57**, 932–41 (1967).
11. Z. Zalevsky, P. García-Martínez and J. García, "Superresolution using gray level coding,"*Opt. Express* **14**, 5178–5182 (2006).
12. Z. Zalevsky, J. Garcia, P. Garcia-Martinez, and C. Ferreira, "Spatial information transmission using orthogonal mutual coherence coding," *Opt. Lett.* **20**, 2837–2839 (2005).
13. V. Mico, J. García, C. Ferreira, D. Sylman, and Z. Zalevsky, "Spatial information transmission using axial temporal coherence coding," *Opt. Lett.* **32**, 736–738 (2007).
14. A. W. Lohmann and D. Paris, "Superresolution for nonbirefringent objects," *J. Opt. Soc. Am.* **3**, 1037–1043 (1964).
15. A. Zlotnik, Z. Zalevsky, and E. Marom, "Superresolution with nonorthogonal polarization coding," *Appl. Opt.* **44**, 3705–3715 (2005)

Part IV
Algorithms for Imaging and Analysis

Chapter 19
Hybrid Imaging Systems for Depth of Focus Extension With or Without Postprocessing

F. Diaz, F. Goudail, B. Loiseaux, and J.-P. Huignard

Abstract We address two different methods to enhance the depth of focus (DOF) according to the application, using amplitude and/or phase masks. The first application consists in increasing the length of the focal line of an optic while preserving its transverse resolution. We propose a novel type of mask inspired from holographic principles that have some advantages compared with the classical binary-phase mask. We then address DOF enhancement in a hybrid imaging system composed of a pupil mask and a digital deconvolution step. We define an optimization criterion based on the final image quality and use it to determine the optimal parameters of phase masks. We show that the parameters may not correspond to point spread functions that are strictly invariant with respect to defocus.

19.1 Introduction

Depth of focus (DOF) enhancement by the use of pupil masks is a topic of great interest. Many types of pupil masks have been proposed with different properties and advantages which make them appropriate for different applications [1, 2]. The key point for choosing the appropriate mask is to define precisely the quality criterion related to the targeted application. The appropriate type of mask and mask parameters are then chosen to optimize this criterion.

In this chapter, we address two different types of applications. The first one consists in increasing the length of the focal line of an optic. This can have application in OCT imaging, for example [2]. For that purpose, the simplest method is probably to use amplitude masks [3]. Continuous phase masks [4] and binary-phase masks (BPMs) [2] can also enhance the DOF by increasing the size of the point

F. Diaz (✉)
Thales Research and Technology, Campus Polytechnique, 1 Avenue Augustin Fresnel, 91767 Palaiseau Cedex, France
and
Laboratoire Charles Fabry de l'Institut d'Optique, CNRS, Université Paris-Sud, Campus Polytechnique, RD 128, 91127 Palaiseau Cedex, France
e-mail: frederic.diaz@thalesgroup.com

T. Fournel and B. Javidi (eds.), *Information Optics and Photonics: Algorithms, Systems, and Applications*, DOI 10.1007/978-1-4419-7380-1_19,

spread function (PSF) in the direction of the optical axis, while maintaining the transverse resolution. In Section 19.2, we compare the performance of the classical BPM and of a novel type of mask inspired from holographic principles.

The second type of application is DOF enhancement in a hybrid imaging system composed of a pupil mask and a digital deconvolution step. Indeed, pupil masks that lead to large DOF enhancement provide strongly blurred images that become sharp only after postprocessing. Such designs allow to relax the constraints on the imaging lenses and to obtain better performance at lower fabrication cost. Using this principle, Cathey and Dowski [1] proposed to include a cubic phase mask that makes the PSF insensitive to defocus and an appropriate postprocessing step to recover the quality of the image. Different types of phase masks have been proposed to enhance the depth of field, such as exponential [5], logarithmic [6], polynomial [7], and rational [8] phase masks. These masks depend on parameters that have to be optimized for a given application. In Section 19.3, we define a quality criterion based on the final (deconvolved) image and use it to determine the optimal parameters of phase masks. We show that the optimal parameters may not correspond to PSFs that are strictly invariant with respect to defocus.

19.2 Uniformizing the Focal Line: The Holographically Generated Complex Mask

The simplest designs to enhance the DOF are probably amplitude masks. Continuous phase masks and BPMs can also be used, and they are usually considered more energy efficient than amplitude masks since they have unit intensity transmission. However, in DOF extension, the relevant parameter is not the total transmitted energy, but the "useful" one is the energy concentrated in the focal line. When this criterion is considered, amplitude/phase (complex) masks may have better performance [9, 10].

We consider in this section that one wants to produce a light distribution as uniform as possible along the focal line. We will define the DOF as the length of the region along the optical axis where the amplitude distribution has fluctuations lower than a given threshold. For a given DOF and allowed fluctuations, our purpose is to determine the mask that maximizes the mean intensity in the DOF region. To reach this goal, we will compare two types of methods: a classical one based on BPM and a new approach based on a holographically generated complex mask (HGCM).

19.2.1 Binary-Phase Mask

The BPM is composed of an array of n concentric rings whose phases are alternatively 0 and π. Their outer radii are denoted as $r_1, r_2, \ldots, r_{n-1}$, and are normalized with respect to the pupil radius.

Assuming a small aperture compared with the focal length, we obtain the following normalized amplitude distribution near the optical axis generated by a lens and a BPM:

$$A(\rho, z) = \frac{\exp(jkz)}{z} \exp\left(\frac{jk\rho^2}{2z}\right) \int_0^R P(\rho_\mathrm{P})\rho_\mathrm{P} J_0\left(\frac{k\rho\rho_\mathrm{P}}{2}\right) \exp\left[\frac{jk\rho_\mathrm{P}^2}{2}\left(\frac{1}{z} - \frac{1}{f}\right)\right] \mathrm{d}\rho_\mathrm{P}, \tag{19.1}$$

where λ is the wavelength, $k = 2\pi/\lambda$, f is the focal length, R is the radius of the lens, ρ and z are the transverse and axial coordinates, respectively, ρ_P is the transverse coordinate in the pupil plane, J_0 is the zero-order Bessel function, and $P(\rho_P)$ is the complex amplitude of the BPM, equal to ± 1.

Using this equation, by performing an exhaustive search, one can find the parameters of the BPM that generate a desired DOF with the highest possible intensity inside the focus region.

Figure 19.1 illustrates the performance of the BPM and shows the PSF of a single lens, a lens with a BPM, and a diaphgramed lens, for a focal length of $f = 160$ mm, an aperture of $D = 16$ mm, at a wavelength of $\lambda = 1,064$ nm. Considering the DOF as the full length at half maximum of the focal line, it is thus possible with the BPM to increase the DOF by a factor of 3 compared with a single lens while maintaining

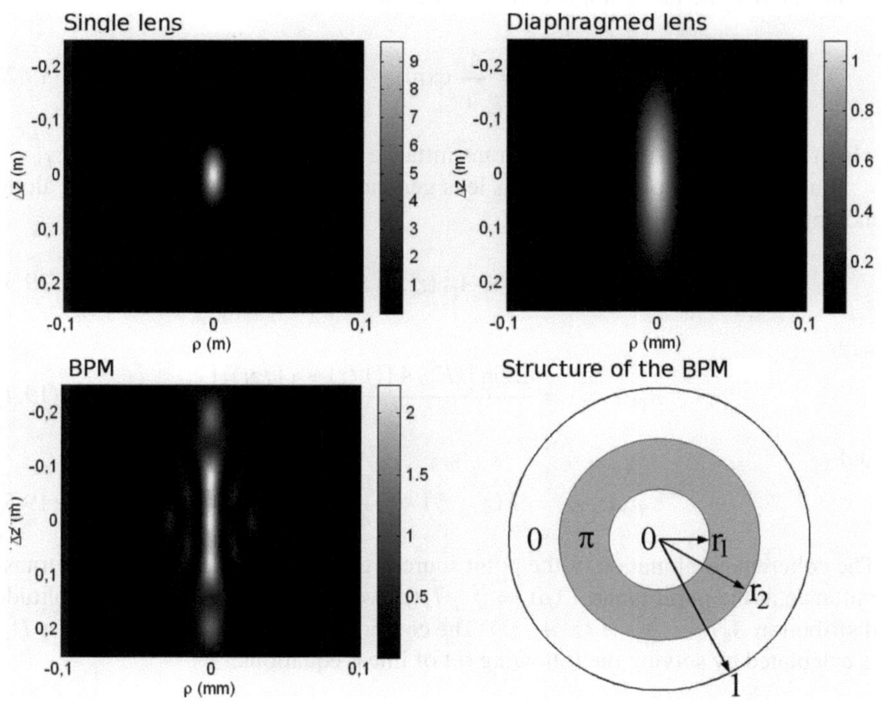

Fig. 19.1 PSF of a single lens, a BPM, and a lens diaphragmed by a factor of $\sqrt{3}$– structure of a BPM

the full width at half maximum. On the contrary, using a diaphragmed lens do not allow to maintain the transverse resolution and involves a loss of energy in the focus region.

19.2.2 Holographically Generated Complex Mask

As an alternative to the BPM, we now propose to design a complex mask with the classical approach used for computer-generated holograms [11]. Such a design first specifies the desired normalized amplitude at chosen points along the optical axis, and then computes the mask that generates this distribution. This makes it possible to "tailor" the shape of the focal line. We call this design as "holographically generated complex mask" (HGCM).

Let us consider n points situated at the coordinates $z_i, i \in [1, n]$ on the optical axis, uniformly distributed throughout the desired DOF region, and set the desired normalized amplitude at each point equal to D_i. In order to obtain low variations inside the DOF region, we first set $D_i = 1$. To calculate the HGCM providing the normalized amplitudes D_i at the coordinates z_i, respectively, we consider these points as elementary radiating point sources that emit waves of amplitudes A_i. Each point source generates complex amplitude at the pupil plane

$$T_i(\rho) = \frac{A_i}{z_i} \exp\left(-\frac{jk\rho^2}{2z_i}\right), \tag{19.2}$$

which can be seen as the complex transmittance of a lens with a focal length z_i.

Illuminated by a plane wave, this lens generates an amplitude distribution along the optical axis equal to:

$$A_z(z, A_i, z_i) = A_i \times A_{\mathrm{pt}}(z, z_i) \times \exp\left(j\varphi_{\mathrm{pt}}(z, z_i)\right) \tag{19.3}$$

with

$$A_{\mathrm{pt}}(z, z_i) = \frac{2\sin\left[kR^2/4\left((1/z) - (1/z_i)\right)\right]}{k(z_i - z)} \tag{19.4}$$

and

$$\varphi_{\mathrm{pt}}(z, z_i) = k(z - z_i) + \frac{kR^2}{4}\left(\frac{1}{z} - \frac{1}{z_i}\right) \tag{19.5}$$

The coherent combination of the point sources generates a complex mask of transmittance at the pupil plane $T(\rho) = \sum_i T_i(\rho)$ which corresponds to the amplitude distribution $A_{z,\mathrm{tot}} = \sum_i A_z(z, A_i, z_i)$. The coefficient A_i so that, $\forall i$, $A_{z,\mathrm{tot}}(z_i) = D_i$, is calculated by solving the following set of linear equations:

$$\mathbf{D} = \mathbf{Z} \times \mathbf{A}, \tag{19.6}$$

where vector **D** contains D_i, vector **A** contains A_i, and **Z** is a $n \times n$ matrix containing the amplitude associated to each point source i, with $A_i = 1$ at the coordinates z_j: $Z_{i,j} = A_z(z_j, 1, z_i)$. The values of A_i are obtained by inverting the matrix **Z**, and the complex transmission of the obtained HGCM is given as follows:

$$T(\rho) = \frac{\sum_i T_i(\rho)}{\max_\rho \left|\sum_i T_i(\rho)\right|} \exp\left(\frac{jk\rho^2}{2f}\right), \tag{19.7}$$

where the phase already introduced by the lens has been removed and the transmittance has normalized so that its maximal modulus is 1. Since we consider spatially incoherent imaging, the interesting value is the intensity distribution along the optical axis:

$$I_z(z) \propto \left|\sum_i A_i \times A_{\mathrm{pt}}(z, z_i) \exp\left(j\varphi_{\mathrm{pt}}(z, z_i)\right)\right|^2. \tag{19.8}$$

While the function $A_{\mathrm{pt}}(z, z_i)$ has relatively slow variations, the $\exp\left(j\varphi_{\mathrm{pt}}(z, z_i)\right)$ phase terms introduce fast and large intensity fluctuations between the point sources. These terms can be removed by choosing point sources satisfying the relation $\varphi_{\mathrm{pt}}(z_i, f) = 2\pi\, p_i$ with p_i integer numbers.

Using this method, we achieved a DOF of 1.6 mm with intensity fluctuation below 1% [11] (Fig. 19.2).

Figure 19.3 shows the PSF of the lens, the HGCM, and the BPM providing the same DOF with the same intensity fluctuation threshold along the optical axis. Their performance are comparable, although the HGCM has an average intensity transmission of 30%.

This lower transmission is compensated by a better concentration of energy in the DOF region, thanks to the contribution of amplitude part of the filter. Moreover, with the HGCM, it is possible to achieve an accurate localization of the DOF

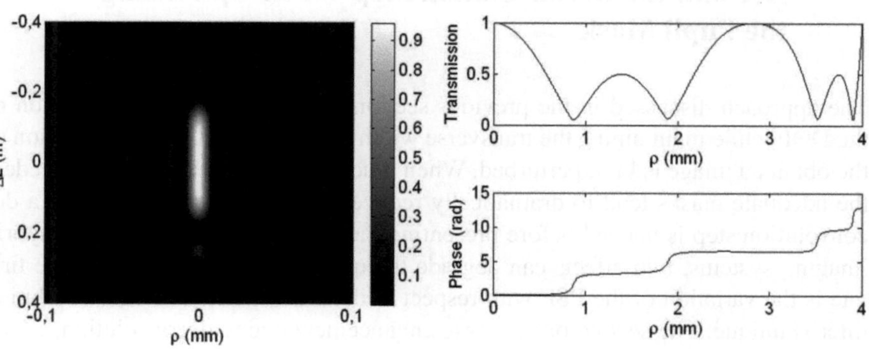

Fig. 19.2 PSF of the HGCM with a DOF of 1.6 mm and fluctuations below 1%, and its corresponding profile

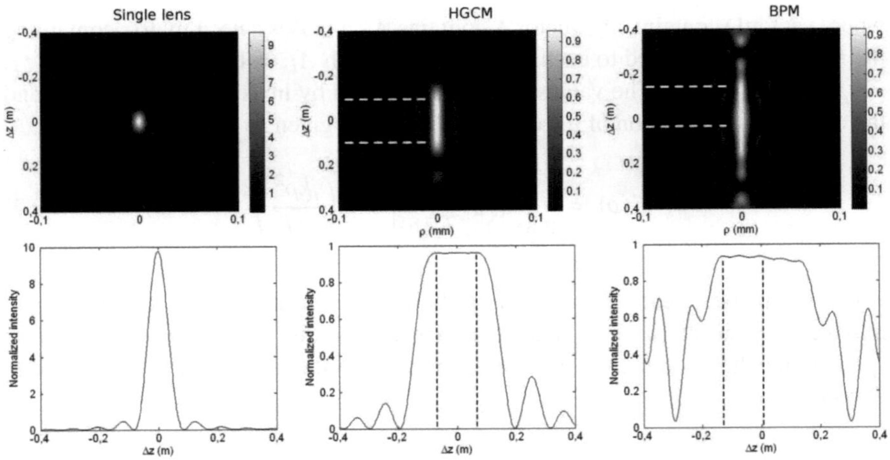

Fig. 19.3 PSF of the lens, the BPM, and the HGCM for a DOF of 1.6 mm and fluctuations below 1%. The DOF region where the variations are below 1% is delimited with the dotted lines

region which is directly linked to the choice of point sources. In our example, the DOF region is centered on the focal point, whereas the DOF region of the BPM is asymmetrically distributed between $z = 158.4$ mm and $z = 160$ mm. Further advantages of the HGCM are the sharper decrease of the intensity at the limit of the DOF region, lower sidelobes, and the better invariance of the transverse spot size with respect to z inside the DOF region. The HGCM is thus a valuable alternative to the BPM for applications when very low fluctuations of the amplitude distribution along the optical axis are required and when the shape of the focal line has to be precisely tailored.

19.3 DOF Extension with Hybrid Imaging System: Taking into Account the Deconvolution Step When Optimizing the Pupil Mask

The approach discussed in the previous section allows for moderate extension of the DOF while maintaining the transverse width of the PSF so that the resolution of the obtained image is little perturbed. When much higher DOF extension is needed, the adequate masks tend to dramatically reduce the resolution. In this case, a deconvolution step is needed before presenting the image to the user. In such hybrid imaging systems, two effects can degrade the quality of the final image. The first one is the variation of the PSF with respect to defocus, whereas the deconvolution filter is unique. The second one is noise enhancement due to deconvolution, which depends on the frequency response of the mask. Indeed, with usual phase masks, the more invariant to defocus the PSF is the more they behave as low-pass filters. In such imaging systems, the relevant quality criterion is the quality of the final image obtained after the deconvolution step. We derive in this section a closed-form expression of such a quality criterion that takes into account the detection noise and the defocus range where the system should provide an optimal image quality. As an example, it will be used to optimize the parameter of a cubic phase mask [1].

19.3.1 Definition of an Image Quality Criterion

We consider incoherent imaging systems. When imaging an object with intensity distribution $O(r)$, the intensity distribution measured $I_\psi(r)$ by the sensor at a defocus ψ is given as follows:

$$I_\psi(r) = h_\psi(r)^* O(r) + n(r), \tag{19.9}$$

where $n(r)$ is the detection noise and $h_\psi(r)$ is the PSF of the optical system for a given defocus ψ which is given as follows:

$$\psi = \frac{\pi R^2}{\lambda}\left(\frac{1}{d_{\mathrm{O}}} + \frac{1}{d_{\mathrm{I}}} - \frac{1}{f}\right), \tag{19.10}$$

where R is the radius of the aperture, f, d_{O}, and d_{I} are, respectively, the focal length, the object distance, and the image sensor plane distance, and the symbol "*" refers to the convolution operation.

It is assumed that an estimator of the object is obtained through a linear deconvolution filter $d(r)$, and has the expression $\hat{O}(r) = d(r) * I(r)$. The mean-squared error (MSE) between the true object and its estimate for a given defocus ψ is thus:

$$\mathrm{MSE}_\psi = \left\langle \left| O(r) - d(r) * I_\psi(r) \right|^2 \right\rangle. \tag{19.11}$$

Assuming that $n(r)$ and $O(r)$ are stationary random processes with power spectral densities $S_{nn}(\nu)$ and $S_{OO}(\nu)$, it can be shown that

$$\mathrm{MSE}_\psi = \int \left| \tilde{d}(\nu)\tilde{h}_\psi(\nu) - 1 \right|^2 S_{OO}(\nu)\mathrm{d}\nu + \int \left| \tilde{d}(\nu) \right|^2 S_{nn}(\nu)\mathrm{d}\nu, \tag{19.12}$$

where the symbol $\sim$ refers to the Fourier transform and ν denotes the spatial frequency. MSE_ψ is the sum of two terms: the first one results from the inadequacy of the deconvolution filter with the actual PSF, since the PSF varies with the defocus, and the second one is the error due to the noise that is enhanced by the deconvolution filter. These two terms are antagonistic, thus minimization of the MSE will be a compromise between them.

Considering the defocus range where the optical system should provide an optimal image quality, the deconvolution filter should be efficient for this whole defocus range $\psi_j \in [0, \psi_{\text{defoc max}}]$. Thus, it is possible to minimize the averaged MSE over the defocus range, which is estimated by averaging MSE_ψ over m values of the defocus that is uniformly distributed in the range $\psi_j \in [0, \psi_{\text{defoc max}}]$:

$$\begin{aligned}\text{MSE}_{\text{mean}} &= \frac{1}{m}\sum_{j=1}^{m}\text{MSE}_{\psi_j},\\ &= \frac{1}{m}\sum_{j=1}^{m}\left[\int\left|\tilde{d}(\nu)\tilde{h}_\psi(\nu)-1\right|^2 S_{OO}(\nu)\mathrm{d}\nu + \int\left|\tilde{d}(\nu)\right|^2 S_{nn}(\nu)\mathrm{d}\nu\right],\\ &= \int S_{OO}(\nu)\left[\frac{1}{m}\sum_{j=1}^{m}\left|\tilde{d}(\nu)\tilde{h}_\psi(\nu)-1\right|^2\right]\mathrm{d}\nu + \int\left|\tilde{d}(\nu)\right|^2 S_{nn}(\nu)\mathrm{d}\nu.\end{aligned} \tag{19.13}$$

To minimize MSE_{mean}, we derivate it with respect to $\tilde{d}(\nu)$ in order to obtain the following filter:

$$\begin{aligned}&\frac{\partial \text{MSE}_{\text{mean}}}{\partial \tilde{d}(\nu)} = 0,\\ &\Leftrightarrow S_{OO}(\nu)\frac{1}{m}\sum_{j=1}^{m}\tilde{h}_{\psi_j}(\nu)\left[\tilde{d}^*(\nu)\tilde{h}^*_{\psi_j}(\nu)-1\right] + Snn(\nu)\tilde{d}^*(\nu) = 0,\\ &\Leftrightarrow S_{OO}(\nu)\frac{1}{m}\tilde{d}(\nu)\sum_{j=1}^{m}\left|\tilde{h}_{\psi_j}(\nu)\right|^2 - S_{OO}(\nu)\frac{1}{m}\sum_{j=1}^{m}\tilde{h}^*_{\psi_j}(\nu) + Snn(\nu)\tilde{d}(\nu) = 0,\\ &\Leftrightarrow \tilde{d}(\nu) = \frac{\sum_{j=1}^{m}\tilde{h}^*_{\psi_j}(\nu)}{(1/m)\sum_{j=1}^{m}\left|\tilde{h}^*_{\psi_j}(\nu)\right|^2 + (S_{nn}(\nu)/S_{OO}(\nu))},\end{aligned} \tag{19.14}$$

which is an averaged Wiener filter.

The proposed image quality criterion is the output signal-to-noise ratio (SNR), which depends on MSE_{mean} calculated with $n > m$ defocus values:

$$\text{SNR}_{\text{out}} = \frac{\int S_{OO}(\nu)\mathrm{d}\nu}{\text{MSE}_{\text{mean}}}. \tag{19.15}$$

The amount of noise in the scene is quantified by the input SNR which is defined as follows:

$$\text{SNR}_{\text{in}}(\text{dB}) = 10\log_{10}\left[\frac{\int S_{OO}(\nu)\mathrm{d}\nu}{\int S_{nn}(\nu)\mathrm{d}\nu}\right]. \tag{19.16}$$

And the SNR at a given defocus is defined as:

$$\text{SNR}_\psi = \frac{\int S_{OO}(\nu)\mathrm{d}\nu}{\text{MSE}_\psi}. \tag{19.17}$$

19.3.2 Example of Application to the Cubic Phase Mask

As an example, we consider the cubic phase mask which is a pure phase mask whose PSF at the focal plane is $h_0(x, y) = |\mathrm{FT}\{P(x, y) \exp[i\phi(x, y)]\}|^2$ with:

$$\phi(x, y) = \alpha \times (x^3 + y^3), \tag{19.18}$$

where x and y are the coordinates of the pupil, normalized to 1, $P(x, y)$ is the pupil function, equal to 1 when $\sqrt{x^2 + y^2} \leq 1$ and 0 otherwise, and α is a parameter that will be optimized according to the SNR criterion, and FT refers to the Fourier transform.

The input SNR is set at $\mathrm{SNR_{in}} = 28\,\mathrm{dB}$ and the maximal defocus value is $\psi_{\mathrm{defoc\,max}} = 9.2$. The object is the classical "Lena" image. The value of $\mathrm{SNR_{out}}$ as a function of α is displayed in Fig. 19.4. It is seen that the optimal phase mask parameter is $\alpha_{\mathrm{opt}} = 10$, leading to $\mathrm{SNR_{out}} = 16.4\,\mathrm{dB}$. With lower α, the response of the optical system to the defocus is not insensitive enough, as shown in Fig. 19.5. It results in a rapid decrease of SNR_ψ with respect to the defocus (Fig. 19.6). With larger α, the response is much insensitive to defocus, leading to a lower variation of SNR_ψ, but the optical system has then a low-pass behavior and the noise is thus enhanced by the deconvolution step. The parameter $\alpha_{\mathrm{opt}} = 10$ is thus a compromise between theses two effects.

The deconvoluted images are displayed in Fig. 19.7 and showed that the criterion $\mathrm{SNR_{out}}$ has an effective correlation with the visual perception. One can notice in Fig. 19.7 the blurred result obtained at $\alpha = 5$ with a defocus $\psi = 9.2$, which corresponds to the decrease of SNR_ψ with respect to the defocus. It results from the

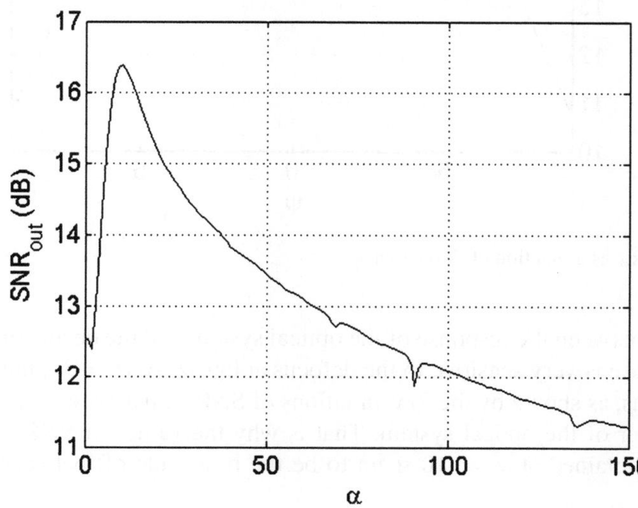

Fig. 19.4 $\mathrm{SNR_{out}}$ as a function of α for Lena

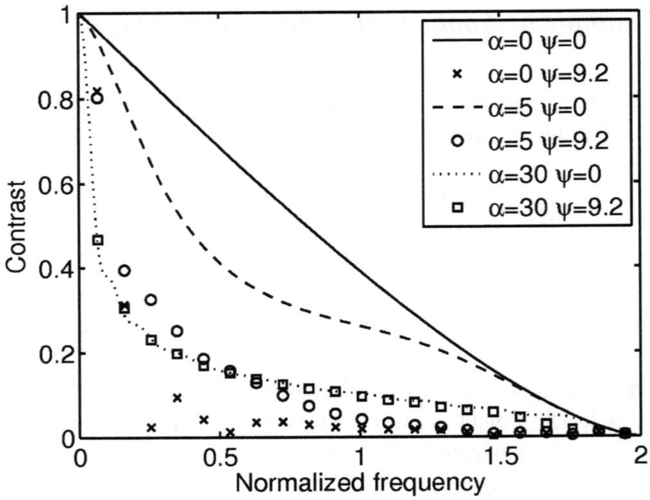

Fig. 19.5 MTF for a lens, and at $\alpha = 5$ and $\alpha = 30$

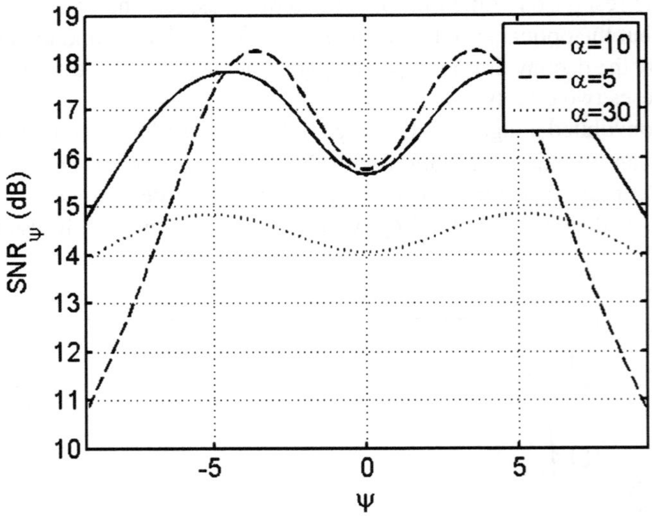

Fig. 19.6 SNR_{ψ} as a function of ψ for Lenan

inadequacy between the response of the optical system and the deconvolution filter, as the response is very sensitive to the defocus at low α. At $\alpha = 30$, the results are homogeneous, as shown by the low variations of SNR_{ψ}, but noisier because of the low-pass filter of the optical system. That is why the values of SNR_{ψ} are small. The results obtained at $\alpha = 10$ seem to be the best trade-off between these two effects.

In consequence, it can be stated that an optimal image quality can be achieved with a nonstrictly invariant PSF.

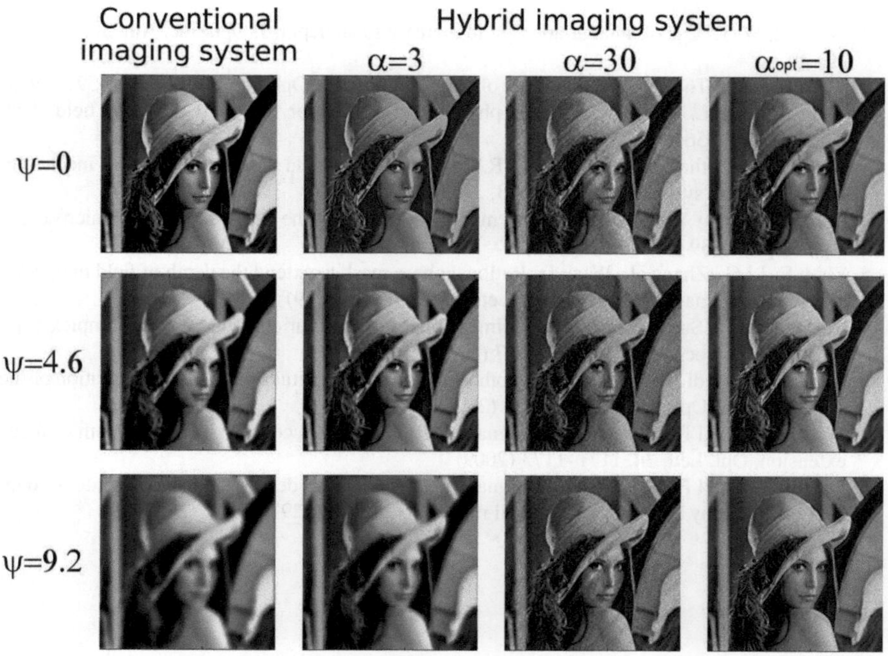

Fig. 19.7 Results obtained with Lena with a conventional imaging system and with a hybrid imaging system at several values of α

19.4 Conclusion

We have discussed two approaches to DOF extension of an optical system, which are adapted to different applications. In the first place, we have addressed the case where a long focal line is needed, while maintaining the transverse resolution. We have proposed a new HGCM which provides better results than the classical binary phase mask, despite its non-negligible absorption, when low fluctuations of intensity along the optical axis are required and when the shape of the focal line has to be precisely tailored. The second application was hybrid imaging systems with a deconvolution step. In this case, we have shown that the optimal shape of the mask results from a trade-off between the insensitivity of the PSF and the defocus, required because the deconvolution filter is unique and the noise enhancement is induced by the deconvolution process.

References

1. Cathey W. T., Dowski E. R.:New paradigm for imaging systems, Appl. Opt. 41, 6080–6092 (2002)
2. Liu L., Diaz F., Wang L., Loiseaux B., Huignard J-P., Sheppard C. J. R., Chen N.: Superresolution along extended depth of focus with binary-phase filters for the Gaussian beam, J. Opt. Soc. Am. A 25, 2095–2101 (2008)

3. Welford, W. T.: Use of annular aperture to increase focal depth, J. Opt. Soc. Am. 50, 749–753 (1960)
4. McLeod, J. H.: The axicon: a new type of optical element, J. Opt. Soc. Am. 44, 592–597 (1953)
5. Yang Q., Liu L., Sun J.:Optimized phase pupil masks for extended depth of field, Opt. Commun. 272, 56–66 (2007)
6. Sherif S. S., Cathey W. T., Dowski E. R.:Phase plate to extend the depth of field of incoherent hybrid imaging systems, Appl. Opt. 43, 2709–2721 (2004)
7. Caron N., Sheng Y.:Polynomial phase masks for extending the depth of field of a microscope, Appl. Opt. 47, 39–43 (2008)
8. Zhou F., Li G., Zhang H., Wang D.:Rational phase mask to extend the depth of field in optical-digital hybrid imaging systems, Opt. Lett. 34, 380–382 (2009)
9. Yun M., Liu L., Sun J., Liu D.:Three-dimensional superresolution by three-zone complex pupil filters, J. Opt. Soc. Am. A 22, 272–277 (2005)
10. Bagheri S., Javidi B.: Extension of depth of field using amplitude and phase modulation of the pupil function, Opt. Lett. 33, 757–759 (2008)
11. Diaz F., Goudail F., Loiseaux L., Huignard J.-P.: Design of a complex filter for depth of focus extension, Opt. Lett. 34, 1171–1173 (2009)
12. Diaz F., Goudail F., Loiseaux L., Huignard J.-P.: Increase in depth of field taking into account deconvolution by optimization of pupil mask, Opt. Lett. 34, 2970–2972 (2009)

Chapter 20
Multispectral Image Pansharpening Based on the Contourlet Transform

Israa Amro and Javier Mateos

Abstract Pansharpening is a technique that fuses the information of a low-resolution multispectral image and a high-resolution panchromatic image, usually remote sensing images, to provide a high-resolution multispectral image. In the literature, this task has been addressed from different points of view being one of the most popular wavelet-based algorithms.

Recently, a new transform, the contourlet transform, has been proposed. This transform combines the advantages of the wavelets transform, with a more efficient directional information representation. The result is a flexible multiscale, multidirection and shift-invariant decomposition that can be efficiently implemented via the a'trous algorithm.

In this chapter, we compare the wavelet-based pansharpening with existing contourlet-based approaches and propose a new pansharpening method based on the contourlet transform. The performance of the contourlet in general, and the proposed method in particular, is assessed numerically and visually for Landsat and SPOT images.

20.1 Introduction

Earth observation satellites provide two different classes of images: a panchromatic image (PAN) with high spatial and low spectral resolutions and a multispectral image (MS) with high spectral and low spatial resolutions. In optical remote sensing, with physical and technological constraints, some satellite sensors supply the spectral bands needed to distinguish features spectrally but not spatially, while other satellite sensors supply the spatial resolution for distinguishing features

I. Amro (✉)
Universidad de Granada, Granada, Spain
and
Al-Quds Open University, Hepron, Palestine
e-mail: iamro@correo.ugr.es; isamro@qou.edu

T. Fournel and B. Javidi (eds.), *Information Optics and Photonics: Algorithms, Systems, and Applications*, DOI 10.1007/978-1-4419-7380-1_20,

spatially but not spectrally. The fusion of high spatial resolution PAN image and the high spectral resolution MS image, providing more comprehensive information [1], is an important issue for many remote sensing and mapping applications. In general, pansharpening algorithms improve the spatial resolution of the MS image while simultaneously retaining its spectral information [2] and provides feature enhancement and improved classification as information from two different images is used effectively [3]. Nowadays pansharpened products are becoming very popular (see, for example, Google Earth [4]) and data providers are offering higher and higher amounts of them at lower and lower costs [5].

In the literature, many pansharpening methods have been proposed for combining PAN with MS image, see [3, 5] for a detailed review. Among them, methods such as Intensity–Hue–Saturation (IHS) [6] and Principal Component Substitution (PCS) [7, 8] provide superior visual high-resolution multispectral images but ignore the requirement of high-quality synthesis of spectral information. More recently, an underlying multiresolution analysis employing the discrete wavelet transform has been used in image fusion. Properties, such as multiresolution, localization, critical sampling, and limited directionality (horizontal, vertical, and diagonal directions) have made the wavelet transform a popular choice for feature extraction, image denoising, and pansharpening. However, wavelets fail to capture the smoothness along the contours [9]. The contourlet transform, an alternative multiresolution approach, provides an efficient directional representation and also efficient in capturing intrinsic geometrical structures of the natural image along the smooth contours [10]. Remote sensing images have presence of natural and man-made objects, e.g., rivers, roads, coastal areas, buildings, which indicate higher geometrical content. Thus, the transformations taking in consideration the geometric structure along with other properties of wavelet transformation will be more useful for pansharpening.

In most of the proposed methods, no explicit physical information about the detection system has been taken into account. However, a new technique proposed in [11] was used to define a wavelet-based fusion method that does incorporate information from the spectral response of the sensor in each band of the low-resolution MS image (LRMS). This prior knowledge is employed in the transformation model that injects spatial detail into the LRMS image. In this chapter, we will briefly describe the wavelet-based pansharpening algorithms and the contourlet-based algorithms sharing the same ideas, propose a new method that cast the WiSpeR method defined in [11] using the contourlet transform and we compare it with some of the most popular methods for pansharpening described in the literature.

The chapter is organized as follows. Section 20.2 provides a short explanation of the contourlet transform. Section 20.3 describes the contourlet-based pansharpening and the proposed algorithm. Experimental results and comparisons are presented in Sect. 20.4 for different datasets and, finally, Sect. 20.5 concludes the chapter.

20.2 Contourlet Transform

Contourlets provide a new system representation for image analysis [10]. The contourlet transform is so-called because of its ability to capture and link the point of discontinuities to form a linear structure (contours). The two-stage process used to derive the contourlet coefficients involves a multiscale transform and a local directional transform. The point of discontinuities and multiscale transformation is obtained via the Laplacian pyramid. The local directional filter bank is used to group these wavelet-like coefficients to obtain a smooth contour. Contourlets provide $2l$ directions at each scale, where l is the number of required orientation. This flexibility of having different numbers of direction at each scale makes contourlets different from other available multiscale and directional image representation [12]. Similar to wavelets, contourlets also have different implementations of the subsampled and nonsubsampled transforms:

1. *Discrete Contourlet Transform (CT)*: The discrete CT is developed in the discrete domain using the fast iterated nonseparable filter banks having an order of N operations for N-pixel images. The transformation stage includes two filter banks: the Laplacian pyramid to generate multiscale decomposition and the directional filter bank (DFB) to reveal directional details at each decomposition level [10] as illustrated in Fig. 20.1a. Similar to the discrete wavelet transform, the discrete contourlet transform is also shift variant.
2. *Nonsubsampled Contourlet Transform (NSCT)*: The NSCT provides a complete shift-invariant and multiscale representation, similar to the redundant wavelet

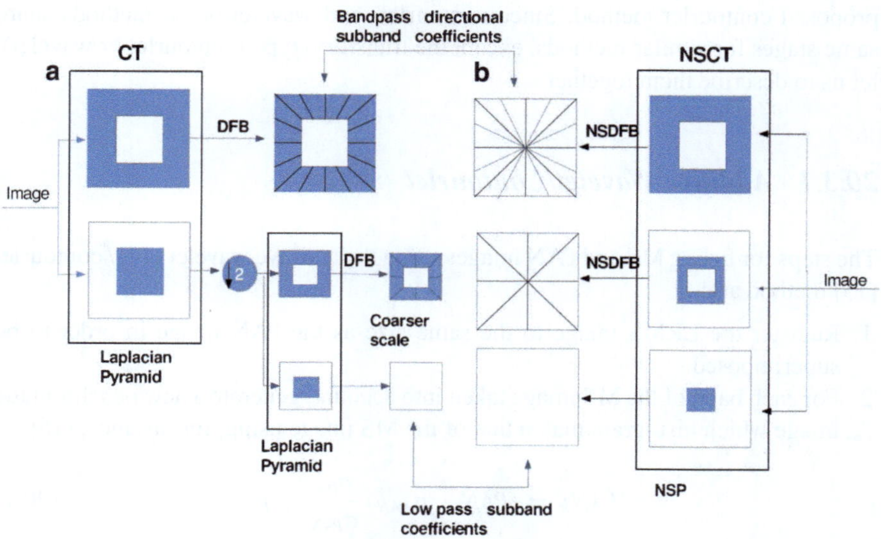

Fig. 20.1 Discrete contourlet transform: (**a**) subsampled contourlet transform, (**b**) nonsubsampled contourlet transform

transform [13], with a fast implementation. The building block of the NSCT is the 2-D two-channel nonsubsampled filter banks (NSFBs). The NSCT is also obtained via a two-stage non shift-invariant process [13] as depicted in Fig. 20.1b. The first part achieves the multiscale property, via the nonsubsampled pyramid (NSP) subband decomposition, while the second part provides directionality information using nonsubsampled directional filter bank (NSDFB). Both stages of the NSCT are constructed to be invertible in order to have an overall invertible system.

20.3 Wavelet- and Contourlet-Based Pansharpening

A number of pansharpening methods using the wavelet and, more recently, the contourlet transform has been proposed. In general, all the transform-based fusion methods consist of three stages. The first stage provides a subband and directional decomposition by the application of the subsampled or nonsubsampled wavelet or contourlet transform to the PAN and MS images. It is followed by the application of various fusion rules onto the transform coefficients. These fusion rules usually comprise, for instance, substituting the original MS coefficient bands by the coefficients of the PAN image or adding the coefficients of the PAN to the coefficients of the original MS bands weighted sometimes, like for the method we propose in this chapter, by a factor related with the contribution of the PAN image to each MS band. The fusion schemes ends with the inverse transform.

This chapter summarizes the most important wavelet-based pansharpening methods, and compares them with the existing contourlet-based ones and the new proposed contourlet method. Since contourlet- and wavelet-based methods share same stages for similar methods, except the transform type (contourlet or wavelet), let us to describe them together.

20.3.1 Additive Wavelet/Contourlet

The steps for fusing MS and PAN images using the additive wavelet [14] / contourlet [15] method are:

1. Register the LRMS image to the same size as the PAN image in order to be superimposed.
2. For each band of the MS image taken into account, generate a new panchromatic image which histogram match that of the MS image using, for instance, [16]

$$PAN_k = (PAN - \mu_{PAN}) \frac{\sigma_{b_k}}{\sigma_{PAN}} + \mu_{b_k}, \tag{20.1}$$

where μ_{PAN}, μ_{b_k} are the mean of the PAN and the MS band b_k, respectively, $k \in B$, and B is the set of bands we are interested in. σ_{PAN} and σ_{b_k} are the standard deviation of PAN and MS band b_k, respectively.

3. Apply the wavelet/contourlet transform to each histogram-matched panchromatic images. Repeat the same transform to each MS band.

$$PAN_k = \bigcup_{i=1}^{n} wPAN_k^i \cup PAN_k^r, \tag{20.2}$$

$$b_k = \bigcup_{i=1}^{n} wb_k^i \cup b_k^r, \quad \forall k \in B, \tag{20.3}$$

where $wPAN_k^i$ and wb_k^i are the wavelet/contourlet coefficients for PAN and MS bands, respectively, PAN_k^r and wb_k^r are the residual (low pass filtered version of original) images of PAN and MS bands, respectively, n is the wavelet/contourlet resolution levels, usually $n = 2$ or 3. The $\cup$ operator means the composition operator that merges the different wavelet/contourlet bands since each band may have different resolution. Note that in the nonsubsampled case, this operator means just adding the different bands.
4. Introduce the details of the panchromatic image into each MS band adding the wavelet/contourlet coefficients of the panchromatic image to those of the MS image

$$b_k^{\text{coef}} = \bigcup_{i=1}^{n} (wPAN_k^i + wb_k^i), \tag{20.4}$$

where b_k^{coef} is the new wavelet/contourlet coefficients of the MS band k.
5. Apply the inverse wavelet/contourlet transform to each MS transformed band

$$b_k^{\text{new}} = b_k^{\text{coef}} \cup b_k^r \tag{20.5}$$

to obtain b_k^{new}, the pansharpened MS band k, $k \in B$. Note that since for the undecimated case, $\sum_{i=1}^{n} wb_k{}^i + b_k^r = b_k$, we do not need to decompose the MS image and we can add b_k to the corresponding PAN coefficients.

20.3.2 *Substitutive Wavelet/Contourlet*

The substitute wavelet/contourlet methods are quite similar to the additive ones but, instead of adding the information of the panchromatic image to each band of the MS image, the pansharpening method simply replaces the MS detail bands with the details obtained by the panchromatic image following these steps for wavelet [17] and contourlet [18] reconstruction:

1. Perform the first 3 steps of the algorithm described in Sect. 20.3.1.
2. For each MS band k, perform the inverse wavelet/contourlet transform to the transformed image formed by the wavelet/contourlet coefficient planes of the

histogram-matched PAN image $wPAN_k^i, i = 1 \ldots n$, and the residual band of the MS image k, that is,

$$b_k^{\text{new}} = \bigcup_{i=1}^{n} wPAN_k^i \cup b_k^r. \tag{20.6}$$

20.3.3 IHS Wavelet/Contourlet

One of the most popular image pansharpening methods are those based on the IHS transformation. The main drawback of these methods is the high distortion of the original spectral information that the resulting MS images present. To avoid this problem, the IHS transformation is followed by the wavelet or contourlet transform to take advantage of the multiresolution property of this transform. Another disadvantage of the IHS-based method, that cannot be solved by these transforms, is that they can only work with three bands due to the IHS transformation.

20.3.3.1 Additive IHS

In order to perform the wavelet [14] and contourlet [19, 20] additive IHS pansharpening, the following steps are followed:

1. Register the LRMS image to the same size as the PAN image in order to be superimposed.
2. Apply the IHS transform to the RGB composition of three MS image, using,

$$\begin{pmatrix} I \\ v1 \\ v2 \end{pmatrix} = \begin{pmatrix} \frac{1}{3} & \frac{1}{3} & \frac{1}{3} \\ \frac{-1}{\sqrt{6}} & \frac{-1}{\sqrt{6}} & \frac{2}{\sqrt{6}} \\ \frac{1}{\sqrt{6}} & \frac{-1}{\sqrt{6}} & 0 \end{pmatrix} \begin{pmatrix} R \\ G \\ B \end{pmatrix}, \tag{20.7}$$

$$H = \tan^{-1}[v2/v1], \quad S = \sqrt{v1^2 + v2^2}.$$

3. Perform histogram matching between the panchromatic image and the intensity component of the IHS image using (20.1) to obtain PAN_I, the PAN image histogram-matched to the I band.
4. Apply wavelet/contourlet decomposition algorithm to the I band of the IHS image and to the histogram-matched PAN one using,

$$I = \bigcup_{i=1}^{n} wI^i \cup I^r, \tag{20.8}$$

$$PAN_I = \bigcup_{i=1}^{n} wPAN_I^i \cup PAN_I^r. \tag{20.9}$$

5. Generate the wavelet/contourlets coefficients of the pansharpened intensity image as the sum of the wavelet/contourlet coefficients of the initial intensity and the histogram-matched PAN image,

$$I^{\text{coef}} = \bigcup_{i=1}^{n} (wPAN_I^i + wI^i). \tag{20.10}$$

6. Apply the inverse wavelet/contourlet transform to reconstruct new intensity image I^{new},

$$I^{\text{new}} = I^{\text{coef}} \cup I^r. \tag{20.11}$$

   Note that, as already happened in the additive wavelet case, since for the undecimated case $\sum_{i=1}^{n} wI^i + I^r = I$, we do not need to decompose the I image.
7. Insert the spatial information of the panchromatic image into the MS one, by applying the inverse IHS transform,

$$\begin{pmatrix} R^{\text{new}} \\ G^{\text{new}} \\ B^{\text{new}} \end{pmatrix} = \begin{pmatrix} 1 & \frac{-1}{\sqrt{6}} & \frac{3}{\sqrt{6}} \\ 1 & \frac{-1}{\sqrt{6}} & \frac{-3}{\sqrt{6}} \\ 1 & \frac{2}{\sqrt{6}} & 0 \end{pmatrix} \begin{pmatrix} I^{\text{new}} \\ v1 \\ v2 \end{pmatrix}. \tag{20.12}$$

20.3.3.2 Substitutive IHS

In order to perform the wavelet [21] and contourlet substitutive IHS pansharpening, the following steps are followed:

1. Perform the first 4 steps of the Sect. 20.3.3.1.
2. Perform the inverse wavelet/contourlet transform to the wavelet/contourlet image formed by substituting the wavelet/contourlet coefficient planes of the intensity image with the corresponding wavelet/contourlet planes of the histogram-matched PAN image,

$$I^{\text{new}} = \bigcup_{i=1}^{n} wPAN_I^i \cup I^r. \tag{20.13}$$

3. Apply the inverse IHS transform using (20.12).

20.3.4 *PCA Wavelet/Contourlet*

The PCA-based method has been popularly used for spectral transformation because the first principal component (PC1) consists of the most variance, making it a suitable choice to replace the PAN component. Like IHS, the main drawback of this method is the high distortion of the original spectral information that the resulting MS images may present. To overcome this problem, Gonzalez et al. [22] proposed a pansharpening method based on the PCA and wavelets methods where only the details of PC1 are replaced by the details of the PAN image.

The steps we need to pansharpen an image using the PCA wavelet [14] and contourlet [12] methods are the following:

1. Register the LRMS image to the same size as the PAN image in order to be superimposed.
2. Apply the PCA transformation to the MS image to obtain the PC1 image.
3. Histogram match the PAN image to the PC1 image.
4. Apply a subsampled or nonsubsampled wavelet or contourlet transformation to the PC1 image and the histogram-matched PAN image.
5. Replace the detail wavelet or contourlet coefficients of PC1 with the detail wavelet or contourlet coefficients of the histogram-matched PAN image.
6. Perform inverse wavelet/contourlet transformation and inverse PCA transformation to obtain a PAN image.

20.3.5 WiSpeR/CiSpeR

The WiSpeR method can be understood as a generalization of different wavelet-based image fusion methods [11]. It uses a modification of the nonsubsampled additive wavelet algorithm where the contribution from the PAN image to each of fused bands depends on a factor generated both from the sensor spectral response and physical properties of the observed object.

The steps for merging MS and PAN images using WiSpeR method are as follows:

1. Register the LRMS image to the same size as the PAN image in order to be superimposed.
2. Generate new panchromatic images, whose histograms match those of each band of the MS image, using (20.1).
3. Perform the n undecimated wavelet planes transform only on the panchromatic images, using (20.2).
4. Calculate the spectral factor λ_k, related to b_k bands [11], where k is the band number.
5. Add the wavelet planes of the panchromatic decomposition to each band of the MS dataset, as the following:

$$b_k^{\text{new}} = b_k + \lambda_k \sum_{i=1}^{n} wPAN_k^i, \tag{20.14}$$

where b_k^{new} is the fused band k, $k \in B$, and B is the number of bands we take into account.

We propose a new contourlet panshapening method, named CiSpeR, that, similar to WiSpeR, depends on a spectral factor to determine the amount of spatial detail of the PAN image that has to be injected into each MS band but it uses the nonsubsampling contourlet transform with necessary filters, and n resolution levels and m directions in each level.

20.4 Experimental Results

The contourlet- and wavelet-based methods mentioned in Sect. 20.3 are evaluated by performing pansharpening on dataset acquired by SPOT 5, and Landsat 7 ETM+ satellites. The MS and PAN images are co-registered for each dataset. The panchromatic images, for both SPOT 5 and Landsat 7 ETM+, are shown in Fig. 20.2.

In the Landsat 7 ETM+ Dataset, we chose a region of interest of the MS image of 256 by 256 pixels with a pixel resolution of 28.5 m, and a region of interest of the PAN image of 512 by 512 pixels, with a pixel resolution of 14.25 m. The MS image consists of six bands from the visible and infrared (IR) region as follows: blue (0.45–0.515 μm), green (0.525–0.605 μm), red (0.63–0.690 μm), Near IR (0.75–0.90 μm), Mid IR (1.55–1.75 μm), and Mid IR (2.09–2.35 μm). The PAN image consists of a single band covering the visible and Near IR (0.52–0.90 μm). The scene, depicted in Fig. 20.3a, was acquired over The Netherlands on May 13, 2000.

The MS image in SPOT 5 dataset covers a region of interest of 256 by 256 pixels with a pixel resolution of 10 m, while the PAN image is 512 by 512 pixels with a pixel resolution of 5 m. The MS image consists of four bands from the visible and infrared region corresponding to green (0.50–0.59 μm), red (0.61–0.68 μm), Near IR (0.78–0.89 μm), Mid IR(1.58–1.75 μm), while the PAN image consists of a single band covering the visible and NIR (0.48–0.71 μm). The scene, depicted in Fig. 20.4a, was acquired over Sevilla (Spain) on February 15, 2003.

Pansharpening results are evaluated visually and numerically using some well-known global quality indexes. The Correlation coefficient (COR) [23] is a

Landsat 7 ETM+ SPOT 5

Fig. 20.2 Panchromatic imagery of the dataset

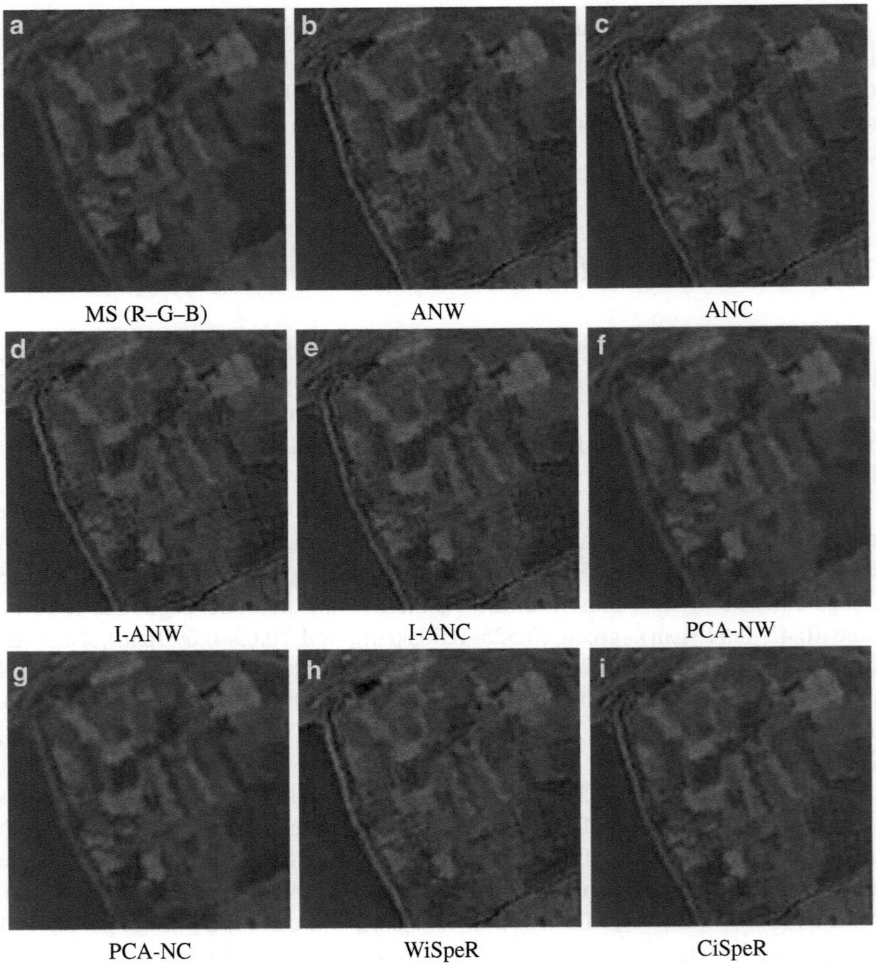

Fig. 20.3 (**a**) Low-resolution image formed from the R–G–B bands of the MS LandSat image. (**b**)–(**i**) Pansharpened images using the methods under study

normalized index that takes values between 0 and 1 (the higher the value the better the quality of the reconstruction) that assess the spatial similarity between each reconstructed multispectral image band and the panchromatic image. Spectral fidelity is assessed by means of the Universal Image Quality Index (UIQI) [24], which also takes values between 0 and 1, indicating a value of 1 the best quality, and the *erreur relative globale adimensionnelle de synthése* (ERGAS) [25] index, whose English translation is relative dimensionless global error in fusion, a global criterion for what the lower the value, the higher the quality of the multispectral image.

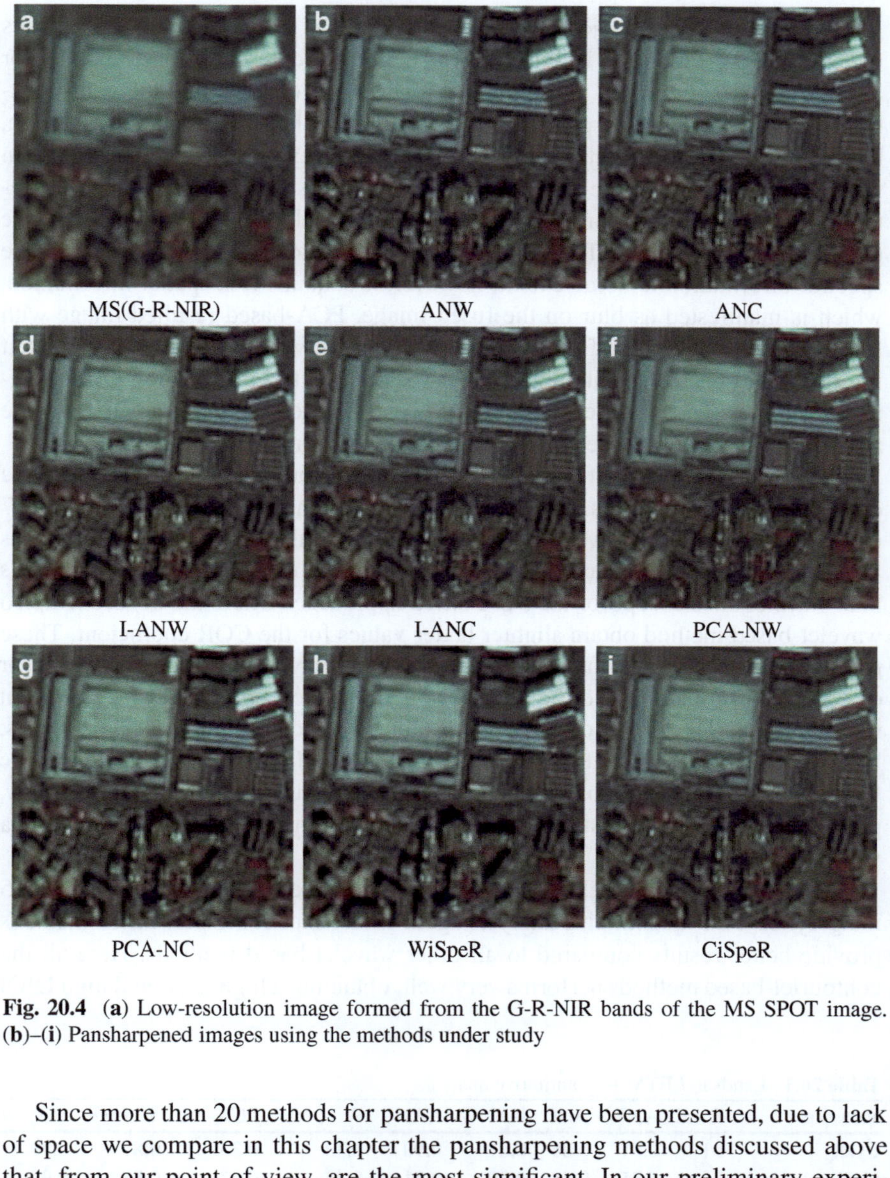

Fig. 20.4 (**a**) Low-resolution image formed from the G-R-NIR bands of the MS SPOT image. (**b**)–(**i**) Pansharpened images using the methods under study

Since more than 20 methods for pansharpening have been presented, due to lack of space we compare in this chapter the pansharpening methods discussed above that, from our point of view, are the most significant. In our preliminary experiments, we have realized that the nonsubsampled decompositions always provide better results than their subsampled counterpart so we are going to center on the nonsubsampled approaches. Also, we realized that, usually, the additive methods performs better than the substitute ones so, in this chapter, we are going to compare the following eight methods: Additive Nonsubsampled Wavelet (ANW), IHS Additive Nonsubsampled Wavelet (I-ANW), PCA Nonsubsampled Wavelet (PCA-NW), Additive Nonsubsampled Contourlet (ANC), IHS Additive Nonsubsampled Contourlet (I-ANC), PCA Nonsubsampled Contourlet (PCA-NC), WiSpeR, and CiSpeR.

The pansharpened images resulted from using the wavelet/contourlet methods under study are presented in Fig. 20.3 for Landsat 7 dataset, and in Fig. 20.4 for SPOT 5 dataset. From the pansharpened images, we observe that the contourlet-based and the new pansharpening method not only enhance the spatial resolution, but also preserve the spectral information of the original MS image better than wavelet ones. Although the IHS-based fusion methods improve the spatial resolution, they induce some color distortion, so these methods cannot be used to preserve the spectral information effectively. The additive contourlet method preserves the spectral information to some extent; however, it contains less spatial information, which is manifested as blur on the fused image. PCA-based resulted image with Landsat does not differ of the original MS image, while with SPOT dataset it improves the spatial resolution and preserves the spectral information of the original MS image. The proposed method, CiSpeR, enhances the images spatially while accurately preserves the spectral information for both imagery datasets.

The visual inspection of the images of Figs. 20.3 and 20.4 agree also with the quantitative analysis results. Table 20.1 shows the quantitative results for Landsat 7 imagery. The highlighted values in the table present the two highest ideal values for each measure. It is clear that the contourlet-based methods, have better results than wavelet ones except for the PCA-based approaches where the nonsubsampled wavelet-based method obtain slighter better values for the COR coefficient. These results are expected since the contourlet is known to have a better representation for directional information, and the nonsubsampled version provides a shift-invariant representation. The figures of merit in this table also indicate that some approaches, like PCA, are spectrally efficient but they did not add many spatial details, while the proposed CiSpeR approach achieves consistent results, spectrally and spatially, providing one of the highest COR values with one of the highest UIQI values and a very low ERGAS value.

Table 20.2 shows the evaluation results with the same quality indexes for SPOT 5 imagery. SPOT 5 quantitative analysis again shows that contourlet-based methods provide better results compared to all other wavelet-based methods. Here all the contourlet-based methods performs very well, obtaining a high COR and high UIQI

Table 20.1 Landsat 7 ETM+ quantitative analysis

Measure	Band	ANW	ANC	I-ANW	I-ANC	PCA-NW	PCA-NC	WiSpeR	CiSpeR
COR	b1	0.81	0.86	**0.87**	**0.90**	0.33	0.33	**0.87**	**0.90**
Ideal value = 1	b2	0.85	0.88	0.89	**0.91**	0.48	0.47	0.88	**0.90**
	b3	**0.90**	**0.90**	0.84	**0.85**	0.41	0.40	0.79	0.80
	b4	**0.95**	**0.95**	–	–	**0.93**	0.91	0.84	0.85
UIQI	b1	0.77	0.84	0.70	0.79	**0.94**	**0.95**	0.57	0.79
Ideal value = 1	b2	0.81	0.87	0.76	0.83	**0.95**	**0.96**	0.66	0.85
	b3	0.80	0.87	0.88	0.91	**0.96**	**0.97**	0.89	0.93
	b4	0.83	0.86	–	–	0.77	**0.88**	**0.88**	**0.93**
ERGAS	–	5.08	3.51	4.25	2.77	3.25	**2.08**	4.72	**2.50**
Ideal = Lowest	–	–	–	–	–	–	–	–	–

Table 20.2 SPOT 5 quantitative analysis

Measure	Band	ANW	ANC	I-ANW	I-ANC	PCA-NW	PCA-NC	WiSpeR	CiSpeR
COR	b1	0.90	0.92	**0.93**	**0.94**	0.83	0.85	0.90	0.92
Ideal value = 1	b2	0.95	**0.96**	0.91	0.93	**0.98**	0.95	0.92	0.94
	b3	0.94	**0.95**	0.93	**0.95**	**0.97**	**0.95**	0.93	**0.95**
	b4	0.93	**0.95**	–	–	**0.95**	**0.96**	0.88	0.92
UIQI	b1	0.86	0.91	0.83	0.88	**0.95**	**0.95**	0.79	**0.92**
Ideal value = 1	b2	0.91	0.94	**0.95**	**0.97**	0.93	**0.95**	0.89	**0.97**
	b3	0.91	**0.95**	0.92	**0.95**	0.93	**0.95**	0.85	**0.96**
	b4	0.82	0.88	–	–	0.88	**0.89**	0.85	**0.94**
ERGAS	–	5.36	3.52	5.40	3.30	3.25	**3.03**	5.27	**2.86**
Ideal = Lowest	–	–	–	–	–	–	–	–	–

with low ERGAS, probably due to the high-resolution of the images, with only 5 m per pixel, that allows a very good representation of the spatial structures using the contourlet transform. For the SPOT image, we can see that PCA give better results than PCA with Landsat. This may be due to the correlation between PC1 and MS bands. While, for the Landsat 7 image, PC1 is very similar to band 4, in SPOT 5 it takes information from the four bands. Again, CiSpeR almost achieves the ideal values in all the spatial and spectral measures.

20.5 Conclusions

In this chapter, wavelet- and contourlet-based pansharpening approaches have been compared and their efficiency to merge Landsat 7 and SPOT images has been evaluated by means of visual and quantitative analysis.

Different image pansharpening methods based on the undecimated wavelet and contourlet transform (Additive, IHS, and PCA) have been experimentally compared. Also the new proposed method CiSpeR was compared with these methods and with WiSpeR. In all methods, contourlet-based pansharpened images present, visually and numerically, better results than those obtained by wavelet for both Landsat and SPOT imagery extracting spatial information from the PAN image missing in the MS image, without modifying its spectral information content.

CiSpeR obtain a very low ERGAS value, smaller than 3, in both imagery, and values very close to the ideal in the other measures. It is a consistent approach that works well spatially and spectrally with different imagery dataset.

Acknowledgements This work has been supported by the Consejería de Innovación, Ciencia y Empresa of the Junta de Andalucía under contract P07-TIC-02698.

References

1. Z. Wang, D. Ziou, C. Armenakis, D. Li, and Q. Li. A comparative analysis of image fusion methods. *IEEE Trans. Geosci.Remote Sens.*, 43(6):1391–1402, 2005.
2. L. Wald. Some terms of reference in data fusion. *IEEE Trans. Geosci. Remote Sens.*, 37(3):1190–1193, 1999.
3. C. Pohl and J. L. Van Genderen. Multi-sensor image fusion in remote sensing: Concepts, methods, and applications. *Int. J. Remote Sens.*, 19(5):823–854, 1998.
4. Google Earth, http://earth.google.com/. *Google Earth Web Site*.
5. L. Alparone, L. Wald, J. Chanussot, P. Gamba, and L. M. Bruce. Comparison of pansharpening algorithms: Outcome of the 2006 GRS-S data-fusion contest. *IEEE Trans. Geosci. Remote Sens.*, 45(10):3012–3021, 2007.
6. P. S. Chavez, Jr and J. A. Bowell. Comparison of the spectral information content of Landsat Thematic Mapper and SPOT for three different sites in the Phoenix, Arizona region. *Photogramm. Eng. Remote Sens.*, 54(12):1699–1708, 1988.
7. J. Zhou, D. L. Civco, and J. A. Silander. A wavelet transform method to merge Landsat TM and SPOT panchromatic data. *Int. J. Remote Sens.*, 19(4):743–757, 1998.
8. P. S. Chavez and A. Y. Kwarteng. Extracting spectral contrast in Landsat thematic mapper image data using selective principal component analysis. *Photogramm. Eng. Remote Sens.*, 55(3):339–348, 1989.
9. V. P. Shah, N. H. Younan, and R. King. Pan-sharpening via the contourlet transform. In *Proc. IEEE Int. Geosci. Remote Sens. Symp. IGARSS 2007*, pages 310–313, 2007.
10. M. N. Do and M. Vetterli. The contourlet transform: An efficient directional multiresolution image representation. *IEEE Trans. Image Process.*, 14(12):2091–2106, 2005.
11. X. Otazu, M. González-Audícana, O. Fors, and J. Núñez. Introduction of sensor spectral response into image fusion methods: Application to wavelet-based methods. *IEEE Trans. Geosci. Remote Sens.*, 43(10):2376–2385, 2005.
12. V. P. Shah, N. H. Younan, and R. L. King. An efficient pan-sharpening method via a combined adaptive PCA approach and contourlets. *IEEE Trans. Geosci. Remote Sens.*, 46(5):1323–1335, 2008.
13. A. L. da Cunha, Jianping Zhou, and M. N. Do. The nonsubsampled contourlet transform: Theory, design, and applications. *IEEE Trans. Image Process.*, 15(10):3089–3101, 2006.
14. M. González-Audícana and X. Otazu. Comparison between Mallat's and the a'trous discrete wavelet transform based algorithms for the fusion of multispectral and panchromatic images. *Int. J. Remote Sens.*, 26(3):595–614, 2005.
15. M. Lillo-Saavedra and C. Gonzalo. Multispectral images fusion by a joint multidirectional and multiresolution representation. *Int. J. Remote Sens.*, 28(18):4065–4079, 2007.
16. W. Dou and Y. Chen. An improved IHS image fusion method with high spectral fidelity. *Int. Archiv. Photogramm., Rem. Sensing Spat. Inform. Sci.*, XXXVII:1253–1256, 2008, part.B7.
17. J. Nuñez, X. Otazu, O. Fors, A. Prades, V. Pala, and R. Arbiol. Multiresolution-based image fusion with additive wavelet decomposition. *IEEE Trans. Geosci. Remote Sens.*, 37(3): 1204–1211, 1999.
18. M. Song, X. Chen, and P. Guo. A fusion method for multispectral and panchromatic images based on HSI and contourlet transformation. In *Proc. 10th Workshop on Image Analysis for Multimedia Interactive Services WIAMIS '09*, pages 77–80, 6–8 May 2009.
19. A. M. ALEjaily, I. A. El Rube, and M. A. Mangoud. Fusion of remote sensing images using contourlet transform. Springer, Berlin, pages 213–218, 2008.
20. X.-H. Yang and L.-C. Jiao. Fusion algorithm for remote sensing images based on nonsubsampled contourlet transform. *Acta Automatica Sinica*, 34(2):274–281, 2008.
21. J. Wu, H. Huang, J. Liu, and J. Tian. Remote sensing image data fusion based on ihs and local deviation of wavelet transformation. In *Proc. IEEE Int. Conf. on Robotics and Biomimetics ROBIO 2004*, pages 564–568, 2004.
22. M. González-Audícana, J. L. Saleta, R. García Catalán, and R. García. Fusion of multispectral and panchromatic images using improved IHS and PCA mergers based on wavelet decomposition. *IEEE Trans. Geosc. Remote Sens.*, 42(6):1291–1298, 2004.

23. V. Vijayaraj, C. G. O'Hara, and N. H. Younan. Quality analysis of pansharpened images. In *Proc. IEEE Int. Geosc. Remote Sens. Symp. IGARSS '04*, volume 1, 2004.
24. Z. Wang and A. C. Bovik. A universal image quality index. *IEEE Signal Process. Lett.*, 9(3):81–84, 2002.
25. L. Wald. Quality of high resolution synthesized images: Is there a simple criterion? In *Proc. Int. Conf. Fusion of Earth Data.*, Nice, France, pages 99–105, 2000.

Chapter 21
Faces Robust Image Hashing by ICA: An Attempt to Optimality

Thierry Fournel and Daniela Coltuc

Abstract This chapter suggests a statistical image analysis approach for image robust hashing. It is designed such that provided hash values are equiprobable and statistically independent, two conditions to have minimum collision probability. The approach is based on Independent Component Analysis (ICA), here used in a configuration known in literature as Architecture II. The statistically independent coefficients issued from image ICA are statistically modeled by the Generalized Gaussian Distribution in order to be quantized such to obtain equiprobable intermediate hash values. A fixed length code is used for binarizing the quantized coefficients. The tests done on face images from FERET database confirm the obtention of intermediate hash values that are equiprobable and independent. The image hashing method is resistant to attacks by jpeg compression, median filtering and additive Gaussian noise.

21.1 Motivation

With the advent of digital technology, the number of applications using digital images has been continuously enlarging. In many cases, images must be archived in databases, requiring large storing volumes and efficient techniques for searching. Often, query images are numerically different from the similar ones in the database, because of compression, image enhancement, filtering and other non-malicious attacks. The retrieval task is not easy in these conditions.

Robust visual hashing may give solutions for such problems. By hashing, images are mapped to short binary strings, called hash values, that can be used for indexing the images in databases. Robust visual hashing provides hash values that are

---

T. Fournel (✉)
Université de Lyon / CNRS, UMR 5516, Laboratoire Hubert Curien/Université de Saint-Etienne, Jean Monnet, 18 rue du Prof. B. Lauras, 42023 Saint-Etienne, France
e-mail: fournel@univ-st-etienne.fr

T. Fournel and B. Javidi (eds.), *Information Optics and Photonics: Algorithms, Systems, and Applications*, DOI 10.1007/978-1-4419-7380-1_21,

invariant for visually similar images but highly sensitive to dissimilarities, visible image modifications. In this way, the retrieval can be successfully performed by simply computing the hash value.

In image processing area, robust image hashing may be placed next to compression and recognition. The idea of compression is present in robust hashing due to the requirement of getting short representations for the image [4]. In image compression, this is obtained by quantization and entropy coding, after statistically modeling the image via transforms (DCT, wavelet, etc.). A series of robust hashing methods uses common compression techniques for obtaining hash values [7, 9]. Opposite to compression, the quantization is much stronger in image hashing since the purpose is not anymore image reconstruction, but image recognition.

Besides its compression behavior, image hashing requires robustness and discriminative power through invariant descriptors [6] as in image recognition but invariance being restricted to unperceptual distortions.

In this chapter, we present an ICA (Independent Component Analysis) based hash function, which is capable to provide independent and uniformly distributed hash values, two compulsory conditions for having minimum collision probability. We focus on the obtention of an intermediate hash value with these properties. The application regards the indexation of images representing human faces.

21.2 ICA in the Context of Robust Image Hashing

Let $\mathscr{I}$ denote a class of images representing, in the case of our application, human faces (Fig. 21.1). *Hash function* is the generic name given to a sequence of operations that transform an image $I \in \mathscr{I}$ into a short binary string, $\mathscr{H}(\mathscr{I})$, called *hash value*. Ideally, a hash function should provide [5]:

1. Hash values with uniform distribution:

$$P[\mathscr{H}(I)] = \frac{1}{2^N}, \tag{21.1}$$

   where I denotes any image $I \in \mathscr{I}$, N is the length of the binary string.
2. Pairwise independence for the hash values of visually different images:

$$P\,[\mathscr{H}(I) = a | \mathscr{H}(I_{\text{dis}}) = b] = P\,[\mathscr{H}(I) = a], \tag{21.2}$$

   where I and I_{dis} denote dissimilar images.
3. Invariance for the hash value of visually similar images:

$$P\,[\mathscr{H}(I) = \mathscr{H}(I_{\text{sim}})] = 1, \tag{21.3}$$

   where I and I_{sim} denote similar images.

Fig. 21.1 Samples from the training set, consisting of 396 images from FERET database. They are frontal views of various persons, with neutral expression

The first two items are general conditions for hashing applications. They minimize the collision probability, that is, the probability of obtaining identical hash values for dissimilar images. The third item is specific to image hashing, where differences between visually similar images must not be sensed by the hash functions. For this reason, image hashing is also called *robust hashing*. The robustness is obtained by using features, which are low sensitive to such differences and by stabilizing the hash values through methods like channel decoding or vector quantization [5, 9].

Our approach focuses on the obtention of a robust hash function satisfying properties (21.1) and (21.2). It is founded on ICA, a mathematical tool able to decompose an image into a series of statistically independent coefficients, which represent a good premise for obtaining the above-mentioned properties.

By ICA, a signal is represented by a basis adapted to signal statistics. This basis is derived by optimizing an independence criterion [3]. There are two possible architectures for performing image ICA: one, where the basis elements are statistically independent and the other one, where the resulting coefficients are statistically independent [8]. Our approach is using the second architecture. According to this one, a signal I is represented by

$$I = \sum_{j=1}^{J} s_j \mathbf{b_j}, \tag{21.4}$$

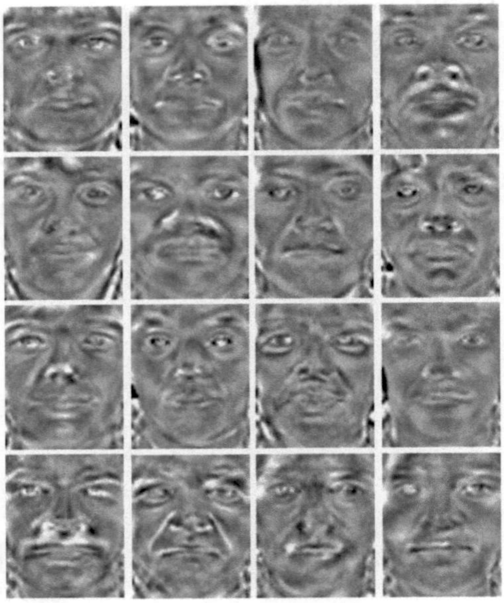

Fig. 21.2 Vector basis, resized as images, reveal primitive faces with characteristics from the training set

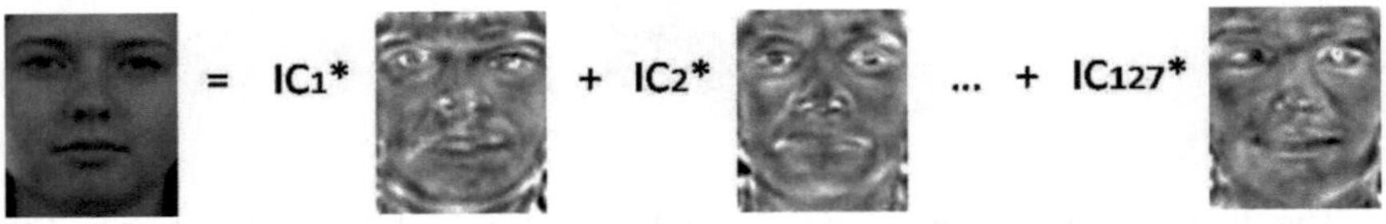

Fig. 21.3 New image represented by ICA basis

where $\mathbf{b_j}$ are the basis vectors and s_j the coefficients, also called Independent Components (ICs). As ICA is designed for 1D signals, images I have to be serialized in order to be represented by (21.4).

The basis vectors $\mathbf{b_j}$ are learned by applying ICA to a training set consisting in images from the same class as image I. The resulted basis has the class characteristics. For instance, the basis vectors obtained by training on face images reveal a set of primitive faces when resized as image I (Fig. 21.2). They represent only the face zone because the images were registered and cropped before training, as explained in Sect. 21.3.

Any new face image belonging to the class, registered and cropped in the same way, may be represented by this basis (Fig. 21.3). The resulted ICs contain image features. Assembled in a vector, they constitute a fingerprint, also named a signature, which is the starting point for our image hash value.

Since the ICs issued from ICA are practically continuous random variables, they have to be quantized and binarized in order to obtain a binary hash value. When the coefficients of distributions are known, it is possible to design quantizers that provide equiprobable quantized values for each IC. In this favorable case, the intermediate hash values are also equiprobable due to ICs independence (the signature probability is equal to the product of ICs probabilities). For instance, J ICs quantized on L equiprobable levels and binarized by using a fixed length code provide intermediate hash values with the probability:

$$P[\mathcal{H}(I)] = \frac{1}{2^{nJ}}, \tag{21.5}$$

where n is the length of shortest binary code able to represent L levels. The demand (21.1) for an ideal hash function is thus satisfied.

It is interesting to note that such intermediate hash values have maximum entropy. Indeed, due to ICA, they are memoryless sources of information and, by appropriate quantization, their alphabet is made of equiprobable symbols. Consequently, both conditions for maximum entropy are satisfied. In our example, each intermediate hash value carries $J \times \log_2(L)$ bits of information.

The use of ICA creates conditions for satisfying also the second demand for an ideal hash function. Indeed, ICs independence is manifested by a sparse matrix of signatures [3]:

$$S = \left[\mathbf{s}^{(1)}\mathbf{s}^{(2)}\ldots\mathbf{s}^{(j)}\ldots\mathbf{s}^{(J)}\right], \tag{21.6}$$

where $\mathbf{s}^{(j)}$ are images ICs signatures represented as column vectors. Sparsity becomes more obvious as the number of ICs increases (ICA algorithms allow the number of ICs as input). For a sufficient number of ICs, the signatures become independent due to the sparsity of S. The experimental results in Sect. 21.3 show a distribution of the normalized Hamming distance between the intermediate hash values, symmetrical and well centered on 50%, which enforces our assertion.

21.2.1 ICs Quantization and Binarization

The experiments on face images have shown ICs histograms that may be modeled by a Generalized Gaussian Distribution (GGD) expressed as

$$p(x) = \frac{\beta}{2\alpha\Gamma\left(\frac{1}{\beta}\right)}\exp - \left(\frac{|x|}{\alpha}\right)^{\beta}, \tag{21.7}$$

where α and β are two parameters determining the distribution width and shape, respectively. From a theoretical point of view, as ICA generates ICs with variance equal to 1, their distribution has a single degree of freedom. In fact, the following relationship exists between α and β:

$$\alpha^2 = \frac{\Gamma\left(\frac{1}{\beta}\right)}{\Gamma\left(\frac{3}{\beta}\right)}. \tag{21.8}$$

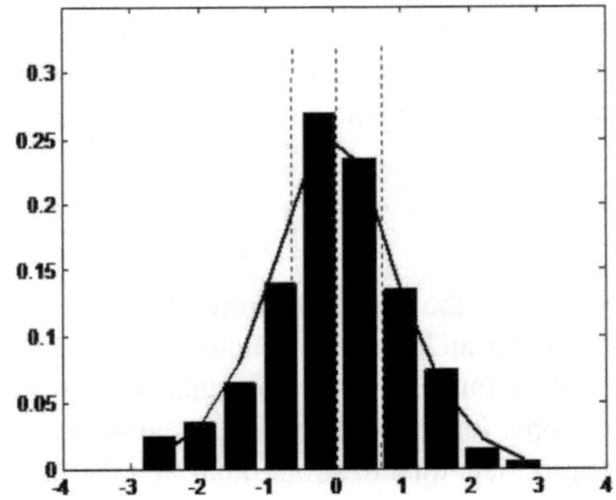

Fig. 21.4 Histogram of first IC and its GGD model. With *dotted lines*, the decision levels for $L = 4$

Consequently, it is sufficient to estimate the parameters β from samples in order to have the complete analytical expressions of IC distribution. The knowledge of the distribution allows the design of quantizers providing L equiprobable quantization levels. The decision levels d_j are found by solving the equations:

$$\int_{d_{j-1}}^{d_j} p(x)\mathrm{d}x = \frac{1}{L}. \tag{21.9}$$

Figure 21.4 shows, for example, the histogram of the first IC. The GGD model of each IC is derived by estimating parameter β on the samples obtained by projecting a new training set on the corresponding basis vector.

Without special constraints, FastICA – the algorithm used in our experiments for learning the basis vectors $\mathbf{b_j}$ – extracts ICs with various entropies. Consequently, by projecting the test images on this basis, various β parameters are obtained for ICs distributions. It is, therefore, necessary to design a quantizer for each IC in order to have equiprobable quantization levels. However, it would be possible to resemble i.i.d. ICs by constraining FastICA to stop when the IC's entropy reaches a predefined threshold. In this case, the quantization task would become much easier since the same quantizer can be used for all ICs. The disadvantage is that, by stopping at a fixed entropy, the ICs independence may be affected.

It is worth to be noted that maximum entropy signatures are obtained only if all ICs are quantized with the same number of levels. In this case, a unique fixed length code may be used for binarizing the entire signature. In order to have intermediate hash values as short as possible, it is necessary to use for L a power of 2.

21.2.2 Noise Reduction

In robust image hashing, any difference between two visually similar images may be considered as noise. By using an ICA approach, one disposes of two mechanisms for noise reduction: the Principal Component Analysis (PCA), usually performed in the first stage of ICA algorithm and ICs selection. In both cases, the noise reduction is obtained by coefficient removal, the difference consisting in the criterion used for it. For PCA, a variance criterion is needed, since it is based on the second-order statistics, while for ICA a fourth-order statistics measure, like negentropy or Kurtosis is appropriate. The removal of PCA coefficients, performed in the learning stage, reduces the number of basis vectors and, consequently, the number of ICs. If needed, a supplementary reduction of ICs may be done after image projecting on the learned basis. Noise reduction by coefficient removal enhances the robustness of the hash value against a series of attacks.

21.3 Experimental Results

Our approach has been tested on a set of face images from FERET database [2]. The ICA basis was learned on 396 gray level images of 384×256 pixels, representing frontal views of different persons, all of them with neutral expression (Fig. 21.1). For tests, it was used another set consisting of 200 images with the same characteristics. All the images were normalized in order to have zero mean and variance equal to one, registered, cropped and resized to 60×50 pixels. The registration was done by using the eyes and mouth coordinates provided by FERET databases. By averaging the training images coordinates, a reference set has been obtained. All the faces, for training or tests, were registered to the reference set. An example is shown in Fig. 21.5.

The learning set was analyzed by FastICA software [1]. For 60×50 pixels images, FastICA can extract at maximum 3,000 ICs. The ICs number has been reduced to $J = 128$, by discarding PCA coefficients carrying about 2% of signal energy. The learned basis consisted of 128 vectors that, resized as 60×50 pixels images, reveal a set of primitive faces (Fig. 21.2). By projecting the test images on this basis, a signature consisting of 128 ICs is obtained for each of them. Each IC histogram has been modeled by the GGD, obtained for variance one and β estimated on signatures samples. The parameter β of the 128 ICs is in the range of $[0.8, 2.8]$ (Fig. 21.6).

By using estimated GGD models, a four-level quantizer was designed for each IC, such to have equiprobable quantized values. Finally, the quantized ICs were binarized by using a 2-bits code. An intermediate hash value of 256 bits has been obtained thus for each test image. The first-order entropy of these intermediate hash values is 0.99 ± 0.01, very close to 1 (Fig. 21.7). This result confirms the property of maximum entropy and, implicitly, the fulfilment of condition (21.1).

Fig. 21.5 Face image after registration, crop and resizing

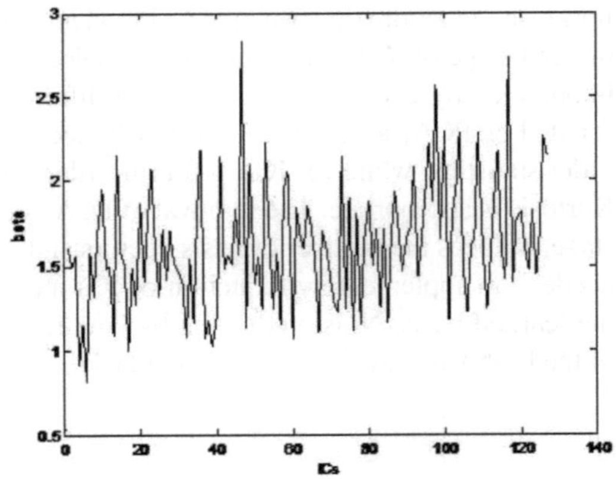

Fig. 21.6 Estimated β for 128 ICs

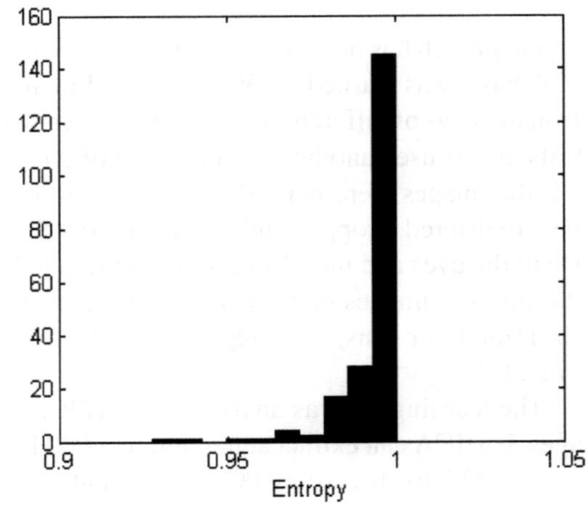

Fig. 21.7 Intermediate hash values entropy

Our hash function have been tested against jpeg compression, median filtering, additive gaussian white noise (AGWN) and image enhancement by histogram equalization. The face images were attacked before preprocessing by registration, crop and resizing. The hash values were compared by using the Hamming distance. Figure 21.8 shows the Bit Error Rate (BER) distribution for jpeg compression with Quality factor 50. The BER of visually similar images is close to zero while that for different images is about 50%. The mean of similar images BERs is 5.8 and the standard deviation 1.8; the different images BERs have a mean of 49.9 and a standard deviation of 3.4. The two BER histograms remain separated until compression with Quality factor 9. This show that, by an appropriate quantization, it is possible to obtain final hash values with residual collision probability for jpeg attacks in this range. The BER mean and standard deviation for maximum attacks intensity allowing this property are given in Table 21.1.

No matter how strong the attack, the BER distribution for visually different images remains symmetrical and centered on 50%. This result is coherent with the

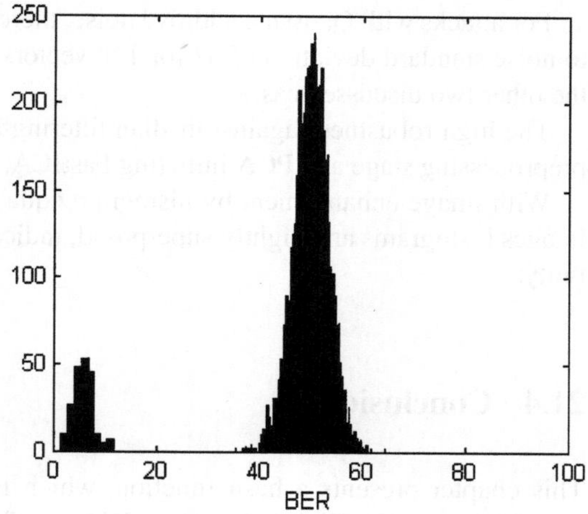

Fig. 21.8 BER distribution for jpeg compression with Quality factor 50. *Left*: histogram of similar images BERs. *Right*: histogram of different images BERs

Table 21.1 BER mean and standard deviation for maximum attack intensity allowing disjoint histograms for similar and different faces

		Similar faces		Different faces	
	Attack	Mean	Standard deviation	Mean	Standard deviation
128 ICs	JPEG $Q = 9$	22.1	4.2	50.1	3.5
$L = 4$	Median 13 × 13	23.5	4.7	49.9	3.4
256 bits	AGWN $\sigma = 62$	17.7	4.4	49.9	3.3
	Image enhancement	18.0	5.7	49.9	3.4
128 ICs	JPEG $Q = 10$	25.4	3.9	49.4	2.8
$L = 8$	Median 12 × 12	27.5	4.4	49.2	2.8
384 bits	AGWN $\sigma = 67$	17.4	4.7	49.0	2.7
	Image enhancement	22.1	6.2	49.2	2.8
64 ICs	JPEG $Q = 7$	19.1	4.0	50.1	4.8
$L = 4$	Median 11 × 11	15.8	4.8	50.1	4.9
128 bits	AGWN $\sigma = 67$	9.7	3.6	49.8	4.8
	Image enhancement	15.6	6.0	49.9	4.8

property of statistical independence for the hash values. The attack intensity mainly modifies the distribution of similar images BER.

By increasing the number of quantization levels to $L = 8$, the Quality factor is also increasing. In this case, the dimension of the hash values increases by 50% since a 3-bits code is needed for binarization.

For ICA basis with 64 vectors, there is no histograms superposition up to a Quality factor 6, when $L = 4$. The hash values are 128 bits long in this case. For jpeg attacks at least, this is the best choice.

The number of ICs and quantization levels have a rather low influence in attacks with median filtering where no histograms superposition appears up to a filtering window around 12 × 12.

For attacks with Gaussian additive noise, there is no histograms superposition up to noise standard deviation of 67 for 128 vectors basis and $L = 4$ and up to 62 for the other two discussed cases.

The high robustness against median filtering and noise is due to resizing in the preprocessing stage and PCA initiating FastICA, both of them low-pass filtering.

With image enhancement by histogram equalization, the similar and dissimilar images histograms are slightly superposed, indicating an increased collision probability.

21.4 Conclusion

This chapter presents a hash function, which is optimal in the sense of properties (21.1) and (21.2). It is based on ICA, configured such to provide statistically independent coefficients for image representation with ICA basis. ICA was completed with a non-uniform quantization and a fixed length binary coding in order to gain equiprobability of the intermediate hash values. The tests on face images have confirmed the targeted properties. In a future work, the hash function will be more extensively tested, also in a print / scan process. It will be completed by a vector quantization stage in order to stabilize the hash values as required in image indexation.

Acknowledgement The second author acknowledges financial support of UEFISCU Romania, Grant 610/2008.

References

1. FastICA package for Matlab. Available at http://www.cis.hut.fi/projects/ica/fastica/
2. FERET Database. Available at http://www.itl.nist.gov/iad/humanid/feret/
3. Hyvarinen A., Karhunen J., Oja E. (2001) Independent Component Analysis. Wiley, New York
4. McCharty E., Balado F., Silvestre G. C. M., Hurley N. J. (2004) A framework for soft hashing and its applications to robust image hashing. In *Proc. IEEE Conf. on Image. Processing* (I: 397–400)
5. Monga V., Benerjee A., Evans B. L. (2006) A clustering based approach to perceptual image hashing. IEEE Transactions on Information Forensics and Security, vol. 1, no. 1, pp. 68–79
6. Monga V., Evans B. L. (2006) Perceptual image hashing via feature points: performance evaluation and trade-offs. IEEE Transactions on Image Processing, vol. 15, no. 11, pp. 3452–3465
7. Skrepth C. J., Uhl A. (2002) Robust hash functions for visual data: an experimental comparison. In F. J. Perales et al. (editors), Springer Lecture Notes on Computer Science, vol. 2652, pp. 986–993. Springer, Berlin
8. Stewart Bartlett M., Movellan J. R., Sejnowski T. J. (2002) Face recognition by independent component analysis. IEEE Transactions on Neural Networks, vol. 13, no. 6, pp. 1450–1464
9. Venkatesan R., Koon S. -M., Jakubowsky M. H., Moulin P. (2000) Robust image hashing. In *Proc. IEEE Conf. on Image. Processing* (III: 664–666)

Chapter 22
Minkowski Metrics: Its Interaction and Complementarity with Euclidean Metrics

Jean-Marie Becker and Michel Goeb

Abstract A certain set of objects has a Minkowski space structure when these objects can be represented by coordinates $(x_0, x_1, \ldots, x_n)$ with $(x_1^2 + \cdots + x_n^2) - x_0^2 \leq 0$. This chapter considers three situations where this structure and its associated tools bring a new light. It first introduces a well-known such space, having its roots in Minkowski's work, the space–time representation. Then, it considers a 3D representation space, S_2^+, for 2×2 symmetrical positive definite matrices, the determinant of these matrices providing this space with a Minkowski structure. An application to a representation of discrete curves is detailed. The last example is another 3D representation space, denoted Ω_2, for plane circles in the Euclidean space $\mathbb{R}^2$. The natural Minkowski quadratic form for a circle in Ω_2 is the square of its radius, for a convenient set of coordinates. An application is given to a new measure of circularity for a finite set of points.

22.1 Introduction

The norm N in a Euclidean space is defined by $\|x\|^2 = N(x)^2 = \sum_{k=1}^{n} x_k^2$ with respect to a certain basis $e_1, e_2, \ldots, e_n$, where $x = \sum_{k=1}^{n} x_k e_k$ with associated dot product $b(x, y) = \sum_{k=1}^{n} x_k y_k$. A Minkowski space can be considered as an extension of the Euclidean space structure. It is structured by a quadratic form that can be expressed under the following form:

$$q(x) = q(x_0, \ldots, x_{n-1}, x_n) \quad \text{with} \quad (x_1^2 + \cdots + x_n^2) - x_0^2 = \|x\|^2 - x_0^2 \quad (22.1)$$

J.-M. Becker (✉)
CPE Lyon, 43 Bd du 11 Nov. 1918, 69616 Villeurbanne Cedex, France
and
Université de Lyon / CNRS, UMR 5516, Laboratoire Hubert Curien / Université de Saint-Etienne, Jean Monnet, 18 rue du Prof. B. Lauras, 42023 Saint-Etienne, France
e-mail: becker@cpe.fr

T. Fournel and B. Javidi (eds.), *Information Optics and Photonics: Algorithms, Systems, and Applications*, DOI 10.1007/978-1-4419-7380-1_22,

in a convenient $(n+1)$-dimensional basis. $\sqrt{q(x)}$ will thus be called a "pseudo-norm", only defined inside the "cone" with equation $(x_1^2+\cdots+x_n^2)=x_0^2$. $q(x)$ is naturally associated with a "dot product" defined by $\sum_{k=1}^{n} x_k y_k - x_0 y_0$. The hyperplane with equation $x_0 = 0$ and, more generally, all affine hyperplanes with equations $x_0 =$ const. have a natural Euclidean structure, by taking $\|r(x)\|$ for its norm instead of $\sqrt{q(x)}$.

Minkowski spaces occur quite naturally in several cases. This chapter demonstrates the interest of this structure using three different examples. The first one, the space–time "continuum" is mainly given for its historical interest. The second example is motivated by a work of the first author to provide a mathematically rigorous approach to curve digitalization on a digital grid. The third one is a kind of zoom on one of the most innovative part of the Ph. D. thesis of the second author. It settles the study of plane circles in a specific Minkowski space; moreover, it introduces a certain canonical measure that interacts with this Minkowski structure; an application is given to the circularity assessment of a set of points.

22.2 The Space–Time Historical Minkowski Space

It is well known that Einstein has a debt towards some mathematicians; prominent among them stands Minkowski, who has been his professor in the Technical University of Zurich. Let us quote Minkowski himself, in a talk he gave in a congress in 1908: "The views of space and time which I wish to lay before you have sprung from the soil of experimental physics, and therein lies their strength. They are radical. Henceforth space by itself, and time by itself, are doomed to fade away into mere shadows, and only a kind of union of the two will preserve an independent reality".

Instead of developing very much this issue, which can be found elsewhere, let us open some tracks that we will find, under other disguises, in the two other examples.

According to the famous Klein classification, a specific geometry is defined by a specific group of transformations preserving a certain property.

For example, Euclidean 2D space is mainly structured by rotations that preserve the norm of a vector, or in a more sophisticated way, that preserve quadratic form $q(x,y) = x^2 + y^2$.

Let us recall that a plane rotation may be described by its matrix:

$$R_\theta = \begin{pmatrix} \cos(\theta) & -\sin(\theta) \\ \sin(\theta) & \cos(\theta) \end{pmatrix}. \tag{22.2}$$

These matrices form a group with $R_{\theta_2} \circ R_{\theta_1} = R_{\theta_1+\theta_2}$.

The space–time equivalent of both quadratic form q and its group of motions [formula (22.2)], in its simplest form, that is, on the real line is as follows. Considering that a certain object stands at $x = 0$ when $t = 0$, we would like to characterize those "events" that can be posterior to this initial situation. By "event" we mean the association of a place x and a time t. The space–time trajectory of this moving

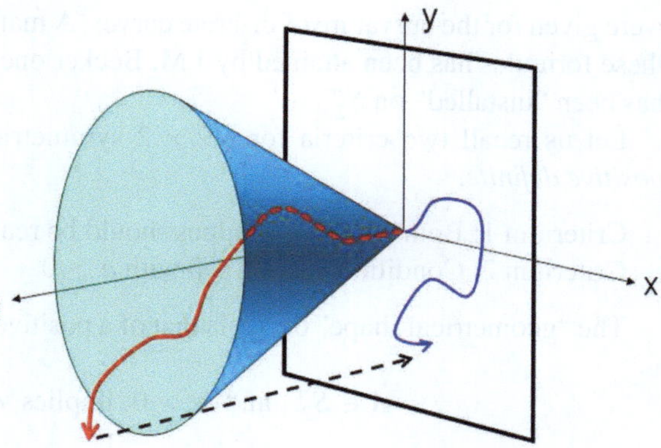

Fig. 22.1 The space–time trajectory of any point lies inside the light-cone. The projection onto x–y plane is the ordinary spatial trajectory

object is a certain curve $x = x(t)$ (x as a function of t), with the constraint that the speed of the object cannot be larger than c, the speed of light. Expressed with formulas:

$$\frac{x}{t} \leq c \quad \text{or} \quad r(t, x) = x^2 - c^2 t^2 \leq 0. \tag{22.3}$$

The left-hand side of the second inequality in (22.3) is the fundamental quadratic form for space–time. It gives a Minkowski structure to the space $\mathbb{R}_+ \times \mathbb{R}$ of events. If one takes a system of units such that $c = 1$, the quadratic form becomes $r(t, x) = x^2 - t^2$. In this case, the associated group is made of the following matrices:

$$H_u = \begin{pmatrix} \cosh(u) & \sinh(u) \\ \sinh(u) & \cosh(u) \end{pmatrix}. \tag{22.4}$$

The group structure arising from the fact that $H_{u_1} \circ H_{u_2} = H_{u_1+u_2}$.

Clearly, if, instead of moving on a line, the object is free to move in the (Euclidean) plane $\mathbb{R}^2$, the space of events would become the 3D space $\mathbb{R}_+ \times \mathbb{R}^2$. The quadratic form attached to this space would be, naturally, $q(t, x, y) = x^2 + y^2 - c^2 t^2$ conditioned by $q(t, x, y) \leq 0$. This constraint means that the space–time trajectory of the moving object should always remain inside a certain cone as illustrated in Fig. 22.1. The generatrices of this cone are the space–time trajectories of photons issued from the origin. This is why this cone is called the light-cone.

22.3 The Minkowski Space S_2^+

The symmetrical positive definite matrices are among the most important tools in linear algebra. Let us denote their set by S_2^+ in the case 2×2, the only one that will be considered here. The personal interest of the first author for these objects, a part of his Ph. D. thesis [2], aroused from [1]. In this paper, analytical formulas were given for the curvature of discrete curves. A mathematically sound approach to these formulas has been attained by J.M. Becker once a Minkowski space structure has been "installed" on S_2^+.

Let us recall two criteria for a 2×2 symmetrical matrix $A = \begin{pmatrix} a & b \\ b & c \end{pmatrix}$ to be *positive definite*:

Criterium 1: Both of its eigenvalues should be real and > 0.
Criterium 2: Condition $det(A) > 0$ with $a > 0$.

The "geometrical shape" of S_2^+ is that of a positive cone, which, precisely, means

$$\begin{aligned} &A \in S_2^+ \text{ and } \alpha > 0 \text{ implies } \alpha A \in S_2^+; \\ &A, B \in S_2^+ \text{ implies } \frac{1}{2}(A + B) \in S_2^+. \end{aligned} \tag{22.5}$$

Remark: If, in criteria 1 and 2 above, signs $>$ are allowed to be signs $\geq$, one gets the *closed* cone set of "symmetrical positive *semi*definite" matrices.

The generic matrix $A \in S_2^+$ can be written $A = x_0\sigma_0 + x_1\sigma_1 + x_2\sigma_2$ with

$$\sigma_0 = \frac{1}{2}\begin{pmatrix} 1 & 0 \\ 0 & 1 \end{pmatrix}; \quad \sigma_1 = \frac{1}{2}\begin{pmatrix} 1 & 0 \\ 0 & -1 \end{pmatrix}; \quad \sigma_2 = \frac{1}{2}\begin{pmatrix} 0 & 1 \\ 1 & 0 \end{pmatrix}$$

with the linear change of coordinates: $x_0 = a + c = \text{trace}(A)$, $x_1 = a - c$, $x_2 = 2b$. $\sigma_0, \sigma_1, \sigma_2$ are called "Pauli matrices" (some authors give this name to slightly different matrices). Notice that neither σ_1 nor σ_2 belong to S_2^+.

Basis $(\sigma_0, \sigma_1, \sigma_2)$ is the adapted basis for the Minkowski space structure of S_2^+, using a quadratic proportional to the determinant function, in the following manner:

$$\begin{aligned} q(A) &= -4\det(A) = 4(b^2 - 4ac) \\ &= -(a + c)^2 + (a - c)^2 + (2b)^2 = -x_0^2 + x_1^2 + x_2^2. \end{aligned} \tag{22.6}$$

In other words, $A \in S_2^+$ if and only if $q(a, b, c) < 0$. One may wonder what is the transformation group preserving q. It is constituted by the following transformations: $\begin{pmatrix} a & b \\ b & c \end{pmatrix} \overset{C_\alpha}{\rightarrow} \begin{pmatrix} a' & b' \\ b' & c' \end{pmatrix}$ defined by

$$\underbrace{\begin{pmatrix} \cos\alpha & +\sin\alpha \\ -\sin\alpha & \cos\alpha \end{pmatrix}}_{\text{Rotation } R_{(-\alpha)}} \begin{pmatrix} a & b \\ b & c \end{pmatrix} \underbrace{\begin{pmatrix} \cos\alpha & -\sin\alpha \\ +\sin\alpha & \cos\alpha \end{pmatrix}}_{\text{Rotation } R_{(\alpha)}} = \begin{pmatrix} a' & b' \\ b' & c' \end{pmatrix}. \tag{22.7}$$

Transformation C_α indeed preserves quadratic form q; the proof is immediate by taking determinants on both sides of formula (22.7).

It is important to give C_α a linear representation. A simple computation of the different $C_\alpha(\sigma_k)$ $k = 0, 1, 2$ gives the following matrix for C_α with respect to basis $(\sigma_0, \sigma_1, \sigma_2)$

$$\begin{pmatrix} 1 & 0 & 0 \\ 0 & \cos 2\alpha & -\sin 2\alpha \\ 0 & \sin 2\alpha & \cos 2\alpha \end{pmatrix}. \tag{22.8}$$

Thus C_α can be considered as a rotation matrix with angle 2α with respect to the last two coordinates.

Let us now consider the affine plane $x_0 = 1$, denoted T_1. It is the set of matrices A with trace$(A) = 1$. T_1 is a convenient "set of representatives". Indeed, due to the cone structure, all matrices $A \in S_2^+$, up to the "scaling" $A \to (1/x_0)A$, can be represented in this plane, or more precisely in the unit disk of T_1, by a matrix with trace 1 ("central projection" onto the unit disk). The restriction of the pseudo-norm to T_1 provides this plane with a Euclidean structure; the "Euclidean norm" in T_1 can be written under different forms

$$\|r(A)\| = x_1^2 + x_2^2 = \sqrt{(a-c)^2 + (2b)^2} = \sqrt{1 - 4\det(A)} = 2\lambda - 1, \tag{22.9}$$

where $0 < \lambda < 1$ is the largest eigenvalue of A (once normalized).

In a similar way, the "Euclidean dot product" in T_1 is

$$\begin{aligned} b(A, A') &= x_1 {x'}_1 + x_2 {x'}_2 = (a-c)(a'-c') + 4bb' \\ &= \|r(A)\| \, \|r(A')\| \cos(2\Delta\theta). \end{aligned} \tag{22.10}$$

The last expression is noticeable: The Euclidean form of the dot-product is retrieved.

Let us conclude this part with an application. Let us consider a digital plane curve, defined as a file of points (x_k, y_k). To each value k_0 of k, one can associate the set of all points of this file with indices k such that $k_0 - N \le k \le k_0 + N$, using a fixed N, for example, $N = 2$ in the case of (Fig. 22.2). $p = 2N + 1$ is the size of the "window" that is placed on the discrete curve, providing an extraction of a certain "cloud of points". To such a cloud is attached its so-called *experimental variance matrix* $(1/p)V$, with coefficients

$$v_{11} = \sum_{k=1}^{p} (x_k - \bar{x})^2; \quad v_{22} = \sum_{k=1}^{p} (y_k - \bar{y})^2; \quad v_{12} = v_{21} = \sum_{k=1}^{p} (x_k - \bar{x})(y_k - \bar{y}). \tag{22.11}$$

($\bar{x}$ and $\bar{y}$ are the coordinates of the center of mass of the cloud.) Such a matrix is positive (semi)definite (immediate proof using Cauchy–Schwarz inequality).

Let us now consider all possible positions of the p points window on the digital curve. Each position generates a certain point in the unit circle, obtained by central projection of the associated variance matrix $(1/p)V$ onto the unit disk of T_1 as explained before. In this way, any discrete curve in the geometrical plane can be associated to a unique discrete transformed curve in the disk, for a given window size $p = 2N + 1$ (see Fig. 22.2).

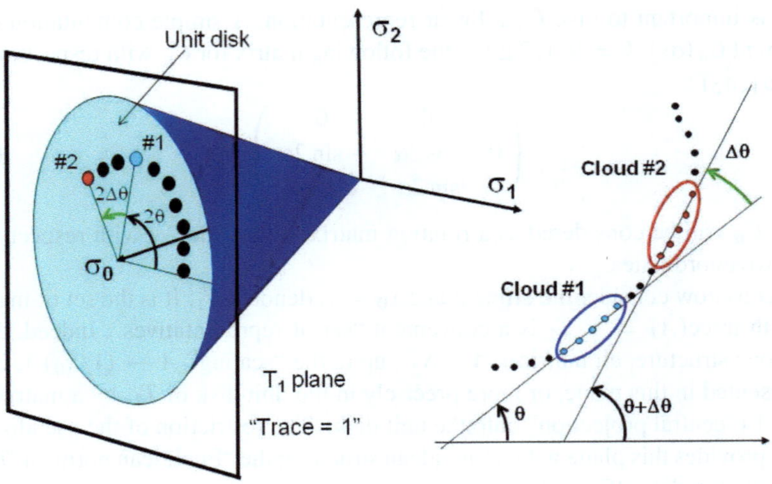

Fig. 22.2 S_2^+, represented as a cone in the $(\sigma_1, \sigma_2, \sigma_3)$ basis, sliced by plane T_1 giving rise to a certain unit disk. If a digital curve is given in the geometrical plane (subfigure on the *right*), there is matching "digital curve" of matrices in the disk. This association relies on a certain window's width

Multiscale contour analysis benefits from the above correspondence between the initial curve and the transformed one. Indeed, the finer the granularity of the initial curve, the closer the transformed curve will be to the unit circle. Thus, the evolution of the different transformed curves provides hints about the most convenient scale for the study of a digital curve.

22.4 The Minkowski Space Ω_2 of Plane Circles

(A general reference for this part is [3]).

We are going to study the set Ω_2 of circles in the Euclidean plane. Such circles depend on three parameters: (a,b), the coordinates of their center and R, their radius, which could be called the circle's "coordinates". There is a better coordinates' choice. It is (a,b,c) coordinates, with a parameter c that we are going to recall.

By comparing the two forms of the equation of a circle with center (a,b) and radius R:

$$(x-a)^2 + (y-b)^2 = R^2,$$
$$x^2 + y^2 - 2ax - 2by + c = 0, \tag{22.12}$$

we can define an "almost" (see below) quadratic form in Ω_2 by

$$q(a,b,c) = R^2 = a^2 + b^2 - c > 0. \tag{22.13}$$

Ω_2 is thus $\mathbb{R}^3$ minus the interior of paraboloid Π with equation:

$$c = a^2 + b^2. \tag{22.14}$$

In other words, Π represents the set of circles with null radius.

There are two differences with the previous Minkowski spaces.

First of all, relationship $c = a^2 + b^2 - R^2$ indicates that, if a, b, and R are lengthes (from the point of view of dimensional analysis), c is to be expressed as *the square of* a length. This is why, having scaled the third axis according to square roots, a paraboloid has taken the place of a cone.

Another apparent difference is that it is the exterior of the paraboloid that is used instead of its interior. But this is merely a question of sign: in the present case, the requirement for the quadratic form R^2 is to be positive instead of negative in the prior cases. Now, let us enrich the structure of Ω_2 by providing it with a measure.

Let us first recall what is the precise meaning of this word in mathematics, and, in a second step, why we need a measure.

A measure deals with "measurable sets" of "objects" (here the objects are circles), to which it attributes a positive or zero value (intuitively, this value is a kind of "size"). Just as triangular inequality is fundamental for distances, the main properties of a measure are its positivity and its additivity for disjoint sets of objects:

$$\forall i, j \quad S_i \cap S_j = \emptyset \Rightarrow \mu(S_1 \cup S_2 \cup \cdots) = \mu(S_1) + \mu(S_2) + \cdots . \tag{22.15}$$

In a somewhat simplified approach shared by most mathematicians, measure theory is nothing more than general integration theory. Indeed, we will represent our measure by an integral (see below).

Why are we interested in measures? Because they may be a valuable substitute to distances. Indeed, assessing a proximity is usually done using a distance or, more generally, a metric. However distances may not be the most adequate tools. For example, for a given "object", it is sometimes "immediate neighbors" that are known. Taking into account whole *sets* of objects instead of dealing with two objects at a time may be a key to a better modelization.

More precisely, we have established (see [3]) that a natural measure in Ω_2 for a (measurable) set of circles S (with parameters a, b, c, and R) is given by the so-called *Stoka* measure

$$\mu(S) = \iiint_S \frac{\mathrm{d}a\ \mathrm{d}b\ \mathrm{d}c}{R^4} = \iiint_S \frac{\mathrm{d}a\ \mathrm{d}b\ \mathrm{d}c}{(a^2 + b^2 - c)^2}.$$

This measure is clearly invariant with respect to the quadratic form of Ω_2. Moreover, it is invariant with respect to other metric operations [in the Euclidean substructure of plane $(a, b, 0)$]: invariance to translations, dilations, and even inversions.

There is an important result that explains the central role of paraboloid Π: The circle with coordinates (a, b, c) includes (resp. excludes) point $M(\alpha, \beta)$ if and only if $(\alpha - a)^2 + (\beta - b)^2 < R^2$ (resp. $> R^2$) which gives $2a\alpha + 2b\beta - (\alpha^2 + \beta^2) > c$. (resp. $< c$). This constraint is naturally interpreted in Ω_2 in the following way: $M(\alpha, \beta)$ in the upper (resp. lower) half-space defined by a certain plane.

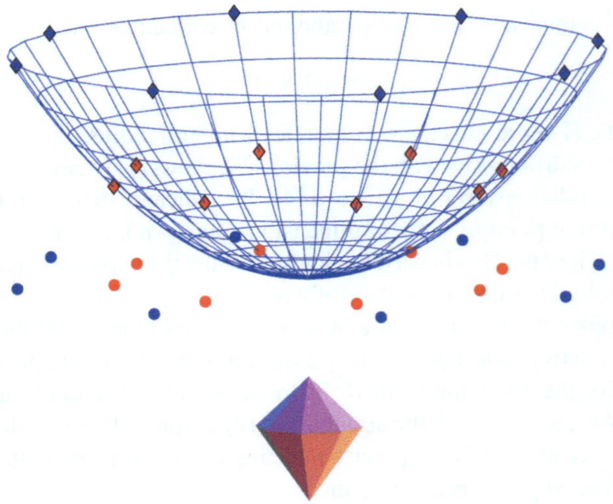

Fig. 22.3 In the $c = 0$ plane of Ω_2 (assimilated to the Euclidean plane, with coordinates $(a, b, 0)$), a double set of points $P = P_i, i = 1 \ldots 8$ and $Q = Q_j, j = 1 \ldots 8$ (disks) which are vertices of regular octagons; their "lifted" versions $P'_i, i = 1 \ldots 8$ and $Q'_j, j = 1 \ldots 8$ on paraboloid Π are represented by *diamonds*; the facets of polyhedron $S(P, Q)$ are the tangent planes to Π in points P'_i and Q'_j

Moreover, this plane has a simple connection with coordinates (α, β). It is the tangent plane $P_\Pi(M)$ to paraboloid Π at point $(\alpha, \beta, \alpha^2 + \beta^2)$ obtained by lifting up vertically M onto Π.

Let us now consider the set $S(P, Q)$ of circles containing a certain finite set of points (P_i) and excluding another finite set (Q_j). Using the result above, it is not difficult to see that the set of all such circles has a polyhedral shape in Ω_2 (see Fig. 22.3).

As it is shown in [3], the Stoka measure of polyhedron $S(P, Q)$ is rather easy to compute.

Remark: we have used the $c = 0$ plane as a copy of the Euclidean space. Plane $c = -1$ (resp. $c = 1$) can be used to modelize the spherical (resp. hyperbolic) non-Euclidean geometries.

Let us terminate this part with an application.

An important metrological issue is to obtain a mathematically based circularity test for a "cloud" of points, that is, a finite set of points P_i.

Here is such a test, a direct consequence of the theory that has been developed:

- Compute the center of gravity G of the cloud.
- Taking G as an origin, dilate (homothetize) the first cloud by a fixed factor (say, e.g., 1.1), into a second cloud Q_j.
- Accept the cloud as circular if the Stoka measure of the set of separating circles $S(P, Q)$ (see Fig. 22.4) is under a certain threshold. Otherwise reject.

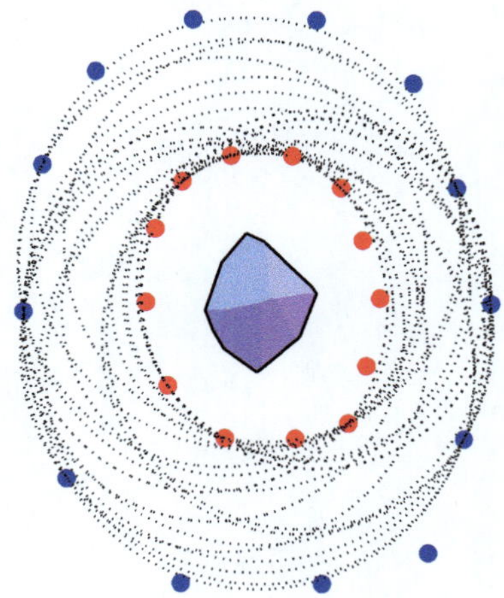

Fig. 22.4 Two "clouds" of points, interior (P) and exterior (Q), and some of their separating circles. The exterior set (Q) has been obtained by a dilation of the interior "cloud" (P). The polygon is the set of all centers of separating circles. This polygon is evidently the projection on the Euclidean plane with coordinates $(a, b, 0)$ of polyhedron $S(P, Q)$

22.5 Conclusion

The three given examples have shown that Minkowski space structure can be very helpful for obtaining through projections (central or linear) with the underlying Euclidean (or non-Euclidean...) space(s), using the x_0 = const. hyperplanes as "blackboards" for the interaction between them.

References

1. Anderson I.M. and Bezdek J.C. (1984) "Curvature and tangential deflection of discrete arcs: a theory based on the commutator of scatter matrix pairs and its application to vertex detection in planar shape data". IEEE Trans. PAMI 6:27–40
2. Becker J.M. (1991) "Contribution à l'analyse d'images par marches quadrantales", Ph. D. Thesis, University of Saint-Etienne, France
3. Goeb M. (2008) "Modèles Géométriques et Mesures de cercles contraints", Ph. D. Thesis, University of Saint-Etienne, France

MIX
Papier aus verantwortungsvollen Quellen
Paper from responsible sources
FSC® C105338
FSC
www.fsc.org

If you have any concerns about our products,
you can contact us on
ProductSafety@springernature.com

In case Publisher is established outside the EU,
the EU authorized representative is:
Springer Nature Customer Service Center GmbH
Europaplatz 3, 69115 Heidelberg, Germany

Printed by Libri Plureos GmbH
in Hamburg, Germany